# Mycobacteria

## I BASIC ASPECTS

# Mycobacteria

## I BASIC ASPECTS

CHAPMAN & HALL MEDICAL MICROBIOLOGY SERIES

*Edited by*

Pattisapu R. J. GANGADHARAM, PH.D.

*Professor of Medicine, Microbiology, and Pathology*
*Director of Mycobacteriology Research,*
*University of Illinois at Chicago College of Medicine, Chicago, Illinois*

P. Anthony JENKINS, PH.D.

*formerly Director, Public Health Laboratory Service,*
*University Hospital of Wales, United Kingdom*

CHAPMAN & HALL

ITP International Thomson Publishing
Thomson Science
New York • Albany • Bonn • Boston • Cincinnati • Detroit • London • Madrid • Melbourne
Mexico City • Pacific Grove • Paris • San Francisco • Singapore • Tokyo • Toronto • Washington

Cover design: Trudi Gershenov



Printed in the United States of America

For more information, contact:

Chapman & Hall
115 Fifth Avenue
New York, NY 10003

Chapman & Hall
2-6 Boundary Row
London SE1 8HN
England

Thomas Nelson Australia
102 Dodds Street
South Melbourne, 3205
Victoria, Australia

Chapman & Hall GmbH
Postfach 100 263
D-69442 Weinheim
Germany

International Thomson Editores
Campos Eliseos 385, Piso 7
Col. Polanco
11560 Mexico D. F.
Mexico

International Thomson Publishing - Japan
Hirakawacho-cho Kyowa Building, 3F
1-2-1 Hirakawacho-cho
Chiyoda-ku, 102 Tokyo
Japan

International Thomson Publishing Asia
221 Henderson Road #05-10
Henderson Building
Singapore 0315





1 2 3 4 5 6 7 8 9 10 XXX 01 00 99 98

**Library of Congress Cataloging-in-Publication Data**

Mycobacteria / edited by P.R.J. Gangadharam, P.A. Jenkins.
p. cm. -- (Chapman & Hall medical microbiology series)
Includes bibliographical references and index.
Contents: v. 1. Basic aspects — v. 2. Chemotherapy.
ISBN 0-412-05451-5 (v. 1: alk. paper). ISBN 0-412-05441-8 (v. 2 : alk. paper)
1. Mycobacterial diseases. I. Gangadharam, Pattisapu Rama Jogi,
II. Jenkins, P.A., (P. Anthony) III. Series
{DNLM: 1. Tuberculosis, Pulmonary. 2. Mycobacterium Infections.
3. Mycobacterium. WF 300 M995 1997}
QR201.M96M93 1997
616'.01474--dc21
DNLM/DLC
for Library of Congress
97-3970
CIP

**British Library Cataloguing in Pubication Data available**

To order this or any other Chapman & Hall book, please contact **International Thomson Publishing, 7625 Empire Drive, Florence, KY 41042.** Phone: (606) 525-6600 or 1-800-842-3636. Fax: (606) 525-7778. e-mail: order@chaphall.com.

For a complete listing of Chapman & Hall's titles, send your request to **Chapman & Hall, Dept. BC, 115 Fifth Avenue, New York, NY 10003.**

To My Parents,
Mr. Pattisapu Venkata Joga Rao
Mrs. Pattisapu Kameswaramma

Who, despite their great fear that I may become a serious victim to this dreadful disease, inspired me to continue research in this field.

Pattisapu Rama Jogi Gangadharam

To,

The Staff of the Tuberculosis Reference Laboratory (1959–1977) and the Mycobacterium Reference Unit (1977–1995) in the Public Health Laboratory, Cardiff who contributed so much to the quality of clinical mycobacteriology in the United Kingdom.

P. Anthony Jenkins

# Contents

# Preface

Tuberculosis had been recognized many centuries before Robert Koch first isolated the tubercle bacillus in 1882. However, it is Koch's discovery which initiated a period of intensive scientific activity that resulted in a variety of methods for detecting the tubercle bacilli by microscopy, by animal inoculation, and by in vitro culture. However, the disease remained a scourge for many more decades and it was not until after the 1940s that effective treatment regimens became available. Not withstanding the development of these regimens, problems have been encountered, especially in their implementation in developing countries, resulting in multidrug resistance, which has become a global problem in the treatment of tuberculosis. In this volume, we have attempted to bring together, with the help of many colleagues who are recognized in their speciality areas, the current state of our knowledge on the basic biology of pathogenic mycobacteria in this volume. The chemotherapy of tuberculosis is elaborately discussed in the other volume.

As the clinical aspects and treatment of the disease became better understood, parallel development took place on general biology of mycobacteria. The "rough, tough, buff" colonial characteristics of tubercle bacilli as described by Koch were paramount until Timpe and Runyon, in 1954, described their four groups of "atypical" mycobacteria. This led to a more intensive study of mycobacteria other than tubercle bacilli, and the number of species grew rapidly. The taxonomic status of some of these species was, however, uncertain and it is largely due to the efforts of the International Working Group of Mycobacterial Taxonomy (IWGMT) that major progress has been made in this area. This group, led by Larry Wayne, has published a number of cooperative studies, where computer-assisted numerical analysis has been used to determine the relationships among a wide variety of mycobacterial species and groups.

Developments in the understanding of the pathology and immunology of mycobacterial infection were slow until comparatively recently, hindered by the lack of techniques to investigate the mechanisms at the genetic level. Over the last decade, this deficiency has been remedied and major advances have been made in unraveling the sequence of events that take place in mycobacterial infection. This upsurge of interest has also been stimulated by the advent of the Acquired Immune Deficiency Syndrome (AIDS) and the concommitent development of multidrug-resistant (MDR) tuberculosis. Despite the many advances that have been made, major problems remain. Rapid methods for the diagnosis of infection and the detection of drug resistance are not routinely available. BCG vaccination has been claimed a success in some countries but not others, and the reasons for this are still not known. Can more effective vaccines be artificially constructed?

We have invited a wide range of international experts to present "state-of-the-art" reviews on each of the important basic aspects of mycobacteria. Coverage and critical analysis of the progress made in recent years have been given greater importance. Because rapid progress is taking place in some fields, it is inevitable that any text can soon become outdated. Every effort has been made to keep such occurrences to a minimum. It is also inevitable that some author's opinion will contradict those of others, as well as some overlap between chapters will occur. Rather than try to avoid this, the editors feel that it is more profitable for readers to have available both differing and convergent viewpoints.

The editors are grateful to the many authors who, despite their own heavy professional commitments, spared their time, efforts, and energies to submit their chapters on time. The editors are conscious of the time factor and the delays that arise in a multiauthored publication. To those who submitted manuscripts promptly, we offer our apologies for these delays. Our special thanks go also to some authors who persevered through illness either of their own or close family members. We sincerely hope that this publication will be an important sourcebook on the basic concepts of mycobacteria.

P.R.J. Gangadharam
P.A. Jenkins

# Acknowledgments

We would like to thank Dr. C.A. Reddy, the Consulting Editor for the Chapman & Hall Medical Microbiology Series, who recognized the importance and timeliness of doing these volumes on mycobacteria, recruited us as the editors, read many of the chapters, and offered constructive suggestions, and has constantly given the needed encouragement to bring these books to fruition. We are also indebted to Mr. Gregory Payne, who was Publisher of Life Sciences at Chapman & Hall and who was invaluable in launching this effort, and to Mr. Henry Flesh (Projects Editor), Ms. Lisa LaMagna (Managing Editor), Ms. Kendall Harris (Editorial Assistant), Barbara Tompkins (Copyeditor), and the able production staff at Chapman & Hall for successfully completing the project.

# Contributors

Burson R. Andersen, Section of Infectious Diseases, Department of Medicine, University of Illinois at Chicago College of Medicine, Chicago, IL. USA.

Frank M. Collins, Mycobacteriology Laboratory, Food and Drug Administration, Bethesda, MD, USA.

Jack T. Crawford, Mycobacteriology laboratory, National Center for Infectious Diseases, Center for Disease Control and Prevention, Atlanta, GA, USA.

Joseph O. Falkinham *111,* Fralin Biotechnology Center, Virginia Polytechnic Institute and State University, Blacksburg, VA, USA.

Pattisapu R.J. Gangadharam, Mycobacteriology Research Laboratories, Department of Medicine, University of Illinois at Chicago College of Medicine, Chicago, IL, USA.

Michael Goodfellow, Department of Microbiology, The Medical School, Framlington Place, Newcastle upon Tyne NE2 4HH, UK.

John M. Grange, Department of Microbiology, National Heart and Lung Institute, Royal Brompton Hospital, Sydney Street, SW3 6NP, UK.

Leonid B. Heifets, Mycobacteriology Laboratory, National Jewish Center for Immunology and Respiratory Medicine and University of Colorado Health Sciences, Denver, CO, USA.

William R. Jarvis, Hospital Infections Program. Centers for Disease Control and Prevention, Atlanta, GA, USA.

Peter A. Jenkins, Formerly, Mycobacteriology Reference Unit, Public Health Laboratory Service, University Hospital of Wales, Health Park, Cardiff, Wales, CF4 4XW13, UK.

Isabel N. de Kantor, Panamerican Institute for food Protection and Zoonoses, INPPAZ, Panamerican Health Organisation (PAHO/WHO), C.C. 44, (1640) Martinez, Buenos Aires, Argentina.

Adalbert Laszlo, National Reference Center for Tuberculosis, Laboratory Center for Disease Control, Ottawa, Ontario, Canada K1A0L2.

John G. Magee. Regional Tuberculosis Reference Center, Regional Public Health Laboratory, Westgate Road, Newcastle upon Tyne NE2 4BE, UK.

Venkata M. Reddy, Mycobacteriology Research Laboratories, Department of Medicine, University of Illinois at Chicago, Chicago IL, USA.

Thomas M. Shinnick, Division of Bacterial and Mycotic Diseases, National Center for Infectious Diseases, Centers for Disease Control and Prevention, Atlanta, GA, USA.

John L. Stanford, Department of Bacteriology, University College London Medical School, London WIP 7LD, UK.

# Biographical Sketch of Dr. P.R.J. Gangadharam

Dr. Pattisapu Rama Jogi Gangadharam was born in Vijayawada, Andhra Pradesh, India. He obtained his B.Sc. (Hons.) and M.Sc. degrees from the Andhra University and his Ph.D. degree from the Bombay University. His research work, which deals almost entirely with mycobacteria and the diseases they cause, spans well over 4 decades, starting from his doctoral work on the chemotherapy of tuberculosis at the Indian Institute of Science, Bangalore, India.

He had been in the first team of international scientists who had established the Tuberculosis Chemotherapy Centre, now called the Tuberculosis Research Centre, at Madras, India, where he participated as a senior member with a major scientific responsibility in many of the controlled clinical trials in tuberculosis. Subsequently, he worked at the Baylor College of Medicine, Houston, Texas as the Director of Mycobacteriology Laboratories at the Harris County Hospital District. This was followed by a move to the National Jewish Hospital, now called the National Jewish Center for Immunology and Respiratory Medicine, Denver, Colorado to become the Director of Mycobacteriology Research and a Professor of Medicine and Clinical Pharmacology at the University of Colorado School of Medicine. Denver, Colorado. Currently, he is working as the Director of Mycobacteriology Research and Professor of Medicine, Microbiology and Pathology at the University of Illinois at Chicago College of Medicine, Chicago, Illinois.

He is an author of around 250 publications and several textbook chapters and editorials. His other single-author book on *Drug Resistance in Mycobacteria* has been a best seller and is widely used all over the world as a reference book on the subject.

Besides participating in several controlled clinical trials, he has made several significant contributions of international recognition. These include the following: urine tests for antimycobacterial drugs to facilitate monitoring compliance in drug intake; serum levels and metabolism of antimycobacterial drugs to guide and monitor clinical response and development of toxicity; discovery and development of animal models for mycobacterial diseases (the beige mouse model discovered by him has been acclaimed as the most valuable animal model for the *Mycobacterium avium* complex); discovery of new antimycobacterial drugs (a drug discovered by him has been named after him); targeted drug delivery using several types of liposomes and sustained drug delivery using several types of biodegradable polymers. Some of his contributions (e.g., peak serum levels) have given the scientific basis for the intermittent chemotherapy of tuberculosis and have been acclaimed by international journals like the *Lancet.*

He has been a recipient of several national and international awards which include the First Robert Koch Centennial Award and the Burroughs Wellcome Award in Microbiology. He is listed in the World's Who is Who from antiquity to the present, and among the Notable Americans of the Bi-centennial Era.

He is a member of several scientific societies, which include The American Thoracic Society, American Society for Microbiology, International Union Against Tuberculosis, Tuberculosis Association of India, and the India Chest Society. He has served on several subcommittees, which include Standards for Tuberculosis Laboratories, of the American Thoracic Society, Safety in Tuberculosis Laboratories of the International Union Against Tuberculosis, and the Scientific Advisory Committee of The American Foundation for AIDS Research.

# Dr. P.A. Jenkins

Dr. P.A. (Tony) Jenkins was born in Cardiff in 1936, took an honors degree in microbiology at the University of Wales and then commenced a study of Farmer's lung (extrinsic allergic alveolitis) for a Ph.D. at the Brompton Hospital under Professor Jack Pepys. He was awarded a Ph.D. in 1964 by the University of London.

In 1965, he was appointed senior scientist at the Tuberculosis Reference Laboratory (Public Health Laboratory Service) in Cardiff and eventually became head of the laboratory in 1977 when the previous director, Dr. Joe Marks, retired. The laboratory, which was renamed the Mycobacterium Reference Unit, remained in Cardiff until Dr. Jenkins retired after 30 year involvement in mycobacteriological diagnosis and research. He has authored or coauthored over 100 papers and articles and presented papers at conferences throughout the world. He was secretary and chairman of the Bacteriology and Immunology Committee of the IUATLD and founder secretary of the European Society for Mycobacteriology. In 1990, he was made president of the Thoracic Society of Wales.

On his small holding in West Wales he now raises pigs and sheep but also keeps in touch with developments in the mycobacterial field and maintains his contributions to the literature.

# 1

# Taxonomy of Mycobacteria

*Michael Goodfellow and John G. Magee*

## 1. Introduction

*Mycobacterium* (Gr. n. *myces* a fungus; Gr. neut. dim. n. *bakterion* a small rod; M.L. neut. n. *Mycobacterium* a fungus rodlet) is undeniably one of the most clinically important and intensively studied of bacterial taxa. Tuberculosis and leprosy, the most significant diseases caused by mycobacteria, have been recognized throughout recorded times. The taxonomic history of the genus *Mycobacterium* is intricate and difficult to disentangle from that of related taxa, notably *Corynebacterium, Nocardia,* and *Rhodococcus* (1). Comprehensive reviews on the early taxonomic history of these genera are available (2–7).

The clinical interest in mycobacteria started with the work of Koch (8) who detected the tubercle bacillus in stained infected tissue and cultivated the organism on inspissated serum medium. From thereon, there was an understandable, but nonetheless clinical bias which promoted a tendency to see different taxonomic kinds of mycobacteria solely in terms of their relationships with *M. tuberculosis.* This taxonomic skew was reflected in the common use of the term "atypical mycobacteria" for strains which could not be identified as either *M. bovis* or *M. tuberculosis.* It was not until the mid-fifties that a much needed classification was introduced for mycobacteria (9–11). In Runyon's classification, mycobacteria, apart from those in the *M. tuberculosis* complex and members of noncultivable taxa, were divided into four overtly artificial groups based on growth rates and pigmentational properties. In recent years, Runyon's groups have been superseded by a more natural classification, which is the subject of the present chapter.

## 2. Circumscription of the Genus

The genus *Mycobacterium* was proposed by Lehmann and Neumann (12) to include the tubercle and leprosy bacilli, organisms which had previously been classified as *Bacterium leprae* (13) and *Bacterium tuberculosis* (14). Subsequently, several hundred mycobacterial species were described, but only 41 of these were included in the Approved Lists of Bacterial Names (15). The number of validly described species of *Mycobacterium* currently stands at 71 (Table 1.1). With the exception of *M. leprae,* which has yet to be cultivated in vitro, mycobacteria can be assigned to two broad taxonomic groups based primarily on the growth rates of members of individual species (16,17). Rapidly growing strains include members of species which, under optimal nutrient and temperature regimes, produce, from dilute inocula, grossly visible colonies on solid media within 7 days, whereas their slowly growing counterparts need at least a week to form colonies under comparable conditions (18). Rapidly growing mycobacteria, which are common saprophytes in natural habitats, have received much less attention than the clinically more relevant slow growers.

Lehmann and Neumann, like many who followed, saw mycobacteria as aerobic, asporogenous rods which were usually acid-alcohol-fast at some stage in the growth cycle. These features were considered sufficient to distinguish mycobacteria from strains of *Corynebacterium diphtheriae* and *Nocardia asteroides*, the causal agents of diphtheria and nocardiosis, respectively. Corynebacteria were seen as non-acid-fast organisms, whereas nocardiae were defined as actinomycetes which produced a substrate mycelium that fragmented into irregular elements. These "differential properties" were to hold sway in the systematics of corynebacteria, mycobacteria, and nocardiae for years to come (5–7). It eventually became apparent that the dependency placed on a few morphological and staining characters was such that elements of the genera *Corynebacterium, Mycobacterium,* and *Nocardia* were virtually interchangeable (1).

It was evident by the early 1950s that better taxonomic criteria were needed for the classification of mycobacteria and related actinomycetes. The introduction and application of a number of chemotaxonomic techniques proved to be the spur for a reappraisal of actinomycete systematics. The determination of wall sugar and amino acid composition (19) and peptidoglycan structure (20) led to the assignment of actinomycetes to several wall groups (chemotypes) and peptidoglycan types (21). Simple wall composition analyses provided the first unambiguous evidence of a close relationship among corynebacteria, mycobacteria, and nocardiae (22); all such strains contain major amounts of *meso*-diaminopimelic acid (*meso*-$A_2$pm), arabinose, and galactose; that is, they have a wall chemotype IV *sensu* Lechevalier and Lechevalier (19), an $A1_\gamma$ peptidoglycan (20) and muramic acid residues that are *N*-glycolated rather than *N*-acetylated as in most other *meso*-$A_2$pm-containing bacteria (23).

Wall chemotype IV actinomycetes classified in the genera *Corynebacterium, Dietzia, Gordona, Mycobacterium, Nocardia, Rhodococcus,* and *Tsukamurella* contain mycolic acids (24–27), that is, high-molecular-weight (30–90 carbon atoms), long-chain, 3-hydroxy fatty acids with an aliphatic side chain at position 2

Table 1.1 Validly published species of *Mycobacterium*

| Species | Authors | Reference | Type strain[a] |
|---|---|---|---|
| **A. Fast-growing mycobacteria** | | | |
| **Pathogenic species** | | | |
| *M. abscessus* | (Moore and Frerich 1953) Kusunoki and Ezaki 1992 | 139, 252 | ATCC 19977 |
| *M. chelonae* | Bergey et al. 1923 | 141 | ATCC 35752 |
| *M. farcinogenes*[b] | Chamoiseau 1973 | 265 | NCTC 10955 |
| *M. fortuitum* subsp. *acetamidolyticum* | Tsukamura et al. 1986 | 109 | ATCC 35931 |
| *M. fortuitum* subsp. *fortuitum* | Da Costa Cruz 1938 | 266 | ATCC 6841 |
| *M. mucogenicum* | Springer et al. 1995 | 251 | ATCC 49650 |
| *M. peregrinum* | Kusunoki and Ezaki 1992 | 139 | ATCC 14467 |
| *M. porcinum*[b] | Tsukamura et al. 1983 | 240 | ATCC 33776 |
| *M. senegalense*[b] | (Chamoiseau 1973) Chamoiseau 1979 | 265, 267 | ATCC 35796 |
| **Nonpathogenic species** | | | |
| *M. agri* | Tsukamura 1981 | 334 | ATCC 27406 |
| *M. aichiense* | Tsukamura et al. 1981 | 264 | ATCC 27280 |
| *M. alvei* | Ausina et al. 1992 | 98 | CIP 103464 |
| *M. aurum* | Tsukamura 1966 | 310 | ATCC 23366 |
| *M. austroafricanum* | Tsukamura et al. 1983 | 304 | ATCC 33464 |
| *M. brumae* | Luquin et al. 1993 | 106 | CIP 103465 |
| *M. chitae* | Tsukamura 1967 | 313 | ATCC 19627 |
| *M. chlorophenolicum* | (Apajalahti et al. 1984) Häggblom et al. 1994 | 100, 263 | DSM 43826 |
| *M. chubuense* | Tsukamura et al. 1981 | 264 | ATCC 27278 |
| *M. confluentis* | Kirschner et al. 1992 | 104 | DSM 44017 |
| *M. diernhoferi* | Tsukamura et al. 1983 | 304 | ATCC 19340 |
| *M. duvalii* | Stanford and Gunthorpe 1971 | 322 | ATCC 43910 |
| *M. fallax* | Lévy-Frébault et al. 1983 | 107 | ATCC 35219 |
| *M. flavescens* | Bojalil et al. 1962 | 119 | ATCC 14474 |
| *M. gadium* | Casal and Calero 1974 | 321 | ATCC 27726 |
| *M. gilvum* | Stanford and Gunthorpe 1971 | 322 | ATCC 43909 |
| *M. komossense* | Kazda and Müller 1979 | 323 | ATCC 33013 |
| *M. madagascariense* | Kazda et al. 1992 | 102 | ATCC 49865 |
| *M. moriokaense* | Tsukamura et al. 1986 | 109 | ATCC 43059 |
| *M. neoaurum* | Tsukamura 1972 | 305 | ATCC 25795 |
| *M. obuense* | Tsukamura et al. 1981 | 264 | ATCC 27023 |

Table 1.1 (*continued*)

| Species | Authors | Reference | Type strain[a] |
|---|---|---|---|
| *M. parafortuitum* | Tsukamura 1966 | 339 | ATCC 19686 |
| *M. phlei* | Lehmann and Neumann 1899 | 316 | ATCC 19249 |
| *M. poriferae* | Padgitt and Moshier 1987 | 108 | ATCC 35087 |
| *M. pulveris* | Tsukamura et al. 1983 | 335 | ATCC 35154 |
| *M. rhodesiae* | Tsukamura et al. 1981 | 264 | ATCC 27024 |
| *M. smegmatis* | (Trevisan 1889) Lehmann and Neumann 1899 | 314, 316 | ATCC 19420 |
| *M. sphagni* | Kazda 1980 | 324 | ATCC 33027 |
| *M. thermoresistibile* | Tsukamura 1966 | 310 | ATCC 19527 |
| *M. tokaiense* | Tsukamura et al. 1981 | 264 | ATCC 27282 |
| *M. vaccae* | Bönicke and Juhasz 1964 | 311 | ATCC 15483 |
| **B. Slow-growing mycobacteria** | | | |
| **Pathogenic species** | | | |
| *M. africanum* | Castets et al. 1969 | 157 | ATCC 25420 |
| *M. asiaticum* | Weiszfeiler et al. 1971 | 193 | ATCC 25276 |
| *M. avium* subsp. *avium* | (Chester 1901) Thorel et al. 1990 | 140, 174 | ATCC 25291 |
| *M. avium* subsp. *paratuberculosis* | (Bergey et al. 1923) Thorel et al. 1990 | 141, 174 | ATCC 19698 |
| *M. avium* subsp. *silvaticum*[b] | Thorel et al. 1990 | 174 | CIP 103317 |
| *M. bovis* | Karlsen and Lessel 1970 | 158 | ATCC 19210 |
| *M. branderi* | Koukila-Kähkölä et al. 1995 | 105 | ATCC 51789 |
| *M. celatum* | Butler et al. 1993 | 99 | ATCC 51131 |
| *M. conspicuum* | Springer et al. 1995 | 221 | DSM 44136 |
| *M. genavense* | Böttger et al. 1993 | 69 | ATCC 51234 |
| *M. haemophilum* | Sompolinsky et al. 1978 | 214 | ATCC 29548 |
| *M. intermedium* | Meier et al. 1993 | 70 | DSM 44049 |
| *M. interjectum* | Springer et al. 1993 | 71 | DSM 44064 |
| *M. intracellulare* | (Cuttino and McCabe 1949) Runyon 1965 | 175, 176 | ATCC 13950 |
| *M. kansasii* | Hauduroy 1955 | 198 | ATCC 12478 |
| *M. lepraemurium*[b] | Marchaux and Sorel 1912 | 177 | None |
| *M. malmoense* | Schröder and Juhlin 1977 | 212 | ATCC 29571 |
| *M. marinum* | Aronson 1926 | 159 | ATCC 927 |
| *M. microti*[b] | Reed 1957 in Breed et al. 1957 | 160 | ATCC 19422 |
| *M. scrofulaceum* | Prissick and Masson 1956 | 228 | ATCC 19981 |
| *M. shimoidei* | Tsukamura 1982 | 230 | ATCC 27962 |

Table 1.1 (*continued*)

| Species | Authors | Reference | Type strain[a] |
|---|---|---|---|
| *M. simiae* | Karasseva et al. 1965 | 194 | ATCC 25275 |
| *M. szulgai* | Marks et al. 1972 | 213 | ATCC 35799 |
| *M. tuberculosis* | (Zopf 1883) Lehmann and Neumann 1896 | 12, 14 | ATCC 27294 |
| *M. ulcerans* | MacCallum et al. 1950 | 161 | ATCC 19423 |
| *M. xenopi* | Schwabacher 1959 | 231 | ATCC 19250 |
| **Non-pathogenic species** | | | |
| *M. cookii* | Kazda et al. 1990 | 101 | ATCC 49103 |
| *M. gastri* | Wayne 1966 | 197 | ATCC 15754 |
| *M. gordonae* | Bojalil et al. 1962 | 119 | ATCC 14470 |
| *M. hiberniae* | Kazda et al. 1993 | 103 | ATCC 49874 |
| *M. nonchromogenicum* | Tsukamura 1965 | 205 | ATCC 23067 |
| *M. terrae* | Wayne 1966 | 197 | ATCC 15755 |
| *M. triviale* | Kubica et al. 1970 | 74 | ATCC 23292 |
| **C. Non-cultivable species** | | | |
| *M. leprae* | (Hansen 1880) Lehmann and Neumann 1896 | 12, 13 | ND |

[a]ATCC, American Type Culture Collection, Rockville, Md., USA.; CIP, Collection National de Cultures de Microorganisms, Paris, France; DSM, Deutsche Sammlung von Mikroorganismen, Braunschweig, Germany; NCTC, National Collection of Type Cultures, Central Public Health Laboratory, London, UK. ND, not determined.
[b]Pathogenic for animals.

(Fig. 1.1). Mycolic acids vary in size and structure, and several techniques of varying degrees of complexity have been developed to recognize the different types (25,26,28). Nonmycobacterial mycolic acids have relatively simple structures and vary in chain length (22–74 carbon atoms) and in the number of cis double bonds (0–5). Mycobacterial mycolates have between 60 and 90 carbon atoms, occur in a variety of structural types (Fig. 1.2), and do not have more than two points of unsaturation (24–26). On pyrolysis, mycolic acid methyl esters yield aldehydes and long-chain fatty acid methyl esters (Fig. 1.3). The mycolic acid methyl esters of mycobacteria yield $C_{22}$ to $C_{26}$ fatty acid methyl esters on pyrolysis whereas those from the other mycolic acid-containing taxa release shorter fatty methyl acid esters (24–26).

Fatty acid, polar lipid, and isoprenoid quinone analyses also provide data of taxonomic value for the classification of mycobacteria and related actinomycetes (21,24,29). Mycolic acid-containing actinomycetes contain relatively simple mixtures of straight-chain and unsaturated fatty acids although 10-methylocta-

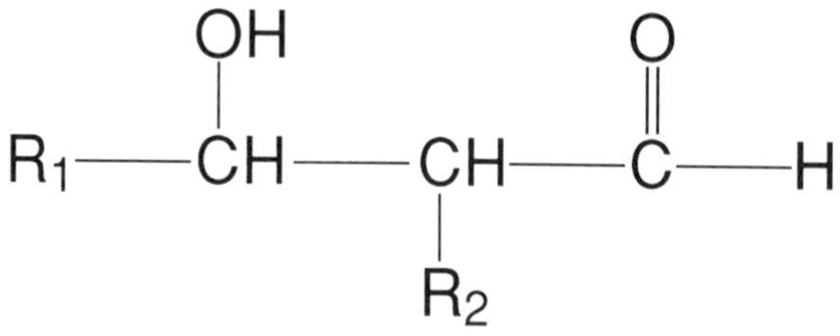

$$R_1-\underset{|}{\overset{OH}{CH}}-\underset{R_2}{CH}-\overset{O}{\overset{\|}{C}}-H$$

Figure 1.1. General formula for mycolic acids. $R_1$ and $R_2$ represent variable side chains.

decanoic (tuberculostearic) acid is also widely encountered. These organisms typically contain diphosphatidylglycerol, phosphatidylethanolamine, phosphatidylinositol, and phosphatidylinositol mannosides although most corynebacteria lack phosphatidylethanolamine. Actinomycetes can be assigned to five phospholipid groups based on semiquantitative analyses of major lipid markers found in whole-organism extracts (30,31).

The menaquinones of wall chemotype IV, mycolic acid-containing actinomycetes fall into a number of distinct patterns (24,29,32). Most animal corynebacteria, gordonae, and mycobacteria have dihydrogenated menaquinones with nine isoprene units, abbreviated as MK9($H_2$), as the main component, but the remaining corynebacteria, dietziae, and rhodococci contain MK8($H_2$) as the predominant isoprenolog. Tsukamurellae ("aurantiaca" strains) are characterized by the presence of fully unsaturated menaquinones with nine isoprene units [MK9; (33)] and nocardiae by hexahydrogenated menaquinones with eight isoprene units, the latter two of which are cyclized (34,35).

The ability to resist decolorization by acidic ethanol following staining with basic fuchsin is a characteristic feature of mycobacteria and some closely related actinomycetes. The acid-alcohol properties of these bacteria could be related to the lipid barrier of the wall mycolyl-arabinogalactan hindering the penetration of the acid (3,36,37). This means that differences in the degree of acid-alcohol-fastness between different organisms might reflect variations in the chemical nature of their walls.

A tighter definition of the genus *Mycobacterium* can be given in light of present knowledge. Mycobacteria are aerobic to microaerophilic actinomycetes which usually form slightly curved or straight, nonmotile rods [(0.2–0.6) × (1.0–10) $\mu$m]. Branching and myceliallike growth may take place with fragmentation into rods or coccoid elements. Cells are acid-alcohol-fast at some stages of growth, and are usually considered gram positive although they are not readily stained by Gram's method. Mycobacteria do not form capsules, conidia, or endospores, rarely exhibit visible aerial hyphae, are catalase positive, produce acid from sugars oxidatively, and, with the exception of strains which do not grow in vitro, can be

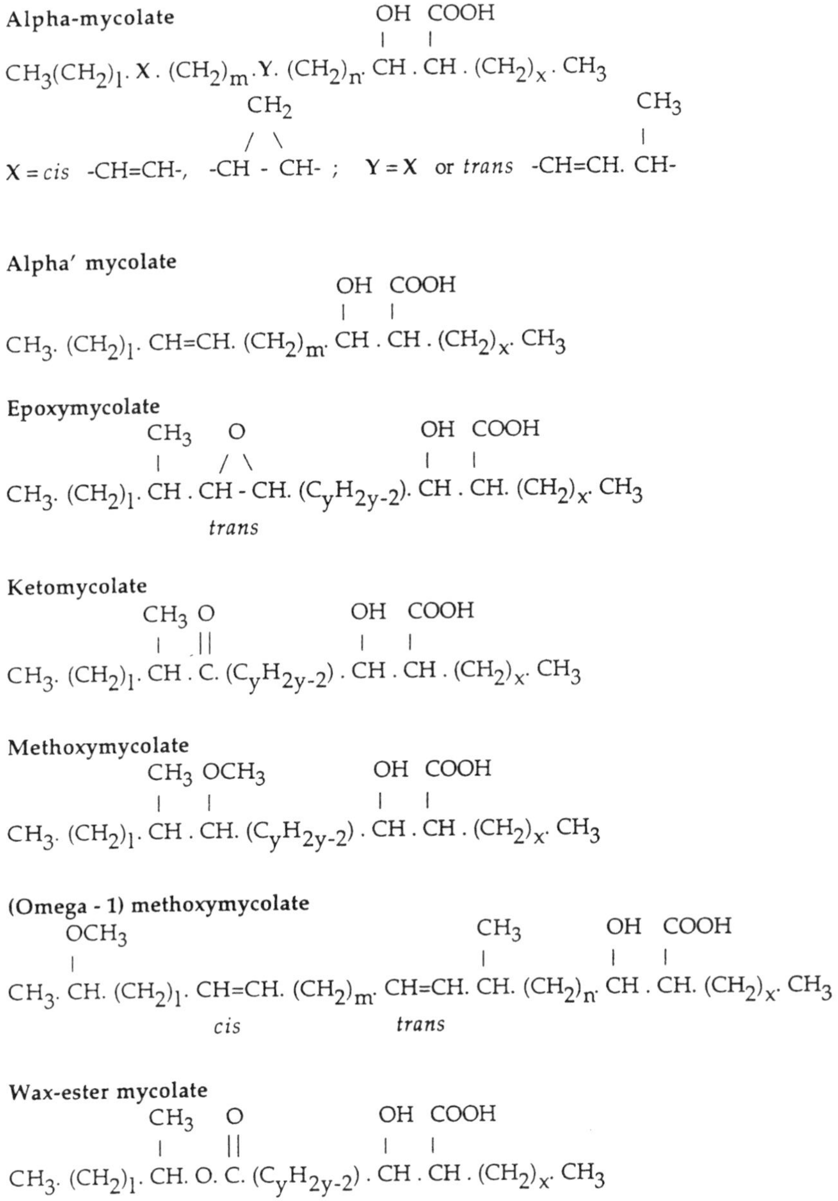

**Alpha-mycolate**

$$\begin{array}{l} \qquad\qquad\qquad\qquad\qquad\qquad\quad OH \;\; COOH \\ \qquad\qquad\qquad\qquad\qquad\qquad\quad\; | \qquad | \\ CH_3(CH_2)_l . X . (CH_2)_m . Y . (CH_2)_n . CH . CH . (CH_2)_x . CH_3 \end{array}$$

$$\begin{array}{l} \qquad\qquad\quad CH_2 \qquad\qquad\qquad\qquad\qquad CH_3 \\ \qquad\qquad\quad / \;\; \backslash \qquad\qquad\qquad\qquad\qquad\quad | \\ X = cis\ \text{-CH=CH-},\ \text{-CH - CH-} ;\quad Y = X \text{ or } trans\ \text{-CH=CH. CH-} \end{array}$$

**Alpha′ mycolate**

$$\begin{array}{l} \qquad\qquad\qquad\qquad\qquad OH \;\; COOH \\ \qquad\qquad\qquad\qquad\qquad\; | \qquad | \\ CH_3 . (CH_2)_l . CH=CH . (CH_2)_m . CH . CH . (CH_2)_x . CH_3 \end{array}$$

**Epoxymycolate**

$$\begin{array}{l} \qquad\quad CH_3 \quad O \qquad\qquad\qquad\quad OH \;\; COOH \\ \qquad\quad\; | \qquad / \; \backslash \qquad\qquad\qquad\quad | \qquad | \\ CH_3 . (CH_2)_l . CH . CH - CH . (C_yH_{2y-2}) . CH . CH . (CH_2)_x . CH_3 \\ \qquad\qquad\qquad\quad trans \end{array}$$

**Ketomycolate**

$$\begin{array}{l} \qquad\quad CH_3 \; O \qquad\qquad\quad OH \;\; COOH \\ \qquad\quad\; | \quad\; || \qquad\qquad\quad\; | \qquad | \\ CH_3 . (CH_2)_l . CH . C . (C_yH_{2y-2}) . CH . CH . (CH_2)_x . CH_3 \end{array}$$

**Methoxymycolate**

$$\begin{array}{l} \qquad\quad CH_3 \; OCH_3 \qquad\quad OH \;\; COOH \\ \qquad\quad\; | \qquad | \qquad\qquad\quad | \qquad | \\ CH_3 . (CH_2)_l . CH . CH . (C_yH_{2y-2}) . CH . CH . (CH_2)_x . CH_3 \end{array}$$

**(Omega - 1) methoxymycolate**

$$\begin{array}{l} \quad OCH_3 \qquad\qquad\qquad\qquad\qquad\qquad CH_3 \qquad\quad OH \;\; COOH \\ \quad\; | \qquad\qquad\qquad\qquad\qquad\qquad\qquad\; | \qquad\qquad\; | \qquad | \\ CH_3 . CH . (CH_2)_l . CH=CH . (CH_2)_m . CH=CH . CH . (CH_2)_n . CH . CH . (CH_2)_x . CH_3 \\ \qquad\qquad\qquad\quad cis \qquad\qquad\qquad trans \end{array}$$

**Wax-ester mycolate**

$$\begin{array}{l} \qquad\quad CH_3 \quad O \qquad\qquad\qquad OH \;\; COOH \\ \qquad\quad\; | \qquad\; || \qquad\qquad\qquad\; | \qquad | \\ CH_3 . (CH_2)_l . CH . O . C . (C_yH_{2y-2}) . CH . CH . (CH_2)_x . CH_3 \end{array}$$

Figure 1.2. Structural types of mycolic acids found in mycobacteria. The values of l, m, n, x, and y are in the range 11 to 35 carbons, all mycolates have a limited range of homologues. The units "($C_yH_{2y-2}$)" incorporate either *cis* double bonds or cyclopropane rings or *trans* double bonds or cyclopropane rings with an adjacent methyl branch in the central part of the chain.

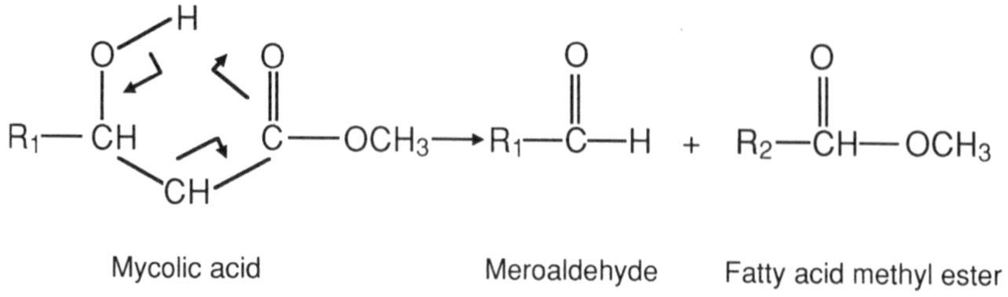

Figure 1.3. Pyrolysis of mycolic acid methyl esters.

divided into rapid and slow growers. Many strains form whitish to cream-colored colonies, but the presence of carotenoid pigments in some strains, notably rapid growers, leads to the formation of bright yellow- or orange-colored colonies. In some cases, the pigments are only produced in response to light (photochromogenic species), but most members of pigmented species also form these pigments in the dark (scotochromogenic species). Diffusible pigments are rarely found.

The wall peptidoglycan contains *N*-glycolated muramic acid, major amounts of *meso*-diaminopimelic acid, arabinose, and galactose, and is the A1$_\gamma$ murein type. Mycobacteria contain major proportions of straight-chain saturated and unsaturated fatty acids with 10-methyloctadecanoic acid; diphosphatidylglycerol, phosphatidylethanolamine, phosphatidylinositol, and phosphatidylinositol mannosides as major polar lipids; dihydrogenated menaquinones with nine isoprene units as the predominant isoprenolog, and mycolic acids with 60–90 carbon atoms and up to 2 double bonds. The fatty acids released on pyrolysis gas chromatography of mycolic acid esters have between 22 and 26 carbon atoms. The genus is also distinguished by characteristic antigenic patterns. The guanine (G) plus cytosine (C) ratio of the DNA ranges from 61 to 71 mole%.

The genus encompasses saprophytes, as well as facultative and obligate pathogens.

The type strain of the genus is *Mycobacterium tuberculosis* (Zopf, 1883 [14]) Lehmann and Neumann, 1896, 363 (12).

One of the responsibilities of the taxonomic subcommittees of the International Committee on Systematic Bacteriology of the International Union of Microbiological Societies is to recommend minimal standards for the publication of new species (38). This practice is intended to prevent the literature becoming cluttered with inadequately described species, as was the case prior to the publication of the *Approved Lists of Bacterial Names* (15). Such standards need to include tests for the establishment of generic identity and for the diagnosis of species. Currently, minimal standards for assigning presumptive mycobacteria to the genus *Mycobacterium* involve the demonstration of (a) acid-alcohol-fastness, (b) the presence of mycolic acids containing 60–90 carbons, which are cleaved to $C_{22}$ to $C_{26}$ fatty acid methyl esters by pyrolysis, and (c) a mole G + C ratio between 61% and 71% (18).

## 3. Suprageneric Relationships

The sequences of genes that code for ribosomal (r) RNA are relatively highly conserved and, hence, can be used to establish taxonomic relationships at the suprageneric level. Two fundamental premises underline this approach, namely that lateral gene transfer has not occurred between 16S rDNA genes and that the extent of evolution or dissimilarity between 16S rRNA sequences of a given pair of organisms is representative of the variation shown by the corresponding genomes. The good congruence found between phylogenies based on 16S rRNA and those derived from studies on alternative molecules, such as ATPase subunits, elongation factors, 23S rRNA, and RNA polymerases, lends substance to this latter point (39). It also seems likely that lateral gene transfer between 16S rDNA genes will be rare, as this gene is responsible for the maintenance of functional and tertiary structural consistency (40).

Analyses of 16S rRNA sequence data show that mycolic acid-containing actinomycetes form a well-defined clade within the evolutionary radiation occupied by actinomycetes (41–45). Mycobacteria form a distinct phyletic line within this clade and have a phylogenetic depth comparable to that of other mycolic acid-containing taxa (Fig. 1.4). These genera, which can be distinguished from one another using a combination of chemical and morphological properties (Table 1.2), are currently assigned to three families, namely Corynebacteriaceae Lehmann and Neumann 1907 (46) (*Corynebacterium* and *Dietzia*), Mycobacteriaceae Chester 1897 (47) (*Mycobacterium*), and Nocardiaceae Castellani and Chalmers 1917 (48) (*Gordona, Nocardia, Rhodococcus,* and *Tsukamurella*).

The taxonomic status of the families Corynebacteriaceae, Mycobacteriaceae, and Nocardiaceae need to be reassessed in light of marked improvements in the classification of mycolic acid-containing actinomycetes. The latter can be assigned to two groups based on the discontinuous distribution of certain chemical markers. Corynebacteria and dietziae have short mycolic acids (22–38 carbon atoms) with N-acetylated muramic acid in their peptidoglycan, whereas gordonae, mycobacteria, nocardiae, rhodococci, and tsukamurellae contain relatively long mycolic acids (34–90 carbon atoms) and N-glycolated muramic acid (Table 1.2).

The congruence between the chemical and molecular sequence data provide the kernel of a proposal to assign mycolic acid-containing actinomycetes to two suprageneric taxa, namely the families Corynebacteriaceae and Mycobacteriaceae (61). The family Corynebacteriaceae encompasses the genera *Corynebacterium* and *Dietzia* and the family Mycobacteriaceae, the genera *Gordona, Mycobacte-*

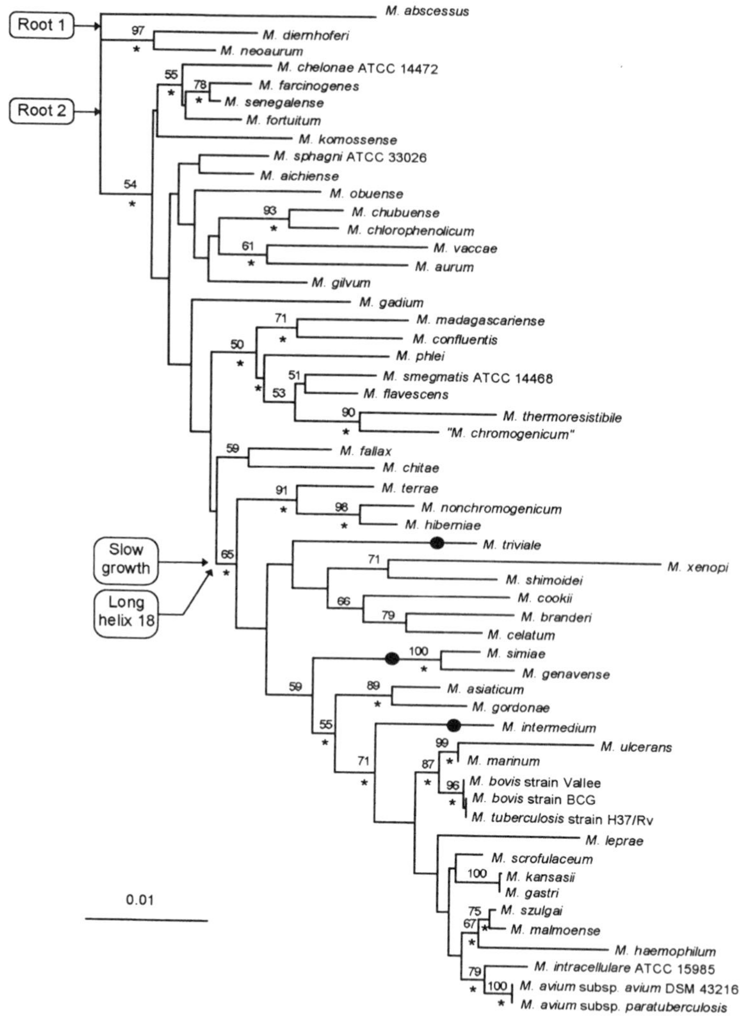


Figure 1.4. An unrooted phylogenetic tree of the genus *Mycobacterium*. The evolutionary distance matrix was calculated according to Jukes and Cantor (49) and the tree generated using the neighbor-joining method (50). The branches with asterisks were ones that were also recovered in the analyses based on the least squares (51) and maximum parsimony methods (52). The bootstrap analysis was carried out using the Jukes and Cantor distance model and the neighbor-joining analysis on 1000 branches. The tree was based on 1227 nucleotide sites. The test strains are type cultures except where the strain information is specified. The scale bar indicates 0.01 substitutions per nucleotide positions. The black circles represent the presence of the short version of helix 18. Root position 1 is based on the use of *Gordona* (6 sequences), *Nocardia pinensis* (1 sequence), and *Rhodococcus* (8 sequences) sequences as outgroups and root position 2 on the use of *Nocardia* (10 sequences) and *Tsukamurella* (2 sequences) sequences.

Table 1.2 Characteristics of genera encompassing wall chemotype IV actinomycetes containing mycolic acids[a]

| | Family | | | | | | |
|---|---|---|---|---|---|---|---|
| | *Corynebacteriaceae* | | *Mycobacteriaceae* | | *Nocardiaceae* | | |
| Characteristics of genus | *Corynebacterium* | *Dietzia* | *Myobacterium* | *Gordona* | *Nocardia* | *Rhodococcus* | *Tsukamurella* |
| Cell morphology | Straight to slightly curved rods which reproduce by snapping division; club-shaped elements may also be formed | Rods and cocci absent | Slightly curved or straight rods, sometimes branching filaments that fragments into rods and coccold elements | Rods and cocci | Substrate mycelium, fragments into rods and cocci | Rods to extensively branched mycelium; the latter fragment into irregular rods and cocci | Straight to slightly curved rods occur singly, in pairs, or in masses |
| Aerial hyphae | Absent | Absent | Usually absent[b] | Absent | Present | Absent | Absent |
| Appearance of visible colonies | 1–2 days | 1–3 days | 2–40 days | 1–3 days | 1–5 days | 1–3 days | 1–3 days |
| Degree of acid-fastness (not necessarily also alcohol fastness) | Sometimes weakly acid-fast | Often partially acid-fast | Usually strongly acid-fast | Often partially acid-fast | Often partially acid-fast | Often partially acid-fast | Weak to strongly acid-fast |
| Strictly aerobic | – | + | + | + | + | + | + |
| Arylsulphatase produced | – | – | + | – | (–) | – | – |
| Peptidoglycan type[c] | $A1_\gamma$ | $A1_\gamma$ | $A1_\gamma$ | $A1_\gamma$ | $A1_\gamma$ | $A1_\gamma$ | $A1_\gamma$ |
| Acyl group of muramic acid[d] | *N*-acetylated | *N*-acetylated | *N*-glycolated | *N*-glycolated | *N*-glycolated | *N*-glycolated | *N*-glycolated |
| Fatty acid types[e] | S, U, (T)[f] | S, U, T | S, U, T[g] | S, U, T | S, U, T | S, U, T | S, U, T |
| Mycolic acid type[h] | Single spot | Single spot | Multiple spots | Single spot | Single spot | Single spot | Two spots |
| Overall size (number of carbons) | 22–38 | 34–38 | 60–90 | 46–66 | 48–60 | 34–52 | 64–78 |
| Number of double bonds | 0–2 | ND | 1–3[i] | 1–4 | 0–3 | 0–4 | 1–6 |
| Fatty esters released on pyrolysis (number of carbons) | 8–18 | ND | 22–26 | 16–18 | 12–18 | 12–16 | 20–22 |

Table 1.2 (*continued*)

| Characteristics of genus | Family | | | | | | |
|---|---|---|---|---|---|---|---|
| | *Corynebacteriaceae* | | *Mycobacteriaceae* | *Gordona* | *Nocardiaceae* | | *Tsukamurella* |
| | *Corynebacterium* | *Dietzia* | *Myobacterium* | | *Nocardia* | *Rhodococcus* | |
| Phospholipid type[j] | 1 | © | 2 | 2 | 2 | 2 | 2 |
| Predominant menaquinone(s)[k] | MK-8($H_2$), -9($H_2$) | MK-8($H_2$) | MK-9 ($H_2$) | MK-9($H_2$) | MK-8 ($H_4$, cycl)[l] | MK-8($H_2$) | MK-9 |
| GC content (mol%) | 51–59 | 63–69 | 62–70 | 63–69 | 64–72 | 67–73 | 67–68 |
| Sensitivity to: | | | | | | | |
| 5-Fluorouracil (20 $\mu$g/ml) | ND | ND | ND | + | − | + | − |
| Lysozyme (50 $\mu$g/ml) | ND | ND | − | + | − | + | − |
| Mitomycin (50 $\mu$g/ml) | ND | ND | ND | + | − | + | − |

[a]Data taken from Goodfellow (53), Boiron et al. (54), Holt et al. (55), and Rainey et al. (27). Symbols: +, positive; −, negative; (−), some strains give positive results.
[b]*Mycobacterium farcinogenes* and *Mycobacterium xenopi* can occasionally produce aerial hyphae.
[c]A, cross-linkage between positions 3 and 4 of adjacent peptide subunits; 1, peptide bridge absent; $\gamma$, *meso*-DAP at position 3 of the tetrapeptide subunits
[d]Acyl group detected using the glycolate test (56):
[e]Abbreviations: S, straight chain; U, monounsaturated; T, tuberculostearic acid (10-methyloctadecanoic) acid; parentheses indicates variable occurrence.
[f]*Corynebacterium bovis* contains tuberculostearic acid (30).
[g]*Mycobacterium gordonae* lacks substantial amounts of tuberculostearic acid (57).
[h]Number of mycolic acid spots on TLC plates.
[i]In mycobacterial mycolic acids, double bonds may be converted to cyclopropane rings; methyl branches and other oxygen functions may be present (58–60).
[j]Phospholipid types: 1, phosphatidylglycerol (variable) and phosphatidylinositol; 2, phosphatidylethanolamine (30,31); © contain an unusual pattern consisting of phosphatidyethanolamine, diphosphatidylglycerol, phosphatidylglycerol, phosphatidylinositol, and phosphatidylinositol mannosides (27).
[k]Abbreviations exemplified by MK-8($H_2$), menaquinones having two of the eight isoprene units hydrogenated.
[l]Nocardiae were originally reported to have predominant amounts of MK-8($H_4$). However, the major component was later shown to correspond to a hexahydrogenated menaquinone with eight isoprene units in which the end two units of the multiprenyl side chain were cyclized (34,35).

*rium, Nocardia, Rhodococcus,* and *Tsukamurella. Corynebacterium amycolatum* (62) and *Turicella otitidis* (63), which lack mycolic acids, fall within the evolutionary radiation encompassed by the family Corynebacteriaceae (44,45). It is also evident from these studies that new taxonomic niches are needed for actinomycetes currently classified as *Nocardia pinensis* and *Tsukamurella wratislaviensis*.

## 4. Intrageneric Relationships

The 16S rRNA similarity studies confirm that the traditional division between fast- and slow-growing mycobacteria is a natural one [Fig. 1.4; (64–67)]. The branching point of the rapidly growing members is close to the root of the mycobacterial subline which shows that these organisms are phylogenetically older than their slow-growing relatives. Most slow growers have an extended 16S rRNA helix at positions 451 to 482 [*Escherichia coli* numbering system; (68)], the exceptions, *M. genavense* (69), *M. intermedium* (70), *M. interjectum* (71), *M. simiae* (65), and *M. triviale* (66) have the short-helix version characteristic of rapidly growing mycobacteria. It is possible that these species have lost the long-helix independently as they are not clustered together in Figure 1.4.

Few phenotypic properties have been found to underpin the division between fast- and slow-growing mycobacteria. However, rapid growers can be separated from slow growers in terms of lipid structure, as shown by differences in resistance to decolorization of bacilli stained with neutral red and treated with alkali (72). This is the basis of the microcolonial neutral red test, the results of which correlate well with growth rates; the slow-growers are more resistant to decolorization. Several other properties show an incomplete correlation with growth rate. Some rapid growers absorb large amounts of iron from an iron-rich medium and develop a rusty color, whereas slow growers do not (73). In addition, many rapid growers can tolerate 5% w/v NaCl in egg medium although only *M. triviale* shares this attribute among the slower growers (74).

An explanation for the differences in the rates of bacterial multiplication between the two mycobacterial groups is still awaited. Possible reasons include differences in key respiratory pathways or energy production, in diffusion of oxygen and nutrients across the cell envelope, in the rate at which certain lipids are synthesized and assimilated into the cell wall, and differences in the number of rRNA (rrn) operons. There is no clear indication from metabolic investigations why *M. leprae* fails to grow in vitro.

There is a broad correlation between the number of ribosomes and the rate of growth of a bacterial strain (76). The number of ribosomes is determined by the production of rRNA, which, in turn, depends on the number of rrn operons, the strength of their promoters, and the efficiency with which the operons are transcribed. It has been shown that fast-growing mycobacteria, such as *M. phlei* and *M. smegmatis*, have two operons (rrnA and rrnB), whereas slow-growing strains, like *M. leprae* and *M. tuberculosis*, have a single rrn operon (76). The rrn operon of slow-growing mycobacteria has the classical structure, namely leader region, 16S rRNA gene, intergene spacer-1, 23S rRNA gene, intergene spacer-2, and 5S rRNA gene (77). There are clear differences between fast- and slow-growing mycobacteria in the sizes of the leader, spacer-1, and spacer-2 regions (78). On the basis of this latter study, it was proposed that the emergence of the slow growers from the main mycobacterial phyletic line was coincident with the deletion of a segment of DNA spanning an rrnB-like operon, leaving an rrn-A-like operon as the sole source of rRNA. The leader and spacer region sequences are highly conserved and could prove to be a useful supplement in establishing phylogenetic relationships between closely related mycobacterial species.

The case for retaining the fast- and slow-growing mycobacteria in a single genus far outweighs grounds for assigning them to two genera, as proposed by Tsukamura (79). The genus *Mycobacterium* is well circumscribed as a result of the application of an impressive array of chemotaxonomic, molecular, and numerical taxonomic methods, most notably in the extensive studies undertaken under the auspices of the International Working Group on Mycobacterial Taxonomy (80–91). Mycobacteria are most accurately characterized by the chemical structure of their mycolic acids (24,26,92) and their antigenic structure (93–95).

Methanolysates of mycobacterial mycolic acids can be resolved into several groups based on the presence or absence of different functional groups in the longer carbon chain of the mycolic acid molecule (Table 1.3). Both one-dimensional (96) and two-dimensional (97) thin-layer chromatographic (TLC) methods have been used for this purpose. The $\alpha$ and $\alpha$ mycolic acids, which lack oxygen functions apart from the 3-hydroxy and carboxy units, are the least polar compounds. In the more polar mycolic acids, the longer chain is substituted with carboxy, epoxy, keto, and methoxy functions. The analysis of mycobacterial mycolic acid methyl esters by TLC provides a sensitive and practical way of establishing mycolic acid patterns, but for many species, additional strains need to be examined to confirm the consistency of such patterns for the characterization of mycobacterial species. Partial congruence is found between the distribution of mycolic acid types and the finer taxonomic relationships between mycobacteria shown in 16S rRNA similarity studies.

Mycobacteria also appear to be unique in producing two different siderophores (exochelins and mycobactins), probably as a response to the lipid-rich nature of the cell envelope. Exochelins are extracellular siderophores, which up to now have received little attention. Two types of siderophore have been described which differ in their solubility in organic solvents (112). In contrast, the highly hydrophobic mycobactins, which are found within the cell envelope, have received much more attention. TLC analysis of mycobactins is straightforward and, based

Table 1.3. Mycobacterial mycolic acid TLC patterns

| Pattern of structural types | Species |
|---|---|
| $\alpha$ only | *M. brumae, M. fallax, M. triviale* |
| $\alpha$, $\alpha'$ | *M. abscessus, M. chelonae* |
| $\alpha$, $\alpha'$, methoxy | *M. agri* |
| $\alpha$, $\alpha'$, epoxy | *M. chitae, M. confluentis, M. farcinogenes, M. fortuitum* subsp. *fortuitum, M. fortuitum subsp. acetamidolyticum, M. peregrinum, M. porcinum, M. senegalense, M. smegmatis* |
| $\alpha$, $\alpha'$, keto | *M. genavense, M. malmoense, M. simiae* |
| $\alpha$, $\alpha'$, keto, wax esters[a] | *M. chubuense, M. diernhoferi*[b], *M. duvalii, M. gilvum, M. obuense, M. parafortuitum, M. shimoidei, M. vaccae* |
| $\alpha$, keto, wax esters[a] | *M. aichiense, M. aurum, M. austroafricanum, M. avium* subsp. *avium, M. avium* subsp. *paratuberculosis, M. avium* subsp. *silvaticum, M. branderi, M. chlorophenolicum, M. conspicuum, M. flavescens, M. gadium, M. hiberniae, M. interjectum, M. intracellulare, M. komossense, M. lepraemurium, M. madagascariense, M. moriokaense, M. nonchromogenicum, M. neoaurum, M. phlei, M. poriferae, M. pulveris, M. rhodesiae, M. scrofulaceum, M. sphagni, M. terrae, M. tokaiense, M. xenopi* |
| $\alpha$, keto, | *M. leprae* |
| $\alpha$, keto, methoxy | *M. africanum, M. asiaticum, M. bovis, M. bovis BCG*[b], *M. celatum, M. gastri, M. gordonae, M. haemophilum, M. kansasii, M. marinum, M. microti, M. szulgai, M. tuberculosis, M. ulcerans* |
| $\alpha$, $\alpha'$, keto, methoxy | *M. thermoresisibile* |
| $\alpha$, $\alpha'$, wax esters# | *M. cookii* |
| $\alpha$, $\omega$-1 methoxy | *M. alvei* |

*Source:* Adapted from Dobson *et al.* (60) to include data from Ausina *et al.* (98), Böttger *et al.* (69), Butler *et al.* (99), Häggblom *et al.* (100), Kazda *et al.* (101–103), Kirschner *et al.* (104), Koukila-Kähkölä *et al.* (105), Luquin *et al.* (106), Lévy-Frébault *et al.* (107), Meier *et al.* (70), Padgitt and Moshier (108), Springer *et al.* (71), Tsukamura *et al.* (109), Valero-Guillam *et al.* (110), and Yassin (personal communication) (111).

*Note: M. mucogenicum* does not appear in the table as data on the TLC mycolate pattern is not available.

[a]Wax ester mycolates are composed of $\omega$-carboxymycolates esterified to 2-eicosanol and homolog.

[b]conflicting results between different workers.

on data presented by Hall and Ratledge (113), provides chemotaxonomic information with a high discriminatory power. However, as in the case of mycolic acid patterns, additional strains need to be analyzed to confirm the chemotaxonomic value of mycobactin $R_f$ values.

Initial attempts to clarify the taxonomy of fast- and slow-growing mycobacteria triggered a worldwide interest in mycobacterial systematics which continues undiminished to the present time (114,115). The founding studies started with the work of Gordon and her colleagues (116–118), who examined many rapidly growing mycobacteria for a large number of tests adapted from those commonly used in nonmycobacterial systematics. These early workers stressed the importance of using overall patterns of test response rather than relying on a few "key" properties, thereby anticipating modern numerical taxonomy, in their successful bid to clarify the taxonomy of rapid growers.

In 1962, Bojalil and his colleagues (119) published the first numerical taxonomic study on the genus *Mycobacterium* with mixed results. Their test strains comprised rapid growers, which had been studied primarily by soil microbiologists, and slow growers, which were mainly of interest to clinical microbiologists. Most of their data were based on the biochemical tests Gordon had used for rapid growers and on the properties that underpin Runyon groups (9–11). The test data were analyzed using the Jaccard coefficient which excludes comparisons between strains based on negative matches. This technique was satisfactory for the classification of the rapid growers but not for the slow growers, which gave uniformly negative results in the tests taken from Gordon's work. Consequently, the slow-growing strains were assigned to clusters on the basis of very few properties, primarily those related to pigment formation (color and stimulation of synthesis by exposure to light). This classification was essentially a reaffirmation of the Runyon groups, that is, slow-growing photochromogens (Group 1), slow-growing nonphotochromogens (Group II), and slow-growing nonphotochromogens (Group III).

This apparent impasse led Wayne (120) to assemble a case for developing a taxonomy of slow-growing mycobacteria which was based on tests specifically designed for this purpose, rather than relying on techniques adapted from those used for other genera. He subsequently noted that differences in reaction rates of a variety of tests were not simple reflections of growth rates but that in some cases, the reaction rates were sufficiently characteristic of members of individual species that they could be used for taxonomic purposes (121). As a consequence of this and later studies, different batteries of tests are now used for the classification of rapid- and slow-growing mycobacteria (see Tables 1.4 and 1.5).

Wayne (120) also recognized that the contributions of individual practitioners needed to be coordinated if the taxonomy of mycobacteria were to be clarified. The International Working Group on Mycobacterial Taxonomy (IWGMT) was formed to provide a vehicle for an ongoing series of collaborative studies. The

Table 1.4. Characteristics differentiating slow-growing species of mycobacteria

| | M. africanum | M. bovis | M. marinum | M. microti | M. tuberculosis | M. ulcerans | M. avium subsp. avium | M. avium subsp. paratuberculosis | M. avium subsp. silvaticum | M. intracellulare | M. lepraemurium | M. asiaticum | M. gordonae | M. gastri | M. kansasii | M. hiberniae | N. nonchromogenicum | M. terrae | M. genavense | M. interjectum | M. simiae | M. haemophilum | M. malmoense | M. szulgai | M. branderi | M. celatum | M. conspicuum | M. cookii | M. intermedium | M. leprae | M. scofulaceum | M. shimodei | M. triviale | M. xenopi |
|---|---|---|---|---|---|---|---|---|---|---|---|---|---|---|---|---|---|---|---|---|---|---|---|---|---|---|---|---|---|---|---|---|---|---|
| Clade | | | —A— | | | | | | —B— | | | C | | D | | | E | | | F | | | G | | | | | | —Single membered— | | | | | |
| Growth & Pigmentation Tests | | | | | | | | | | | | | | | | | | | | | | | | | | | | | | | | | | |
| Growth at: 25°C | – | – | V | – | – | + | + | — | — | + | | + | + | + | + | + | + | + | V | – | | + | V | + | + | + | + | V | + | / | + | – | + | – |
| Growth at: 42°C | – | – | – | – | – | – | + | + | + | V | – | | – | – | V | – | V | – | + | – | | – | – | – | + | + | – | – | + | / | – | + | V | + |
| Growth at: 45°C | – | – | – | – | – | – | V | | – | V | – | – | – | – | – | – | – | – | | – | | – | – | – | + | + | – | – | – | / | – | V | – | + |
| Microaerophilic growth | + | + | | | – | + | – | | | – | | | – | | – | | | + | | | | | + | + | | | | | | | + | | | |
| Pigmentation | N | N | P | N | N | N | N | N | N | N | N | P | S | N | P | S | N | N | N | S | P | N | N | S | N | N | S | S | P | / | S | N | N | N |
| Enzyme Tests | | | | | | | | | | | | | | | | | | | | | | | | | | | | | | | | | | |
| Alpha esterase | + | + | – | + | + | – | + | | | + | | + | + | – | – | | – | V | | + | V | | – | V | | | | | + | / | V | | | + |
| Acid phosphatase | V | – | + | – | – | V | – | – | – | – | – | + | V | + | + | + | + | + | | – | – | – | – | + | | + | V | | + | / | – | + | + | – |
| Aryl sulphatase (10 days) | – | V | + | – | – | + | – | – | – | + | – | | V | + | + | + | V | + | – | V | + | – | V | V | | | + | + | + | / | V | – | + | + |
| Aryl sulphatase (14 days) | | | + | | | + | | | | + | – | | | + | + | | + | + | – | | + | | + | V | + | + | + | + | | / | V | – | + | + |
| Catalase (45mm foam) | – | – | V | – | – | – | – | – | – | – | – | + | + | – | + | | + | + | + | | + | – | – | + | – | – | – | + | | / | + | – | + | – |
| Catalase resists 68° | – | – | + | – | – | + | V | + | – | + | – | + | + | – | + | | + | + | + | + | + | – | – | + | | + | – | | + | / | + | + | + | + |
| Niacin | V | – | – | + | + | V | – | – | – | – | – | – | – | – | – | | – | – | – | – | V | – | – | – | – | – | | | | | – | – | – | – |
| Nictonamidase | V | – | + | + | + | – | + | + | | + | – | – | – | + | + | – | + | – | | + | + | + | + | + | + | | – | – | – | / | + | + | – | + |
| Nitrate reduction | V | – | – | V | + | – | – | – | – | – | – | – | – | – | + | + | – | + | – | – | – | – | – | + | | – | – | – | – | / | – | – | + | – |
| Pyrazinamide | V | – | + | + | + | – | + | + | | + | – | – | – | – | – | – | + | – | + | + | + | + | + | + | + | | – | – | – | / | + | + | – | + |
| Tellurite (8 days) | – | – | – | | – | | + | | + | + | | | | – | – | | – | – | | | | – | V | | | | – | | | / | – | | – | – |
| Tellurite (9 days) | | + | – | | + | | + | | + | + | | | | – | – | | + | + | | | | – | + | | | | – | | | / | + | | – | + |
| Tween (10 days) | V | V | + | V | V | – | – | V | – | – | – | + | + | + | + | V | + | + | – | – | – | – | | | – | – | + | – | + | / | – | + | + | – |
| Urease | + | + | + | + | + | – | – | – | – | – | – | – | – | + | + | – | – | – | + | + | + | – | V | + | – | – | – | – | + | / | + | – | – | – |

Table 1.4. Continued

| | M. africanum | M. bovis | M. marinum | M. microti | M. tuberculosis | M. ulcerans | M. avium subsp. avium | M. avium subsp. paratuberculosis | M. avium subsp. silvaticum | M. intracellulare | M. lepraemurium | M. asiaticum | M. gordonae | M. gastri | M. kansasii | M. hiberniae | N. nonchromogenicum |
|---|---|---|---|---|---|---|---|---|---|---|---|---|---|---|---|---|---|
| Clade | | | —A— | | | | | | —B— | | | C | | D | | | E |
| Resistance to: | | | | | | | | | | | | | | | | | |
| p-nitro-benzoic acid | − | − | | − | − | V | + | | + | + | | − | + | − | + | | V |
| Ethambutol (1 mg/l) | − | − | | − | − | + | + | + | + | + | | − | | − | + | V | + |
| Ethambutol (5 mg/l) | − | − | | − | − | V | + | + | + | + | | − | V | − | V | − | − |
| Hydroxylamine (125 mg/l) | − | − | | − | − | + | + | | + | + | | + | + | + | + | | + |
| Hydroxylamine (250 mg/l) | − | − | | − | − | + | + | | − | + | | + | | V | − | | + |
| Hydroxylamine (500 mg/l) | − | − | | − | − | + | V | | − | + | | + | V | − | − | | + |
| Isoniazid (1 mg/l) | − | − | | − | − | + | + | + | + | + | | + | + | − | − | + | + |
| Isoniazid (5 mg/l) | − | − | | − | − | | + | | + | + | | | V | − | − | + | + |
| Isoniazid (10 mg/l) | − | − | | − | − | V | + | | + | + | | V | V | − | − | + | + |
| Oleic acid (250 mg/l) | − | − | | − | − | − | + | | | + | − | + | V | − | − | | V |
| Thiophen-2-carboxylic acid hydrazide (5 mg/l) | − | − | | − | + | + | + | + | + | + | | + | + | + | + | | + |

| | M. terrae | M. genavense | M. interjectum | M. simiae | M. haemophilum | M. malmoense | M. szulgai | M. branderi | M. celatum | M. conspicuum | M. cookii | M. intermedium | M. leprae | M. scofulaceum | M. shimodei | M. triviale | M. xenopi |
|---|---|---|---|---|---|---|---|---|---|---|---|---|---|---|---|---|---|
| Clade | | | F | | | G | | | | | —Single membered— | | | | | | |
| Resistance to: | | | | | | | | | | | | | | | | | |
| p-nitro-benzoic acid | + | | | + | + | + | + | | + | | | | / | + | + | + | V |
| Ethambutol (1 mg/l) | + | + | | + | + | − | | | + | + | − | | / | + | | + | + |
| Ethambutol (5 mg/l) | − | + | | + | + | − | | | | − | − | | / | + | − | − | + |
| Hydroxylamine (125 mg/l) | + | | | + | + | + | | | | | | | / | + | V | | + |
| Hydroxylamine (250 mg/l) | + | | | + | + | + | | | | | | | / | + | − | | + |
| Hydroxylamine (500 mg/l) | + | | | + | − | V | − | | − | | | | / | + | − | | − |
| Isoniazid (1 mg/l) | + | + | | + | + | + | − | | + | + | − | | / | + | | + | − |
| Isoniazid (5 mg/l) | + | | | | | − | | | − | − | − | | / | + | | + | − |
| Isoniazid (10 mg/l) | + | | | V | | − | | | − | − | − | | / | + | | + | − |
| Oleic acid (250 mg/l) | V | | | V | | − | + | | | | | | / | + | | | − |
| Thiophen-2-carboxylic acid hydrazide (5 mg/l) | + | | | + | | + | + | | + | | | | / | + | + | + | + |

*Source:* Data from Goodfellow and Wayne (16), Magee (29), Wayne and Kubica (17) and from references cited for each species in the narrative text.

*Note:* Abbreviations: +, at least 85% strains positive; −, at least 85% strains negative; V, variable. Pigmentation: P, photochromogenic: S, scotochromogenic: N, nonpigmented.

Table 1.5. Characteristics differentiating fast-growing species of mycobacteria

| | M. abscessus | M. chelonae | M. chubuense | M. chlorophenolicum | M. confluentis | M. madagascariense | M. farcinogenes | M. fortuitum subsp. acetamidolyticum | M. fortuitum subsp. fortuitum | M. peregrinum | M. senegalense | M. diernhoferi | M. neoaurum | M. aurum | M. vaccae | M. aichiense | M. chitae | M. fallax | M. flavescens | M. gadium | M. gilvum | M. komossense | M. obuense | M. phlei | M. smegmatis | M. sphagni | M. agri | m. alvei | M. austroafricanum | M. brumae | M. duvalii | M. mucogenicum | M. moriokaense | M. parafortuitum | M. porcinum | M. poriferae | M. pulveris | M. rhodesiae | M. thermoresistibile | M. tokaiense |
|---|---|---|---|---|---|---|---|---|---|---|---|---|---|---|---|---|---|---|---|---|---|---|---|---|---|---|---|---|---|---|---|---|---|---|---|---|---|---|---|---|
| Clade | A | | B | | C | | | —D— | | | | E | | F | | | | | | | | | | | | | | —Single membered— | | | | | | | | | | | | |
| Enzyme Tests | | | | | | | | | | | | | | | | | | | | | | | | | | | | | | | | | | | | | | | | |
| Acetamidase | – | – | – | | + | – | | + | + | + | + | + | + | V | + | – | + | – | – | | – | – | – | V | + | – | V | | – | | – | | – | V | + | – | – | – | – | + |
| Acid phosphatase | + | + | – | + | + | + | | – | + | + | | + | – | – | – | + | + | | – | | V | + | – | + | – | + | – | | – | | – | | | – | – | | + | + | – | – |
| Aryl sulphatase 3 days | + | + | | + | – | + | – | + | + | + | + | – | – | V | V | + | – | – | V | – | – | – | + | – | – | + | V | + | + | – | – | + | + | – | + | – | – | + | – | + |
| Aryl sulphatase 7 days | + | + | + | + | – | + | – | + | + | + | + | + | – | V | + | + | + | | V | V | + | | + | V | + | + | | + | + | | – | + | + | + | + | | V | + | – | + |
| Aryl sulphatase 14 days | – | V | + | – | + | – | + | + | + | + | + | + | + | V | + | – | + | + | + | + | + | – | – | + | + | + | + | + | + | + | + | V | + | + | – | – | + | – | + | – |
| α-Esterase | V | V | + | | + | | | – | V | | | – | – | + | – | – | + | + | + | | – | + | V | + | – | – | + | | V | | – | | + | – | V | | + | – | + | + |
| β-Esterase | + | + | + | | – | | | + | V | | | + | + | + | V | V | + | + | + | | V | + | + | + | – | – | + | | + | | + | | + | V | + | | + | – | + | + |
| Nitrate reductase | V | – | – | | – | – | | – | V | V | | V | + | + | + | – | + | + | + | + | + | + | – | + | + | + | – | + | – | + | – | + | – | + | | | + | + | + | – |
| Tween hydrolysis | V | + | | | | | | | + | + | | + | | | – | | – | | V | | | | | + | – | | | | | | | | | – | | | | | – | |
| Hippurate hydrolysis | V | V | | | | | | | V | | | + | | | + | | | | – | | | | | – | + | | | | | | | + | | | | | | | | |

Table 1.5. *(continued)*

| Clade | M. abscessus | M. chelonae | M. chubuense | M. chlorophenolicum | M. confluentis | M. madagascariense | M. farcinogenes | M. fortuitum subsp. acetamidolyticum | M. fortuitum subsp. fortuitum | M. peregrinum | M. senegalense | M. diernhoferi | M. neoaurum | M. aurum | M. vaccae | M. aichiense | M. chitae | M. fallax | M. flavescens | M. gadium |
|---|---|---|---|---|---|---|---|---|---|---|---|---|---|---|---|---|---|---|---|---|
| | A | | B | | C | | | | —D— | | | E | | F | | | | | | |
| Growth at: | | | | | | | | | | | | | | | | | | | | |
| 42°C | – | – | – | | + | – | | + | V | V | | – | V | – | + | | – | – | V | |
| 45°C | – | – | – | – | – | – | – | – | – | – | – | – | – | – | – | – | – | – | – | – |
| 52°C | – | – | – | – | – | – | – | – | – | – | – | – | – | – | – | – | – | – | – | – |
| Pigmentation | N | N | S | S | N | S | S | N | N | N | S | N | S | S | P | S | N | N | S | S |
| Growth on: | | | | | | | | | | | | | | | | | | | | |
| MacConkey | + | + | | | | – | – | – | + | + | – | – | – | – | – | | – | – | – | |
| 5% w/v NaCl agar | + | – | | + | – | – | – | – | + | + | + | – | | V | + | | + | – | + | + |
| Utilisation tests: | | | | | | | | | | | | | | | | | | | | |
| Xylose | – | – | + | + | | + | | – | – | – | | + | V | + | V | – | – | – | – | |
| Trehalose | V | V | V | + | | + | | – | V | + | | – | + | V | V | V | – | – | – | |
| Mannitol | – | – | V | V | | + | | – | V | + | + | + | + | + | + | + | – | – | V | + |
| Sorbitol | – | – | V | V | | + | | – | – | – | | – | – | – | – | – | – | – | V | |
| Resistance to: | | | | | | | | | | | | | | | | | | | | |
| Hydroxylamine 500 | + | + | – | | | – | – | + | + | + | | – | + | V | – | – | – | + | – | |

| Clade | M. gilvum | M. komossense | M. obuense | M. phlei | M. smegmatis | M. sphagni | M. agri | m. alvei | M. austroafricanum | M. brumae | M. duvalii | M. mucogenicum | M. moriokaense | M. parafortuitum | M. porcinum | M. poriferae | M. pulveris | M. rhodesiae | M. thermoresistibile | M. tokaiense |
|---|---|---|---|---|---|---|---|---|---|---|---|---|---|---|---|---|---|---|---|---|
| | | | | | | | —Single membered— | | | | | | | | | | | | | |
| Growth at: | | | | | | | | | | | | | | | | | | | | |
| 42°C | | | | + | + | | + | – | – | – | | – | + | + | + | – | + | | + | |
| 45°C | – | – | – | + | + | – | + | – | – | – | – | – | – | – | – | – | – | – | + | – |
| 52°C | – | – | – | + | – | – | – | – | – | – | – | – | – | – | – | – | – | – | + | – |
| Pigmentation | S | S | S | S | N | S | N | N | S | N | S | N | N | P | N | S | N | S | S | S |
| Growth on: | | | | | | | | | | | | | | | | | | | | |
| MacConkey | | – | | – | – | – | – | | – | | – | + | | – | | | | | – | |
| 5% w/v NaCl agar | | – | | + | + | – | + | – | + | – | | – | + | + | | | | | + | |
| Utilisation tests: | | | | | | | | | | | | | | | | | | | | |
| Xylose | – | – | – | V | + | – | + | – | + | – | – | | – | V | – | + | – | + | – | + |
| Trehalose | – | + | V | V | + | V | – | + | – | + | – | | V | – | + | + | – | – | – | V |
| Mannitol | + | + | + | + | + | + | – | – | + | – | + | | + | + | + | + | – | + | – | + |
| Sorbitol | – | + | V | V | + | – | – | | – | | – | | + | – | – | + | – | – | – | + |
| Resistance to: | | | | | | | | | | | | | | | | | | | | |
| Hydroxylamine 500 | – | – | – | – | – | – | – | + | – | – | – | | – | – | + | – | – | – | – | – |

*Source:* Data from Goodfellow and Wayne (16), Magee (29), Wayne and Kubica (17) and from references cited for each species in the narrative text.
*Note:* Abbreviations: +, at least 85% strains positive; –, at least 85% strains negative; V, variable pigmentation; P, photochromogenic; S, scotochromogenic; N, nonpigmented.

results of the first full cooperative study, which focused on scotochromogenic slow growers of Runyon Group II, was presented at the 1969 conference of the IWGMT (80). This and all subsequent studies followed a set of guiding principles which have been described in detail elsewhere (16,122–124). In essence, replicate sets of 50–100 strains selected for each study were sent to each of the participants, all of whom agreed to examine them, under code, according to technical procedures that they chose and to forward their data to the coordinator of that study. No attempt was made to dictate which tests should be examined or how they were to be performed. The code was broken once all of the experimental work had been completed.

Most of the strains included in the collaborative studies were representative of validly described *Mycobacterium* species which had been proposed over the previous decade; the pooled data served primarily to underpin the taxonomic status and descriptions of these species. The tests with the greatest discriminatory power to distinguish between species were highlighted, but because each investigator was free to choose individual tests and methods for their performance, it was not possible to assess the interlaboratory reproducibility of the techniques. This point was addressed in a series of cooperative studies designed to determine test reproducibility under rigorously defined conditions; from these studies, a series of tests emerged that were highly reproducible (125,126). Data from all of these cooperative studies were used to generate frequency matrices for computer-assisted identification of slowly growing mycobacteria in clinical laboratories (127,128).

## 5. Toward a Polyphasic Species Concept

The species is usually considered to be the basic unit in biological classification, but the development of a universally accepted species concept for bacteria is proving to be a formidable task (129). Recent bacterial species concepts have tended to reflect the taxonomic methods used to classify individual strains. The dramatic impact wrought by the application of modern taxonomic methods, notably chemotaxonomy, molecular systematics, and numerical taxonomy, on bacterial classification is eloquent testimony to this point (130–132). Technique-driven approaches to the circumscription of bacterial species are reasonably sound in an operational sense but fail to take account of the fact that species are products of evolutionary processes. It is, therefore, the overall pattern of properties shown by cultivable bacteria not the processes which gave rise to them which is currently seen to be paramount in bacterial classification.

Ravin (133,134) recognized three very practical kinds of species: genospecies, which encompass mutually interfertile forms and correspond most closely to the biological species concept; nomenspecies, made up of individuals resembling the nomenclatural type strain; and taxospecies, groups of organisms which share a high proportion of common properties. Genospecies have received little attention from bacterial systematists and should not be confused with genomic species, that is, with organisms which share high DNA : DNA similarity values.

A comprehensive approach to the delineation of bacterial species makes use of both genotypic and phenotypic data, thereby building on the ideas of Ravin (134). This conceptual approach to bacterial classification, known as polyphasic taxonomy, was introduced by Colwell (135) to signify successive or simultaneous taxonomic studies on groups of organisms using appropriate data acquisition systems. By their very nature, polyphasic taxonomic studies can be expected to yield well-defined species, a stable nomenclature, and improved species definitions. In general, recent approaches to mycobacterial classification have had an implicitly polyphasic dimension.

Numerical taxonomic surveys, in particular those conducted by the IWGMT (80–91), have provided the framework for the current classification of mycobacteria at the species level. The primary aim of such studies is to assign individual mycobacterial strains to homogeneous groups or clusters, defined on the basis of overall similarity, which can be equated with taxospecies. The IWGMT cooperators realized that it was essential to evaluate the taxonomic integrity of taxospecies by examining representative strains using independent taxonomic criteria. For this reason, immunologically based tests were usually used to provide independent corroboration of the resultant classification. Additionally, but to a lesser extent, phage typing and chemical analyses served a similar function.

Bacterial species can also be defined in molecular terms. When DNA is dissociated into the single-stranded form and allowed to reassociate in the presence of single-stranded DNA from another organism, hybrid molecules (heteroduplexes) are formed if there is sufficient homology between the respective nucleotide base sequences. The amount of molecular hybrid formed and its stability on heating provides a measure of relatedness. The term genomic species is applied to strains, including the type strain, which share 70% or more DNA : DNA relatedness with a $\Delta T_m$ of 5°C or less [$T_m$ is the melting temperature of the heteroduplex as determined by stepwise denaturation; $\Delta T_m$ is the difference in $T_m$ in °C between the homologous and heterologous hybrids formed under standardized conditions (136)].

It is not always realized that DNA : DNA similarity values, which are sensitive to experimental conditions, do not reflect that actual degree of nucleotide similarity values at the level of the primary structure. Because thermal stabilities of heteroduplexes fall by about 1–1.5°C for every 1% of unpaired bases (137), bacterial DNA heteroduplexes are only formed when strains show DNA relatedness values of more than 50–70%. It has been estimated that organisms which have 70% or more DNA similarity will have at least 96% DNA sequence identity (138).

DNA : DNA relatedness studies have been used to evaluate the taxonomic status of mycobacterial species circumscribed in numerical taxonomic surveys and to provide evidence for establishing new species. Kusunoki and Ezaki (139), for example, used a number of DNA pairing techniques to unravel taxonomic relationships within the *M. fortuitum* complex. The resultant data together with corresponding phenotypic information led them to propose that "*M. peregrinum*," be revived as an independent species and that *M. chelonae* subspecies *abscessus* be raised to species status as *M. abscessus*. In contrast, *M. avium* Chester 1901 (140), *M. paratuberculosis* Bergey et al. 1923 (141), and the wood pigeon mycobacteria have been shown to belong to a single genomic species (142–144). DNA : DNA pairing data have provided powerful evidence for the recognition of *M. alvei* (198), *M. brumae* (106) *M. hiberniae* (103), and *M. interjectum* (71) as distinct species. In general, DNA similarity studies reveal a low degree of relatedness between slowly and rapidly growing mycobacteria (98,123).

The taxonomic relationships of representatives of potentially novel mycobacterial species are increasingly being determined by comparing the nucleotide sequences of their rRNA with corresponding sequences held in rRNA databases. The discovery of unique 16S rRNA sequences has provided valuable information for the recognition of several mycobacterial species, including *M. branderi* (105), *M. confluentis* (104), *M. cookii* (101), *M. intermedium* (70), *M. interjectum* (71), *M. madagascariense* (102), and *M. mucogenicum* (145).

The advantages of using 16S rRNA sequences for the delineation of new species need to be seen within the context of the limitations of the method (129). It is, for instance, not possible to set absolute 16S rRNA similarity values for the delineation of species because of different rates of nucleotide sequence divergence. It is also important to remember that strains belonging to different genomic species can have identical 16S rRNA sequences and that only complete or near-complete nucleotide sequences of the macromolecule allow reliable comparisons with corresponding results held in 16S rRNA databases. Information on 16S rRNA sequences can be accessed through the Internet (*http://www.bdt.org.br/structure/molecular.html*).

Diagnostic bacteriologists require rapid, accurate identification of clinically significant isolates with a minimum of effort and equipment, whatever the type and source of data needed to distinguish species. A new diagnostic mycobacteriology is emerging from improvements in the subgeneric classification of mycobacteria (146,147). Computer-assisted identification methods are available for the analysis of phenotypic data (127,128), whereas nucleic acid probes can be used to identify isolates at different levels of specificity (148–150). Antibody probes for individual proteins, such as T-catalase and $\alpha$ antigen protein, can provide a function similar to that served by RNA probes (151–153). Similarly, chromatographic analyses of fatty acids, including mycolic acids, provide valuable data for diagnostic purposes (154,155).

## 6. Classification of Slow-Growing Mycobacteria

The current classification of slow-growing mycobacteria is mainly based on an extensive series of polyphasic taxonomic studies, notably those carried out under the direction of the IWGMT. The initial IWGMT studies, which involved the application of modern taxonomic techniques to the circumscription and delineation of most of the validly described species of slow-growing mycobacteria (80,81,90), were successful although some strains failed to aggregate with any of the major clusters and were too few in number to permit assignment to new taxa. This problem was addressed in a number of open-ended IWGMT studies in which unusual or rarely encountered strains were thoroughly characterized using both phenotypic and genotypic methods (82–85). This strategy allowed expanded descriptions of several taxa to be drawn, notably *M. asiaticum, M. malmoense, M. shimodei, M. simiae*, and *M. szulgai*.

In the most recent IWGMT study, Wayne et al. (87) set out to address three major taxonomic challenges:

- To investigate the taxonomic status of a selected set of slow-growing mycobacteria that had proved difficult to classify using conventional phenotypic properties (85)
- To determine whether it was possible to establish consistent criteria for determining the depth of branching in 16S rRNA-based phylogenetic trees of slow-growing mycobacteria to help define species in a phylogenetic taxonomic sense
- To evaluate the potential of certain chemotaxonomic and molecular taxonomic methods, including restriction fragment-length polymorphism (RFLP) analysis of polymerase chain reaction (PCR) amplified DNA, for the identification of mycobacteria at species, subspecific, and infrasubspecific levels.

The good overall agreement found among the 16S rRNA sequencing, DNA : DNA relatedness, probe hybridization, antigenic binding, and phenotypic characterization data allowed 51 out of 66 "difficult" test strains to be assigned to known species. It was, therefore, concluded that the confidence placed in the polyphasic taxonomic approach to the delineation of mycobacterial species is fully warranted. However, the collaborators were unable to identify any single 16S rRNA interstrain nucleotide difference value that unequivocally defined species borders, although they were in a position to reaffirm the value of DNA : DNA relatedness studies in the delineation of bacterial species (129,131,136). They were also of the view that DNA : RFLP and multilocus enzyme electrophoretic analyses were useful for epidemiological and ecological studies where tracing individual strains is the primary objective.

The application of molecular systematic techniques, notably 16S rRNA sequencing, allows comparisons to be made between phenetic and phylogenetic classifications. With respect to the slow-growing mycobacteria, good congruence has been found between genotypic and phenotypic data, thereby confirming the taxonomic status of most of the currently recognized species. Indeed, 22 out of the 32 species of slow-growing mycobacteria can be assigned to seven multimembered clades; the taxonomic integrity of which is supported by several treeing algorithms and high bootstrap values (Fig. 1.4). Additional comparative studies are needed to determine the finer supraspecific relationships of the single-membered phyletic lines.

The separation of validly described species of slow-growing mycobacteria on the basis of 16S rDNA sequences is not only important in its own right but also opens up the prospect of developing PCR or oligonucleotide probes for the identification of unknown isolates from clinical material and environmental samples. In addition, the ability to obtain 16S rRNA sequence data from difficult to cultivate and uncultured mycobacteria provides a powerful way of highlighting additional centers of taxonomic variation within the evolutionary radiation formed by slow-growing mycobacteria. Minimal standards have been proposed for the description of slow-growing *Mycobacterium* species (18).

The proposition that nomenclature should reflect genomic relationships (129) and that all preconceived notions should be examined within this context (156) also applies to mycobacteria. Consequently, in the remaining part of this section, taxonomic pen-pictures of individual species of slow-growing mycobacteria are considered within the context of their perceived evolutionary relationships. This approach to classification is particularly valuable in the case of slow-growing mycobacteria, as it allows the distribution of existing diagnostic tests to be seen within the context of an emerging natural taxonomy (Table 1.4). It is, for instance, interesting that for the most part the constituent members of the multimembered clades have a common mycolic acid pattern (Table 1.3).

### *6.1 The* **Mycobacterium tuberculosis** *Clade*

This clade encompasses the following:

*M. africanum* Castets et al. 1969 (157)
*M. bovis* Karlson and Lessel 1970 (158)
*M. marinum* Aronson 1926 (159)
*M. microti* Reed in Breed et al. 1957 (160)
*M. tuberculosis* (Zopf 1883; [14]) Lehmann and Neumann 1896 (12,14)
*M. ulcerans* MacCallum et al. in Fenner 1957 (161)

Members of the *M. tuberculosis* complex, that is, *M. africanum, M. bovis, M. microti,* and *M. tuberculosis,* have many properties in common and form a phyletic line within the evolutionary radiation encompassed by the slow-growing mycobacteria (Fig. 1.4). All members of the complex appear to be obligate pathogens, as they have not been shown to replicate outside the animal or human body. They are also strict mesophiles showing little or no growth below 30°C or above 37°C. Strains assigned to the complex occupy a taxonomic position that is quite separate from that of all other validly described species classified in the genus *Mycobacterium.* The early taxonomic history of the four taxa constituting the complex has been reviewed by Grange (162).

Originally, *M. tuberculosis* included both the human and the bovine tubercle bacillus, but in 1970, the latter was given the separate specific epithet, *M. bovis* (158). Karlsen and Lessel (158) pointed out that the bovine and human strains differed in biochemical and cultural properties and in their virulence for a range of animals. They also correctly stated that the epidemiological differences in infection due to the two types made it important to distinguish between them. Indeed, until the advent of specific chemotherapy, the problem of human mycobacterial disease, other than leprosy, was overwhelmingly one associated with *M. tuberculosis,* which was spread between human beings by the airborne route, and with *M. bovis,* which could also be spread from infected cattle to humans by way of milk. Because of the differences in epidemiology of these infections and the attendant differences in the necessary control measures, a lot of emphasis was placed on finding methods to distinguish between these two groups of organisms. The net result was a weighting in the battery of tests used to characterize slow-growing mycobacteria with the techniques which tended to exaggerate differences between *M. bovis* and *M. tuberculosis* particularly evident in numerical taxonomic studies (127).

The situation outlined above was further complicated by the taxonomic position of the vaccine strain, *M. bovis* BCG (Bacille–Calmette–Guérin). This organism was derived from a nonvirulent varient of *M. bovis* (163) and in the ensuing years has been maintained and used independently in a number of laboratories throughout the world. When a sample of these isolates were reexamined, they formed a cluster which showed no greater matching affinity to *M. bovis* than to *M. tuberculosis* (127). Samples of BCG isolates have also been distinguished from *M. bovis* and *M. tuberculosis* in mycolic acid (164) and pyrolysis mass spectrometric analyses (165). In this latter study, the BCG isolates were found to be heterogeneous and more closely related to laboratory adapted strains of *M. tuberculosis* than to recent isolates of either *M. bovis* or *M. tuberculosis.*

The distinction between *M. bovis* and *M. tuberculosis* becomes even more tenuous when *M. africanum* and *M. microti* are considered. Castets *et al.* (157) proposed *M. africanum* for a group of strains causing human tuberculosis in tropical Africa and having properties intermediate between the human and bovine types. In a numerical taxonomic analysis of *M. africanum* (166), the test organisms were assigned to two main clusters; strains from West Africa (Dakar, Yaounde, and Mauritania) were similar in many respects to the human type, whereas those from Rwanda and Burundi more closely resembled the bovine type. Clinical isolates of *M. africanum* were found to be indistinguishable from those of *M. bovis* on the basis of pyrolysis mass spectrometry (165).

*Mycobacterium microti* was originally derived from a vole (*Microtus agrestis*) by Wells (167). A similar organism was isolated from the Cape hyrax or dassie (*Procavia capensis*), and in common with the vole bacillus, it has a very limited range of susceptible hosts (168). *Mycobacterium microti* has properties which are intermediate between *M. bovis* and *M. tuberculosis* and shows very low virulence in guinea pigs and humans. It has been used in vaccine trials and seems to give a similar measure of protection as that provided by *M. bovis* BCG.

The constituent members of the *M. tuberculosis* complex can be distinguished on the basis of their phenotypic and PyMS profiles, but the balance of taxonomic evidence clearly indicates that *M. africanum, M. bovis,* and *M. microti* should become subjective synonyms of *M. tuberculosis.* Thus, in cross-sensitivity studies using experimentally sensitized animals, sensitins (extracts) of *M. tuberculosis* cannot be separated from those of *M. africanum, M. bovis,* and *M. microti,* although they are readily distinguished from those of other *Mycobacterium* species (169). Similarly, immunodiffusion analysis of bacillary extracts show a pattern of antigens common to all members of the complex (93), whereas immunologic analysis of T-catalase extracted from *M. tuberculosis* indicates structural identity with that of *M. africanum* and *M. bovis,* and marked divergence from that of other *Mycobacterium* species (170). In addition, DNA relatedness values show *M. bovis* and *M. tuberculosis* to be virtually identical (171). Finally, members of all four taxa have common mycolic acid patterns (Table 1.3). Despite epidemiological and ecological considerations, it is only a matter of time before members of the *M. tuberculosis* complex are classified as a single species that encompasses appropriate subspecific taxa.

Members of the *M. tuberculosis* complex share a shallow branch of the phylogenetic tree of slow-growing mycobacteria with *M. marinum* and *M. ulcerans* (Fig. 1.4). The integrity of this phyletic line is supported by several treeing algorithms and a high bootstrap value. It is also interesting that the constituent species of this evolutionary line have a common mycolic acid pattern (Table 1.3) but can be distinguished by a number of phenotypic tests.

*Mycobacterium marinum* is a photochromogenic organism which causes disease of fish and self-limiting granulomatous skin lesions (swimming pool or fish tank granulomas) in humans; its natural habitat appears to be water. On primary isolation, it grows readily at 30°C although not at 37°C, but after several subcultures, it can become adapted to growth at 37°C. Members of the species form a discrete cluster in numerical taxonomic studies and have a unique agglutinating serovar (81,85). They can also be distinguished from other slow-growing mycobacteria by skin testing (172). A large structural divergence between the T-catalase of *M. marinum* and those of *M. avium* and *M. kansasii* has been detected serologically (173). *Mycobacterium marinum* strains have also been classified under the now invalid names "*M. balnei*" and "*M. platypoecilus*."

*Mycobacterium ulcerans,* a nonphotochromogenic organism, represents another species associated with skin lesions. However, it causes a more progressive malignant disease than *M. marinum* and is found in tropical regions, whereas *M. marinum* is common in temperate climates. *Mycobacterium ulcerans* grows at 30°C but not at 37°C and, in contrast, to *M. marinum,* it does not adapt to growth at 37°C on repeated subculture. Immunodiffusion analysis reveals a number of specific antigens which distinguish *M. ulcerans* strains from other slow-growing mycobacteria (93). Organisms described as "*M. buruli*" are now considered to belong to the species *M. ulcerans* (17).

### *6.2. The* Mycobacterium-avium *Clade*

This clade consists of the following:

*M. avium* subspecies *avium* (Chester 1901; [40])
Thorel et al. 1990 (174)
*M. avium* subspecies *paratuberculosis* (Bergey et al. 1923; [141])
Thorel et al. 1990 (174)
*M. avium* subspecies *silvaticum* Thorel et al. 1990 (174)
*M. intracellulare* (Cuttino and McCabe 1949; [175])
Runyon 1965 (176)

Associated species:
*M. lepraemurium* Marchoux and Sorel 1912 (177)

*Mycobacterium avium* was considered to be the avian variety of the "tubercle bacillus" until it became abundantly clear from DNA relatedness (178), numerical taxonomic (121), and catalase immunologic distance measurements (170) that it is a distinct taxon which is readily separable from members of the *M. tuberculosis* complex. In contrast, *M. avium* and *M. intracellulare* are often lumped together as the *M. avium-intracellulare* (MAI) complex. The detection of a third phenotypically similar group, designated cluster 4, which encompasses a number of organisms that do not react with nucleic acid probes for either *M. avium* or *M. intracellulare,* expands the scope of the MAI complex (85). Members of the MAI complex cause pulmonary disease in elderly patients, lymphadenitis in children, and disseminated disease in HIV-infected patients. *Mycobacterium scrofulaceum* is often associated with *M. avium* and *M. intracellulare* in the so-called MAIS complex (179). However, the 16S rRNA tree does not support this association (Fig. 1.4).

*Mycobacterium avium, M. intracellulare, M. lepraemurium, M. paratuberculosis*, cluster 4, and the wood pigeon bacillus are difficult to separate using phenotypic properties (87,90). Nevertheless, *M. avium* and *M. intracellulare* are clearly distinct species based on T-catalase serology (153,180,181) and DNA-relatedness data (182). In contrast, the extensive DNA and phenotypic relatedness found among *M. avium, M. paratuberculosis,* and the wood pigeon bacillus led Thorel et al. (174) to propose that these taxa be classified in one species with the designations *M. avium* subsp. *avium, M. avium* subsp. *paratuberculosis,* and *M. avium* subsp. *silvaticum,* respectively. There is also good evidence that *M. lepraemurium* should be reduced to a subspecies of *M. avium* (114). All members of the *M. avium clade*, including *M. intracellulare*, share a number of group specific antigens (93) and have a common mycolic acid pattern (Table 1.3).

The taxonomic structure of the MAI complex has been clarified, although not resolved, in a series of IWGMT cooperative studies (84–87). It appears that members of the complex are actively evolving, with the separation of the constituent taxa having been relatively recent, as reflected in the shallow branching in the 16S rRNA tree (Fig. 1.4), and with both phenotypic convergence and phenotypic divergence in evidence. The phenotypic diversity and intercluster overlap shown by members of the MAI complex might be a consequence of both subspecific variation in the structure of their chromosomal DNAs, as reflected in the different restriction fragment-length polymorphisms [RFLP (183)] and the presence of plasmids (185,186). Data from RFLP analyses (183) show that there are 3% or fewer base substitutions within the species *M. avium;* the corresponding figure for strains that fall into the *M. intracellulare* DNA homology group is 13%.

Members of the MAIS complex produce similar polar glycopeptidolipid surface antigens that are responsible for very specific seroagglutination of whole cells (187). These agglutination serovars have been the subject of detailed studies with over 30 serovars now recognized. Serotypes 1 to 28 are usually considered to belong either to *M. avium* or *M. intracellulare,* and serovars 41 to 43, to *M. scrofulaceum.* Wayne et al. (188) found that most strains belonging to serovars 1 to 6 and 8 to 11 were identified as *M. avium* using data from independent DNA relatedness (182), Gen-Probe (150), and T-catalase serologic studies (153,180). Similarly, most strains belonging to serovars 7, 12 to 20, and 23 to 25 were identified as *M. intracellulare,* and most belonging to serovars 41 to 43 were identified as *M. scrofulaceum.* However, there was evidence that strains of a given serovar can, on occasion, be placed in different species. Wards et al. (189) found that DNA restriction endonuclease analysis patterns within the *M. avium* complex appeared to reflect the serotypes of strains though the correlation was not perfect. Members of avium serovar 2 are the most virulent and commonest cause of disease in birds.

*Mycobacterium avium* subsp. *paratuberculosis* strains were originally found to be associated with chronic hypertrophic enteritis, or Johne's disease, of cattle by Johne and Frothingham (190) who considered them to be a variant of the "avian tubercle bacillus." The organism also causes enteritis in ruminants and has been implicated in the pathogenesis of Crohn's disease in humans (191). *Mycobacterium avium* subsp. *paratuberculosis* can be distinguished from other members of the *M. avium* complex by its stable mycobactin dependence (174) and the presence of multiple copies of a unique insertion element, IS 900 (183,191,192). Three RFLP types have been recognized in animal isolates of the organism using IS 900 as a probe. Bovine and ovine strains shared the same RFLP types (types A and B), whereas a single caprine isolate had a unique RFLP pattern (type C).

*Mycobacterium avium* subsp. *silvaticum* strains characteristically produce alkaline phosphatase, grow well on media at pH 5.5, are inhibited by cycloserine (50 $\mu$g/ml), do not grow in egg medium, and do not have a requirement for mycobactin. Strains of this subspecies are obligate pathogens for animals and cause tuberculosis in birds, notably cranes and wood pigeons, and paratuberculosis in mammals, especially deer.

*Mycobacterium lepraemurium,* the rat leprosy bacillus, causes indurating and ulcerating disease of the skin and lymph nodes of rats and mice. The organism is not easily cultivated in vitro but does grow on Ogawa egg medium and on agar-based media supplemented with cytochrome c and $\alpha$-ketoglutaric acid.

### *6.3.* **Mycobacterium gordonae** *Clade*

This clade consists of the following:

*M. asiaticum* Weiszfeiler et al. 1971 (193)
*M. gordonae* Bojalil et al. 1962 (119)

The first recorded strains of *M. asiaticum* were isolated from monkeys in 1965, but these organisms were considered to be variants of *M. simiae* (194). It was not until 1971 that Weiszfeiler and his co-workers recognized four of these strains to be members of a novel species and named them *M. asiaticum* (193). There is a convincing evidence from DNA relatedness (195,196), intradermal sensitin testing (195), numerical taxonomic (84) and M- and T-catalase studies (173,196) that the organism forms a good species.

It is important to distinguish between *M. asiaticum* and *M. gordonae* strains in routine diagnostic laboratories because *M. gordonae* is a common contaminant in clinical specimens but is rarely, if ever, clinically significant, whereas *M. asiaticum* is rarely encountered but, when it is isolated, is often the cause of disease in humans. The two organisms form a distinct phyletic line in the 16S rRNA tree (Fig. 1.4), have a common mycolic acid pattern (Table 1.4), and show considerable phenotypic overlap (84). However, *M. asiaticum* is photochromogenic and *M. gordonae* consistently scotochromogenic.

There are also strong grounds for believing that *M. gordonae* forms a good species (17,80,84,146). A nucleic acid probe specific for *M. gordonae* is available commercially (Gen-Probe) and a cross-absorbed antibody probe specific for the

M-catalase of *M. gordonae* has also been described (196). The organism is often known as the "tap water scotochromogen," as it is commonly found in water. Some strains carrying the invalid name "*M. aquae*" belong to the species.

### *6.4. The* **M. kansasii** *Clade*

This clade consists of the following:

*M. gastri* Wayne 1966 (197)
*M. kansasii* Hauduroy 1955 (198)

*Mycobacterium kansasii* can be considered as a distinct species on the basis of antigenic composition, sensitin specificity, phage typing, and numerical taxonomic data (81,197,199). Strains grow well at 37°C and are usually photochromogenic, producing a bright yellow pigment. Scotochromogenic or nonchromogenic variants are occasionally encountered. Continuous incubation of photochromogenic strains in the light leads to the formation of orange-colored crystals due to excess carotene synthesis. The organism can cause chronic human pulmonary disease and generalized infection in HIV-positive patients. It has also been isolated from piped water supplies but is rarely encountered in the natural environment.

There is evidence of genetic diversity among *M. kansasii* strains. Most isolates from Australia, Japan, and South Africa have been reported to react with either of two *M. kansasii*-specific probes (200), whereas those from Belgium and Switzerland reacted less frequently with either probe (201). A probe prepared from a highly repeated DNA element from *M. tuberculosis* highlighted five different RFLP patterns among isolates reacting with a *M. kansasii*-specific probe (201). Distinct RFLP patterns were also demonstrated among isolates that did not react with the *M. kansasii*-specific probe (201,202).

*Mycobacterium gastri* is an uncommonly encountered species that has been isolated from gastric washings, sputum, and soil. It has never been directly associated with disease in humans. This organism is usually nonphotochromogenic, although two photochromogenic strains with antigenic and biochemical properties characteristic of *M. gastri* have been reported (203).

*Mycobacterium gastri* and *M. kansasii* fall into separate numerical taxonomic clusters (81) and differ in agglutination serotypes (197). In contrast, immunodiffusion studies of cell extract antigens have demonstrated a strong similarity between these organisms (93,204). The T-catalase of *M. gastri* also shows a close structural relationship to that of *M. kansasii* (173). These observations, together with the fact that *M. gastri* and *M. kansasii* have identical 16S rRNA sequences (65) and a common mycolic acid pattern, cast doubt on whether these taxa are separate species. However, the distinction between them, whether it is at species or subspecies level, is very important because *M. kansasii* is usually significant when isolated from clinical material, whereas *M. gastri* is rarely, if ever, clinically significant.

### *6.5. The* Mycobacterium nonchromogenicum *Clade*

This clade encompasses the following:

*M. hiberniae* Kazda et al. 1993 (103)
*M. nonchromogenicum* Tsukamura 1965 (205)
*M. terrae* Wayne 1966 (197)

*Mycobacterium nonchromogenicum, M. terrae,* and *M. triviale* are often assigned to the so-called "*M. terrae*" complex for diagnostic purposes (17,124,128), but it is clear from 16S rRNA sequence data that *M. triviale* forms a distinct and relatively deep clade (Fig. 1.4). In contrast, *M. hiberniae, M. nonchromogenicum* and *M. terrae* are phylogenetically close. The constituent members of this group share a common mycolic acid pattern (Table 1.3) but can be distinguished using DNA-relatedness data (103,206).

*Mycobacterium nonchromogenicum* and *M. terrae* encompass nonchromogenic organisms which fall into separate but related clusters in numerical phenetic surveys (74,84,89,90). A tenuous distinction can be drawn between the two organisms by dermal hypersensitivity testing (90), but immunodiffusion analysis does not permit their differentiation (93). Positive reactions for nicotinamidase and pyrazinamidase and negative nitrate reduction provide the most definitive means of distinguishing between *M. nonchromogenicum* and *M. terrae.* Members of these species are commonly encountered in sputum, presumably as extrinsic contaminants from dust and water, but only rarely cause disease in humans.

*Mycobacterium hiberniae* is a scotochromogenic, rose-pink-pigmented, nonpathogenic organism which has been isolated repeatedly from sphagnum vegetation, true moss, and soil in Ireland (103). The organism is easily differentiated from *M. terrae* by its rose-pink pigment and its positive arylsulfatase activity. It can also be distinguished from *M. terrae* due to differences in 16S rRNA sequences, major pyrolysis esters, and DNA : DNA pairing values. There are also marked differences in major pyrolysis esters and DNA-relatedness values between *M. hiberniae* and *M. nonchromogenicum.*

### *6.6. The* Mycobacterium simiae *Clade*

This clade encompasses the following:

*M. genavense* Böttger et al. 1993 (69)
*M. interjectum* Springer et al. 1993 (71)
*M. simiae* Karassova et al. 1965 (194)

*Mycobacterium genavense, M. interjectum,* and *M. simiae* form a relatively deep branch of the mycobacterial phylogenetic tree [Fig. 1.4; (71)]. Two other species, namely *M. intermedium* and *M. triviale,* are sometimes assigned to this group. This broader grouping is characterized by slow growth at the phenotypic level and the simultaneous presence of a molecular signature associated with fast-growing mycobacteria, namely a short helix 18 (65).

*Mycobacterium simiae* is a photochromogenic organism which was originally isolated from lymph nodes of apparently healthy monkeys (194). It is rarely encountered in clinical laboratories but has been isolated from patients with pulmonary disease, nonpulmonary lesions, and environmental sources. Three agglutinating serotypes, labeled simiae 1, simiae 2, and avian 18, have been recognized among strains identified as *M. simiae* (83). In the same study, numerical taxonomic analysis showed *M. simiae* to be composed of a highly homogeneous cluster of serovar 18 strains and a less tightly linked subcluster comprising the other serovars. The taxonomic integrity of *S. simiae* is also supported by chemotaxonomic (87), DNA relatedness (195), and additional serological data (196). Strains previously designated "*M. habana*" (207) are now classified as *M. simiae* (208–210).

*Mycobacterium genavense* is an opportunistic pathogen which frequently causes disseminated infections in patients with AIDS. It grows poorly on conventional media and requires mycobactin for growth. The organism can be distinguished from other slow-growing mycobacteria by its fastidious growth, in particular its preference for broth media and its inability to grow on standard solid media used for isolation of mycobacteria. A characteristic whole-organism fatty acid pattern differentiates *M. genavense* from *M. simiae* (211) although the two organisms have a common mycolic acid pattern (Table 1.3).

*Mycobacterium interjectum* was proposed by Springer et al. (71) to accommodate strains isolated from a lymph node of a child with chronic lymphadenitis. Additional strains misclassified as *M. gordonae* formed a homogeneous group based on DNA-relatedness and numerical taxonomic data (87). Most of the strains also reacted in a plate hybridization assay with reference DNA from *M. interjectum.* Lipid analysis distinguished the *M. interjectum* strains from the type strain of *M. gordonae,* a finding of considerable practical importance. *Mycobacterium interjectum* can be differentiated from *M. simiae* by the presence of *cis*-10-hexadecanoic acid and heptadecanoic acid and the absence of *cis*-11-hexadecanoic acid (71).

### *6.7. The* **Mycobacterium szulgai** *Clade*

This clade encompasses the following:

*M. malmoense* Schröder and Juhlin 1977 (212)
*M. szulgai* Marks et al. 1972 (213)

Associated species:
*M. haemophilum* Sompolinsky et al. 1978 (214)

The 16S rRNA sequence of the type strain of *M. szulgai* differs by only two nucleotides from the *M. malmoense* sequence (87). Nevertheless, DNA-relatedness and RFLP data, as well as lipid patterns, support the separation of *M. szulgai* and *M. malmoense* strains into two distinct species. *Mycobacterium szulgai* has been distinguished from *M. malmoense* and several other slow-growing mycobacteria by specific precipitation of M-catalase (196). *Mycobacterium haemophilum* can also be assigned to the *M. szulgai* clade despite the relatively low bootstrap value shown in Figure 1.4. *Mycobacterium haemophilum* and *M. szulgai* have a common mycolic acid pattern (Table 1.3).

*Mycobacterium szulgai* forms a distinct numerical taxonomic cluster (83,84) and shows a unique TLC pattern of surface lipid antigens (213). The pattern of pigmentation of this species is unusual in that it is scotochromogenic when grown at 37°C but can be photochromogenic when grown at 25°C. The organism is most easily confused with *M. flavescens* when the most commonly used diagnostic tests are applied (82), but its susceptibility to picric acid (0.2% w/v) and sodium chloride (5% w/v), and resistance to hydroxylamine hydrochloride (125 $\mu$g/ml) separate it from the latter species. *Mycobacterium szulgai* is an occasional pathogen which principally causes chronic pulmonary disease although disseminated infections have been reported in AIDS patients and immunocompromised individuals.

*Mycobacterium haemophilum* has been shown to be a good species (215–218) which has a novel phenolic glycolipid antigen (219). It requires hemin or ferric ammonium citrate for growth, is biochemically inactive, grows well at 30°C but not at 37°C. The organism is a recently recognized pathogen of immunocomprised patients. Evidence that it causes cervical lymphadenitis in healthy children suggests that the source of *M. haemophilum* is environmental.

*Mycobacterium malmoense* can be distinguished from all other slow-growing mycobacteria on the basis of catalase serology (181), lipid TLC patterns (83), numerical taxonomy (83–85), DNA RFLP patterns (87), and limited DNA-relatedness data (87,206). There is evidence of genetic diversity within the species. Five types of surface glycolipid patterns and two major and three minor rRNA ribotypes have been recognized (220). However, no correlation was found between the glycolipid and ribotype groups.

The organism is isolated relatively frequently in the United Kingdom and some other European countries. However, its frequency could be underestimated, as it grows poorly on conventional laboratory media and the similarity of its biochemical activities to those of other slow-growing nonphotochromogenic species, such as *M. gastri* and *M. nonchromogenicum*. Consequently, negative phosphatase and positive pyrazinamidase reactions are among the most powerful biochemical features that distinguish *M. malmoense* from *M. gastri*. Negative responses for high catalase, acid phosphatase and β-galactosidase separate *M. malmoense* from *M. nonchromogenicum*. These distinctions are important as *M. malmoense* is considered to be a pathogen whereas *M. gastri* and *M. nonchromogenicum* are not.

### *6.8. Single-Membered Clades*

The detailed phylogenetic relationships of the remaining validly described species of slow-growing mycobacteria are not clear, although a loose association is apparent between members of some of these taxa (Fig. 1.4).

*Mycobacterium branderi* was proposed as a new species using growth, biochemical, lipid, and 16S rRNA sequence data (105). The organism, which was isolated from clinical samples in Finland, is nonchromogenic, grows well between 25°C and 45°C, gives a positive arylsulfatase reaction after 14 days and has a unique glycolipid pattern. The two nearest neighbors to *M. branderi* are *M. celatum* and *M. cookii,* although members of these taxa have different mycolic acid patterns (Table 1.3). *Mycobacterium branderi* and *M. cookii* strains are easy to distinguish as the latter, in contrast to the former, are scotochromogenic and do not grow at 37°C.

*Mycobacterium celatum* was proposed to accommodate isolates from bronchial washes, sputum specimens, and blood from HIV-positive and HIV-negative persons (99). In conventional diagnostic tests, the organism most closely resembles members of the *M. avium* complex and *M. xenopi*. The mycolic acid profile of *M. celatum* is qualititatively similar but quantitatively different from that of *M. xenopi* and is clearly distinct from that of *M. avium* and related taxa. *Mycobacterium celatum* strains form two highly related 16S rRNA subgroups, but the members of these are indistinguishable using the minimal standards proposed for the definition of slow-growing *Mycobacterium* species (18). In the diagnostic laboratory, *M. celatum* can be distinguished from *M. xenopi* by colony morphology and growth at 45°C, and from the *M. avium* complex by DNA probes.

*Mycobacterium conspicuum* was isolated repeatedly from immunocompromised patients with disseminated mycobacterial infections (221). The case reports on these patients suggest that *M. conspicuum* has a pathogenic potential similar to those of other nontuberculosis pathogenic bacteria such as *M. avium, M. celatum,* and *M. genavense*. The organism shows a distinct metabolic reaction pattern, has a unique high-performance liquid crystallographic (HPLC) mycolic acid profile, and only grows in liquid culture at 37°C. It also characteristically synthesises two branched fatty acids with 15 and 17 carbon atoms, respectively.

*Mycobacterium cookii* contains nonpathogenic organisms that were repeatedly isolated from sphagnum vegetation and surface water of moors in New Zealand (101). The organisms grows at 22°C and 31°C but not at 37°C and shows catalase, acid phosphatase, and arylsulfatase activities. Representative strains form a distinct numerical taxonomic cluster and have a unique mycolic acid pattern.

*Mycobacterium intermedium* was repeatedly isolated from the sputum of a single patient with pulmonary disease (70). The organism, which was considered to occupy an intermediate position between the rapid and slow growers in the *Mycobacterium* phylogenetic tree, can be distinguished from other slow-growing photochromogenic mycobacteria by its mycolic acid pattern (Table 1.3). *Mycobacterium intermedium* can be confused with *M. asiaticum* but, unlike the latter, it shows eugonic growth, is $\beta$-galactosidase positive and grows at 22°C. It can be distinguished from *M. kansasii* by the results of tests for $\alpha$-esterase, $\beta$-esterase, $\beta$-galactosidase, and nicotinamidase activities.

*Mycobacterium scrofulaceum* forms a distinct species on the basis of numerical taxonomic (80,90), immunodiffusion (93), immunoelectrophoretic (222) and reciprocal dermal skin testing data (80,169,195). The three serovars established within the series by seroagglutination (223,224) correspond to TLC patterns seen with surface lipid antigens (80,225), which are of the alkali stable type (226). Goslee et al. (227) proposed a fourth serovar and designated it serovar 44.

Although scotochromogenic, *M. scrofulaceum* is sometimes assigned to the MAIS complex (179). However, DNA homology data (182), immunologic distance measurements on T-catalase (173), and phenotypic properties (82,86), as well as 16S rRNA sequence data, all provide evidence that the evolutionary distance of *M. scrofulaceum* is sufficient to make it inappropriate to include this species in a complex with *M. avium* and *M. intracellulare.* The organism causes cervical adenitis in children and has been isolated from gastric aspirates, sputum specimens, and the environment (228). *Mycobacterium scrofulaceum* is identical to *M. marianum*, but this name, although the earlier of the two, was discarded because of its orthographic similarity to the valid epithet *M. marinum* (228).

*Mycobacterium shimoidei* was revived as a valid nomenspecies (230) after being omitted from the Approved Lists of Bacterial Names (15). It forms a loose phylogenetic relationship with *M. xenopi* (Fig. 1.4) and is distinct on the basis of numerical taxonomic (82,84) and DNA-relatedness data (206). The organism is thermophilic, growing at 45°C but not at 25°C, and, like *M. malmoense*, it is biochemically unreactive. *Mycobacterium shimodei* can be distinguished from *M. malmoense* on the basis of mycolic acid pattern and acid phosphatase activity. The organism is thought to cause pulmonary disease in humans, although only a few cases of infection have been documented (114).

*Mycobacterium triviale* forms a distinct numerical taxonomic cluster (74,84) and is one of the five slow-growing mycobacteria that display the short 16S rRNA helix. The most distinctive feature of the organism is its ability to grow in the presence of 5% w/v sodium chloride, a property not shared by any other known slow-growing mycobacteria. *Mycobacterium triviale* is also unusual as it only forms $\alpha$-mycolates (Table 1.3). The organism has been isolated from sputum but is not considered a pathogen.

*Mycobacterium xenopi* was first isolated from a cold-blooded animal, the toad *Xenopus laevis* (231), but despite this, it is unique among mycobacteria as it grows only poorly at 37°C preferring 42–45°C. The organism forms a good species on the basis of T-catalase serology (173) and numerical taxonomic data (90). Members of this species have been implicated in chronic pulmonary diseases and nonpulmonary infections in immunocompromised persons but have also been isolated from body secretions in the absence of disease and from environmental sources.

### *6.9. Noncultivable Species*

*Mycobacterium leprae* was first observed microscopically by Hansen (232) in tissue taken from a leprosy patient. Subsequent studies confirmed that *M. leprae* is the causal agent of human leprosy. The organism has not yet been cultured in artificial media despite periodic claims to the contrary. The only methods of propagating *M. leprae* are by injecting leprous material either into the footpads of healthy mice (233), or into the nine-banded armadillo (234), where it causes a generalized infection, allowing the production of large numbers of organisms for biochemical and immunological studies. The detailed taxonomic affinities of *M. leprae* have still to be resolved. The organism forms a distinct phyletic line (Fig. 1.4), produces $\alpha$- and keto-mycolates (Table 1.3) and several species-specific antigens.

Some armadillos used for the experimental propagation of *M. leprae* have been shown to be infected with other slow-growing mycobacteria (235,236). These armadillo-derived mycobacteria (ADM) were assigned to a number of homogeneous groups based on nutritional requirements, mycolic acid patterns, and DNA-relatedness data (237). These organisms are unrelated to *M. leprae* or to any other validly described mycobacterial species and, hence, appear to represent new species (238).

## 7. Classification of Fast-Growing Mycobacteria

Fast growing mycobacteria are ubiquitous in natural habitats but have received less attention than their more clinically significant, slow-growing relatives. The vast majority of human infections attributed to rapidly growing mycobacteria are caused by *M. abscessus, M. chelonae,* and *M. fortuitum* (114,115,239). Rare cases of human disease have been associated with *M. flavescens, M. neoaurum, M. peregrinum,* and *M. thermoresistibile* strains. Other fast-growing mycobacteria, such as those classified as *M. duvalii, M. gilvum, M. rhodesiae,* and *M. vaccae,* are found in clinical material, but there is no evidence that they cause disease. *Mycobacterium farcinogenes* and *M. senegalense,* agents of bovine farcy, are closely related to *M. fortuitum* (Fig. 1.4). The same may prove to be true of *M. porcinum,* which causes lymphadenitis in swine (240).

The first sound strategy for the classification of rapidly growing mycobacteria was developed by Gordon and her colleagues who examined many strains for a balanced set of biochemical properties and delineated species on the basis of overall reaction patterns rather than on the results of a few subjectively chosen tests (116–118,241,242). These celebrated studies established the taxonomic integrity of *M. fortuitum, M. phlei,* and *M. smegmatis* and simultaneously paved the way for a comprehensive series of numerical phenetic investigations that provided a framework for the classification of rapidly growing mycobacteria. The large pools of data required for computer analysis were relatively easy to generate as fast-growing mycobacteria are readily grown and metabolically active.

Rapid-growing mycobacteria have been the subject of three cooperative studies carried out under the aegis of the IWGMT (88,89,91). These investigations were designed to determine the taxonomic status of established and recently described species of fast-growing mycobacteria and their relationships with strains now classified in the genera *Gordona, Nocardia,* and *Rhodococcus* (53). As before, the collaborators examined sets of judiciously chosen strains, pooled their biochemical, cultural, and physiological data for computer analysis and evaluated the resultant numerical classifications within the context of results from independent chemical, immunological, phage susceptibility, and experimental pathogenicity studies.

There have been a plethora of numerical taxonomic analyses carried out on rapidly growing mycobacteria by individual teams of investigators. The products of these numerical analyses have been interpreted within the context of relationships found using other proven taxonomic methods. As with slow-growing mycobacteria, valuable data have been derived from the application of immunologically based tests (89,93,243,244), and, to a lesser extent, bacteriocin (245), chemotaxonomic (24), and phage typing studies (246) have served a similar purpose. Chemical- and molecular-based techniques are increasingly being used to clarify the taxonomy of rapidly growing mycobacteria, notably quantitative fatty acid (100,247,248), DNA : DNA relatedness (98,106,249), and 16S rRNA RFLP analyses (250).

The taxonomic status of most of the validly described species of rapid-growing mycobacteria has been underpinned by the results of chemotaxonomic, DNA : DNA relatedness and 16S rRNA-based investigations. It is evident from the 16S rRNA tree (Fig. 1.4) that most species of rapidly growing mycobacteria are well separated from each other, although in the majority of cases, the distances between the branching points are small. The configuration of the tree is sensitive to the addition of new sequences and to the application of different treeing algorithms. Nevertheless, it is encouraging that the 26 species of rapid-growing mycobacteria can be assigned to 7 multimembered and 12 single-membered phyletic lines [Fig. 1.4; (251)]. It is evident from Figure 1.4 that the taxonomic status of six out of the seven multimembered clades is supported by high bootstrap values and several treeing algorithms. Representatives of 12 fast-growing *Mycobacterium* species have still to feature in 16S rRNA sequencing studies; a partial sequence is available for *M. mucogenicum* (251).

The 16S rRNA data are not only helping to clarify the intrageneric relationships between fast-growing mycobacterial species but also show that some well-established taxonomic groups, such as the *M. fortuitum* and *M. parafortuitum* complexes, are polyphyletic and, hence, in need of revision. As with slow-growing species, it is important that their fast-growing counterparts are considered within the unfolding phylogenetic framework. The distribution of diagnostic tests recommended for the identification of rapidly growing mycobacteria also needs to be seen within this context (Table 1.5). The minimal standards needed for the description of new fast-growing *Mycobacterium* species should be built on this new taxonomic perspective.

### *7.1. The* **Mycobacterium chelonae** *Clade*

This clade encompasses the following:

*M. abscessus* (Moore & Frerich 1953 [252])
Kusunoki & Ezaki 1992 (139)
*M. chelonae* Bergey et al. 1923 (141)

*Mycobacterium chelonei* was proposed by Bergey et al. (141) for Friedmann's (1903) turtle tubercle bacillus. It was subsequently pointed out (253) that the correct name of this organism is *M. chelonae,* given the gender of the binomial name for the turtle, *Chelona corticata.* The correct spelling is used here. Although rarely isolated from sputum, strains of *M. chelonae* cause wound infections and cervical adenitis. Evidence of prosthetic valve endocarditis caused by *M. chelonae* (254) suggests that replacement valves can be contaminated with this organism. *Mycobacterium chelonae* and *M. abscessus* are now seen as distinct, albeit phylogenetically related species (251) although this has not always been apparent.

Initial confusion over the status and taxonomic affinities of *M. abscessus* (252), *M. borstelense* (255), and *M. runyonii* (119) was clarified when Stanford and Beck (256) proposed that all of three species be reduced to synonyms of *M. chelonae.* In subsequent work, this species was considered to consist of two geographical variants, the first accommodated organisms previously classified as *M. abscessus* and *M. runyonii* and the second encompassed *M. borstelense* and *M. chelonae* strains. Prior to this study, *M. abscessus* and *M. runyonii* had been shown to be identical (257). In contrast, *M. borstelense* was seen either as a separate species (258,259) or as a variety of *M. abscessus* (260). Kubica et al. (89) also recognized these subgroups and raised them to subspecific status as *M. chelonae* subsp. *chelonae* and *M. chelonae* subsp. *abscessus*. This proposal was supported by the results of additional studies (91,261).

Preliminary DNA : DNA-relatedness data failed to clarify the relationship between the two *M. chelonae* subspecies. Baess (262) reported that *M. chelonae* subsp. *abscessus* ATCC 19977 and *M. chelonae* subsp. *chelonae* ATCC 19235 were closely related, as the two strains showed 99% DNA homology as determined by spectrophotometric hybridization. However, this finding was not conclusive, as strain ATCC 19235 is not the type strain (T) of *M. chelonae* subsp. *chelonae*. In a more comprehensive DNA : DNA-relatedness study, which included the type strain of each of the subspecies, the reannealing values suggested that the two taxa should be seen as independent species (249). Nevertheless, it was left to Kusunoki and Ezaki (139) to formally reestablish *M. abscessus* as a distinct species. Several phenotypic properties are available to distinguish between *M. abscessus* and *M. chelonae* (17,139), although members of these taxa share a unique mycolic acid pattern (Table 1.3).

### *7.2. The* **Mycobacterium chubuense** *Clade*

This clade encompasses the following:

*M. chlorophenolicum* (Apajalahti et al. 1986 [263])
Häggblom et al. 1994 (100)
*M. chubuense* Tsukamura et al. 1981 (264)

*Mycobacterium chlorophenolicum* and *M. chubuense* are schotochromogenic and form a distinct phyletic line (Fig. 1.4) but have different mycolic acid patterns (Table 1.3). *Mycobacterium chubuense,* which encompasses soil isolates, forms a distinct species on the basis of numerical taxonomic data (264). *Mycobacterium chlorophenolicum,* which was initially misclassified as *Rhodococcus chlorophenolicus* (263), contains chlorophenol-degrading strains independently isolated from enrichment cultures prepared with chlorophenol-contaminated soil and sludge collected from different locations in Finland. In contrast to *M. chubuense, M. chlorophenolicum* strains are acid phosphatase positive (Table 1.5).

### *7.3. The* **Mycobacterium confluentis** *Clade*

This clade encompasses the following:

*M. confluentis* Kirscher et al. 1992 (104)
*M. madagascariense* Kazda et al. 1992 (102)

*Mycobacterium confluentis* and *M. madagascariense* show a low level of differences in their rRNA but have different mycolic acid patterns (Table 1.3). Kirschner et al. (104) proposed *M. confluentis* for nonpigmented mycobacteria isolated from sputum of a healthy male. The organism grows well between 22°C and 41°C and can be distinguished from other fast-growing mycobacteria by its susceptibility to antitubercular drugs. *Mycobacterium madagascariense* was proposed by Kazda et al. (102) for scotochromogenic mycobacteria isolated from sphagnum vegetation in Madagascar. The organism grows between 22°C and 31°C but not at 37°C, metabolizes iron, and shows putrescine oxidase activity.

### *7.4. The* **Mycobacterium fortuitum** *Clade*

This clade encompasses the following:

*M. farcinogenes* Chamoiseau 1973 (265)
*M. fortuitum* subsp. *acetamidolyticum* Tsukamura et al. 1986 (109)
*M. fortuitum* subsp. *fortuitum* da Costa Cruz 1938 (266)
*M. peregrinum* Kusunoki and Ezaki 1992 (139)
*M. senegalense* (Chamoiseau 1973 [265]) Chamoiseau 1979 (267)

*Mycobacterium fortuitum* was proposed by da Costa Cruz (266) for a strain isolated from an injection-site abscess. Stanford and Gunthorpe (268) subsequently found the organism to be identical to Kuster's frog tubercle bacilli which Bergey et al. (141) had raised to species status as *M. ranae.* Runyon (269) challenged this binomial as a *nomen ambiguum* and, for stability in nomenclature, requested conservation of the epithet *fortuitum;* the Judicial Commission of the International Committee of Systematic Bacteriology of the International Association of Microbiological Societies ruled in support of this recommendation in 1974 [Opinion 51; (270)]. The taxonomic status of *M. fortuitum* is supported by an abundance of data, notably by the results of electrophoretic analysis of enzymes and proteins (271–273), and numerical taxonomic (89,274), immunodiffusion, and immunoelectrophoretic data (93,275,276). Other organisms which belong to this species include *M. giae* (277) and *M. minetti* (278).

*Mycobacterium fortuitum* is the correct name for a well-described species which is composed of several subspecies and biovars. The status and description of the biovars has been considered in detail (279); there is also evidence that *M. fortuitum* biovars can be distinguished by sequence differences in their 16S rRNA genes (280). *Mycobacterium fortuitum* subsp. *acetamidolyticum* was proposed by Tsukamura et al. (109) for a strain isolated from the sputum of a patient with lung disease. The organism was readily distinguished from all other mycobacteria on the basis of numerical taxonomic data but showed 94% DNA : DNA relatedness with the type strain of *M. fortuitum* subsp. *fortuitum.* It is also readily distinguished from all other fast-growing mycobacteria by its ability to use acetamide and L-glutamate as simultaneous carbon and nitrogen sources.

The overwhelming majority of human infections attributed to rapidly growing mycobacteria are caused by *M. fortuitum* and other species assigned to the so-called *M. fortuitum* complex (114,115,281,282). This broad and artificial assemblage has been a catchall for *M. abscessus, M. chelonae, M. fortuitum, M. mucogenicum*, and *M. peregrinum*. Members of all of these taxa are rapidly growing, nonphotochromogenic mycobacteria which give a postive 3-day arylsulfatase test, grow on MacConkey agar lacking crystal violet, degrade periodic acid–Schiff, and are resistant to 500 $\mu$g/ml hydroxylamine hydrochloride (17).

It is common practice in many diagnostic laboratories to identify isolates only to the level of the *M. fortuitum* complex. Nevertheless, it is good practice to distinguish between members of the different species, as they cause different diseases and have different drug susceptibilities (283–287). The constituent species of the *M. fortuitum* complex can be distinguished from one another using cultural and biochemical tests (Table 1.5), as well as by chemical, serological, and molecular systematic criteria. There is evidence that the environment is the source of rapidly growing mycobacteria that infect patients (114,115). The development of DNA fingerprinting technology, especially pulse-field gel electrophoresis, has facilitated studies on the epidemiology of clinical diseases caused by these organisms (284).

*Mycobacterium peregrinum,* first proposed by Bojalil et al. (119), has had an unsettled taxonomic history. The organism was considered by some to be synonymous with *M. fortuitum* (89,268,289,290) and, hence, was omitted from the Approved Lists of Bacterial Names (15). In contrast, data from isoelectric focusing of $\beta$-lactamase (288), DNA : DNA relatedness (249,262), and serological analyses (291) supported the reintroduction of the taxon as an independent species. *Mycobacterium peregrinum* was formally reestablished as a distinct species in 1992 by Kusunoki and Ezaki (139). The organism is a relatively rare cause of disease but does so in the same setting as other members of the *M. fortuitum* complex.

Lipid data led several investigators to the view that actinomycetes causing bovine farcy in eastern and western Africa belonged to the genus *Mycobacterium* (265,267,292–294). Chamoiseau (265) proposed the name *M. farcinogenes* for these bovine isolates and recognized two subspecies, *tchadense* and *senegalanese*. It later became apparent that strains from Chad differed from those isolated from zebu cattle in Senegal. Thus, the latter were relatively fast growing, showed broader and more intense amidase and glucolytic activities, contained a characteristic mycoside C′ (295), caused a more generalized peritonitis in guinea pigs (267) but did not show a relatively high degree of DNA relatedness with strains from Chad. On the basis of these and other differences, Chamoiseau (267) proposed the latter be classified as *M. farcinogenes* and their more rapidly growing counterparts as *M. senegalense*. Members of each of these species differ from all other known mycobacteria, as they are found in lesions of the lymphatic system or the parenchyma of zebu cattle and present a stable mycelium, a positive malonamidase test and a distinctive pathogenicity for guinea pigs (267).

*Mycobacterium farcinogenes* and *Mycobacterium senagalense* form a distinct evolutionary clade together with *M. fortuitum* and *M. peregrinum* (Fig. 1.4). In addition, members of all four species have a common mycolic acid pattern (Table 1.3). DNA relatedness (262,296), chemotaxonomic (297–300), numerical taxonomic (301) and serological data (302,303) underpin the relationship with *M. fortuitum* but also confirm the distinction between *M. farcinogenes* and *M. senegalense*. Despite the improvements in the taxonomy of all of these taxa, better diagnostic methods are needed to separate *M. farcinogenes* and *M. senegalense* strains from one another and from other members of the *M. fortuitum* complex.

### *7.5. The* **Mycobacterium neoaurum** *Clade*

This clade encompasses the following:

*Mycobacterium diernhoferi* Tsukamura et al. 1983 (304)
*Mycobacterium neoaurum* Tsukamura 1972 (305)

Saito et al. (91) found that *M. aurum, M. diernhoferi, M. neoaurum*, and *M. parafortuitum* strains shared many phenotypic characters and concluded that these taxa should not be separated, but assigned to a grouping of convenience, the "*M. parafortuitum* complex," pending the acquisition of additional taxonomic data. *Mycobacterium aurum* and *M. neoaurum* were subsequently lumped together in an aggregate taxon equated with the *M. parafortuitum* complex (306). Comparative 16S rRNA sequence studies show quite unequivocally that *M. diernhoferi* and *M. neoaurum* form a deep-rooted clade which is sharply separated from a corresponding phyletic line composed of *M. aurum* and *M. vaccae* (Fig. 1.4).

*Mycobacterium neoaurum* forms a distinct numerical taxonomic cluster (264,304). Little is known about this organism, although there is evidence that it can cause bacteremia (307). *Mycobacterium diernhoferi* was originally described by Boñicke and Juhasz (308) for nonphotochromogenic bacteria isolated from soil in a cattle field. The organism, which was omitted from the Approved Lists of Bacterial Names (15), was redescribed mainly on the basis of numerical taxonomic data (304). Serological data help underpin the taxonomic integrity of this species (89,91,309). *Mycobacterium diernhoferi* is not known to be associated with human disease.

### *7.6. The* **Mycobacterium vaccae** *Clade*

This clade encompasses the following:

*Mycobacterium aurum* Tsukamura 1966 (310)
*Mycobacterium vaccae* Bönicke and Juhasz 1964 (311)

*Mycobacterium aurum* and *M. vaccae* form a recognizable clade, albeit one supported by a relatively low bootstrap value (Fig. 1.4). The status of *M. aurum* as a distinct species is not only supported by 16S rRNA data but also by the results of DNA relatedness (262) and numerical taxonomic studies (91,310). The organism, which is found in soil, is occasionally seen in the sputum of patients, but is not related to disease.

*Mycobacterium vaccae* was proposed by Boñicke and Juhasz (311) for mycobacteria associated with milk or cattle or their environs. The original confusion over the taxonomic status of the species (89,91) was resolved by the results of DNA relatedness experiments (262). *Mycobacterium aurum* and *M. vaccae* cannot be distinguished on the basis of agglutination or bacteriocin typing (245,312) but they do have different mycolic acid patterns (Table 1.3).

### *7.7. Single-Membered Clades*

Nearly full 16S rRNA sequences are available for an additional 12 validly described mycobacterial species, although the exact position of these organisms in the mycobacterial tree has still to be ascertained (Fig. 1.4). Nine of these species are scotochromogenic; the remaining species produce nonpigmented colonies.

#### NONPIGMENTED SPECIES

*Mycobacterium chitae* and *M. fallax* show a loose phylogenetic relationship in the mycobacterial tree generated using the neighbor-joining method (Fig. 1.4) but have markedly different mycolic acid patterns (Table 1.3). Tsukamura (313) proposed *M. chitae* for a number of nonpathogenic mycobacteria isolated from soil. There is evidence that these organisms form a homogeneous numerical taxonomic cluster and a distinct serotype (91,313). *Mycobacterium fallax* was proposed by Lev́y-Frebault et al. (107) for strains isolated from river water and sputum. Members of this species form a distinct numerical taxonomic cluster and have a simple mycolic acid pattern (Table 1.3).

*Mycobacterium smegmatis* was the name given by Trevisan (314) to the Smegma Bacillus of Alvarez and Tavel (315); the organism was redescribed by Lehmann and Neumann 1899 (316). Gordon and her colleagues (116,118) showed the taxon to be a good species, a conclusion subsequently borne out by the results of numerical taxonomic (89,257,310), DNA relatedness (262,296), immunodiffusion and immunological studies (275,276,317,318). In addition, species-specific sensitins (319) and phages (89,246) have been detected. *Mycobacterium smegmatis* is usually associated with secretions of the normal genitalia and with soft lesions associated with accidental or surgical trauma. The clinical conditions caused by these organisms have been considered in detail (285,320). The 16S rRNA sequence data suggest a relatively close relationship between *M. smegmatis* and *M. confluentis, M. flavescens, M. madagascariense, M. phlei*, and *M. thermoresistibile* (Fig. 1.4). *Mycobacterium phlei* and *M. smegmatis* have also been found to share a large number of ribosomal precipitinogens (244).

## SCOTOCHROMOGENIC SPECIES

This artificial grouping includes several nonpathogenic species, namely *M. aichiense* Tsukamura et al. 1981 (264), *M. gadium* Casal and Calero (321), *M. gilvum* Stanford and Gunthorpe 1971 (322), *M. komossense* Kazda and Müller 1979 (323), *M. obuense* Tsukamura et al. 1981 (264); *M. phlei* Lehmann and Neumann 1899 (316), and *M. sphagni* Kazda 1980 (324), all of which form either yellow or orange colonies. The remaining two members of the group, *M. flavescens* Bojalil et al. 1962 (119) and *M. thermoresistibile* Tsukamura 1966 (310), have occasionally been implicated as agents of human disease (114). There is some evidence from the mycobacterial tree that *M. flavescens, M. phlei, M. smegmatis,* and *M. thermoresistibile,* all of which encompass thermotolerant strains, might be related (Fig. 1.4).

*Mycobacterium phlei* and *M. thermoresistibile* are the only two mycobacteria that can grow up to 52°C and which remain viable after heating at 60°C for 4 h. The former, the first described species of rapidly growing, scotochromogenic mycobacteria, was proposed for the Timothy Bacillus or Grass Bacillus I of Moëller (325,326). The distinctiveness of this species is supported by catalase and esterase patterns (272), sensitin skin testing (319), phage and bacteriocin typing (245,246), and by the results of numerical taxonomic (89,274,310), immunofluorescence, immunodiffusion, and immunoelectrophoretic studies (275,276,317,318,327). *Mycobacterium thermoresistibile* not only forms a homogeneous numerical taxonomic cluster (91,264,310) but has a unique mycolic acid pattern (Table 1.3). Members of this taxon grow slowly on primary isolation and therefore can be confused with slow-growing scotochromogenic mycobacteria, but colonies appear in 3–5 days when subcultured onto egg media (17).

*Mycobacterium flavescens* was proposed by Bojalil et al. (119) for strains isolated from guinea pigs. The taxonomic integrity of the species is demonstrated by a characteristic bacteriocin pattern (245), species-specific antigens (89,322), and numerical taxonomic data (89,91,264,306) although there is evidence that the taxon is heterogeneous (225,328). Tsukamura and Mizuno (329) recognized two subspecies of *M. flavescens,* one of which contained strains previously labeled *M. gallinarum* (330). *Mycobacterium acapulcensis* Bojalil et al. (119) was found to be identical to *M. flavescens* (331).

Stanford and Gunthorpe (322) assigned 20 fast-growing, scotochromogenic mycobacteria with similar saccharolytic properties to three groups. One group contained the type strain of *M. flavescens;* the remaining two were allotted species status as *M. duvalii* and *M. gilvum.* The latter, which encompassed isolates from sputum and pleural fluid, is a good species on the basis of antigenic, lipid, and numerical taxonomic data (264,306,322,332).

*Mycobacterium komossense* (323) and *M. sphagni* (324) were proposed for strains isolated from sphagnum vegetation in northern Europe. The distinctiveness of each of these species is supported by characteristic lipid and immunodiffusion patterns, sensitin testing and numerical taxonomic data (264,323,324). Similarly, *M. aichiense* and *M. obuense*, which encompass soil isolates, form well-circumscribed numerical taxonomic clusters (264). In contrast, *M. gadium* (321) was isolated from a sputum specimen of a patient with pulmonary disease.

### *7.8. Additional Species of Rapidly Growing Mycobacteria*

The phylogenetic position of the remaining 13 validly described species in the mycobacterial tree has still to be determined. Members of these taxa can be assigned to two artificial groups on the basis of whether they form pigmented or nonpigmented colonies. Only 2 out of the 13 species, namely *M. mucogenicum* (251) and *M. porcinum* (240), are known to cause disease.

#### NONPIGMENTED STRAINS

This group encompasses seven species. *Mycobacterium agri* was originally described by Tsukamura (333) for strains isolated from alkali-treated soil samples. The species was not included in the Approved Lists of Bacterial Names (15) but was revised by Tsukamura (334) mainly on the basis of numerical taxonomic data. *Mycobacterium alvei* (98) and *M. brumae* (106) were proposed for organisms isolated from soil, water, and human sputum samples in Spain. Members of each of these species form a distinct group based on phenotypic and DNA-relatedness data. *Mycobacterium agri* and *M. alvei* can be distinguished from one another, and from all other mycobacterial species, by their unique mycolic acid patterns (Table 1.3). Similarly, *M. brumae* produces $\alpha$-mycolates which characteristically release 22-carbon atom esters on pyrolysis. In contrast, *M. pulveris,* which was proposed by Tsukamura *et al.* (335) for isolates from house dust, has a fairly standard mycolic acid pattern.

*Mycobacterium mucogenicum* (251) and *M. porcinum* (240) were proposed for organisms seen to be similar to *M. fortuitum.* Indeed, *M. mucogenicum* had previously been assigned to the *M. fortuitum* complex as the *M. chelonae*-like organism (336,337). This taxon forms a distinct group biochemically and has a characteristic mycolic acid profile as determined by high-performance liquid chromatography. Partial 16S rRNA sequence data underpin the relationship to members of the *M. fortuitum* complex. Additional evidence that *M. mucogenicum* and other species of rapidly growing mycobacteria differ genotypically comes from RFLP analyses of PCR amplification products of a 439 base-pair sequence of the 65-kDa heat shock protein. The most common type of diseases caused by *M. mucogenicum* are posttraumatic wound infections and catheter-related sepsis, clinical diseases that are relatively commonly caused by members of other taxonomic groups of rapidly growing mycobacteria. The specific epithet of the organism reflects the highly mucoid character of most clinical isolates on solid media. *Mycobacterium porcinum* strains, which were isolated from submandibular lymph nodes of swine, can be distinguished from *M. fortuitum* as they do not reduce nitrate, show positive succinamidase activity, and use benzoate as a sole source of carbon in the presence of ammmoniacal nitrogen. However, *M. fortuitum* and *M. porcinum* share a common mycolic acid pattern (Table 1.3).

### PIGMENTED STRAINS

Five of the remaining six species contain scotochromogenic mycobacteria that were mainly isolated from environmental sources. *Mycobacterium austroafricanum* (304), a member of the *M. parafortuitum* complex, encompasses strains isolated from water in South Africa; these organisms formed a distinct numerical taxonomic cluster. Similarly, *M. rhodesiae* (264) and *M. tokaiense*, which were originally omitted from the Approved Lists of Bacterial Names (15), form distinct clusters. None of these species have been implicated as agents of diseases, although *M. rhodesiae* strains were isolated in Rhodesia from sputum of patients with pulmonary tuberculosis.

*Mycobacterium duvalii* was proposed by Stanford and Gunthorpe (322) mainly on the basis of detailed immunodiffusion analyses augmented by a series of biochemical tests. The species description was based on four strains isolated from leprous lesions. Two of these strains were isolated by Duval and Wellman (338) and erroneously considered to be *M. leprae*. The taxonomic standing of the species is supported by numerical taxonomic and lipid data (306,332). Even less is known of *M. poriferae* which was isolated from a marine sponge (108). This organism was considered to be related to *M. parafortuitum* but can be distinguished from the latter on the basis of mycolic acid composition (Table 1.3).

*Mycobacterium parafortuitum* was proposed by Tsukamura (339) for soil isolates which resembled *M. fortuitum* in their rapid growth, utilization of organic acids, and pattern of amidase activity. However, these strains differed from *M. fortuitum* in showing negative reactions in the 3-day arylsulphate test, in salicylate and PAS degradation, in their inability to use nitrate as a sole nitrogen source, and by their capacity to form acid from arabinose, inositol, mannitol, and xylose. Members of the taxon show marked increases in pigmentation after exposure to light and form a distinct DNA-relatedness group (262).

## 8. Conclusions and Future Perspectivess

The crucial point in the development of a taxonomic system for mycobacteria was the application of an integrated or polyphasic taxonomic approach which allowed the delineation of species using complementary genotypic and phenotypic data. This approach was exemplified in the impressive series of collaborative studies carried out under the *aegis* of the International Working Group on Mycobacterial Taxonomy. It is clear from such studies that the genus *Mycobacterium* is well circumscribed and encompasses many well-defined species. It is particularly impressive that the taxonomy of strains previously lumped into "taxonomic complexes" is being clarified. The improved mycobacterial taxonomy has facilitated the recognition of many new species, several of which contain previously unrecognized pathogens, and provides a sound base for the development of improved practical techniques for the identification of unknown mycobacteria.

The spectacular advances made in mycobacterial classification not only raise a number of new questions but also challenge the rationale behind existing strategies. The focus in recent years has been on the generation of good quality taxonomic information, notably numerical taxonomic, DNA relatedness, and 16S rRNA sequence data. There is now a need to pay greater attention to the development of appropriate data-handling techniques and the management of taxonomic databases, especially since the acquisition of genotypic and phenotypic data is becoming increasingly automated and, hence, relatively straightforward (340,341). It is also clear that advances in mycobacterial systematics will, as with other genera, depend on gaining rapid access to taxonomic information through international taxonomic networks (340).

Additional perspectives for future work include the need to:

- Highlight additional chemical and phenotypic markers for the subgeneric classification of mycobacteria.
- Determine the effects on genetic instability on the phenotypic expression of mycobacteria.
- Develop "user-friendly" software tools for the comparison of polyphasic taxonomic data; cross-checking data is critical to sound taxonomic work.
- Unravel the subspecific structure of mycobacterial species which encompass human and animal pathogens.
- Devise new selective media for determining the numbers, types, and activities of mycobacteria in natural habitats.
- Discover the full extent of mycobacterial species diversity.

It is clear that although much has been accomplished, much remains to be done!

**Acknowledgment**

The authors are grateful to Dr. Jongsik Chun for providing the phylogenetic tree (Fig. 1.3).

**References**

1. Bousfield IJ, Goodfellow M (1976) The "*rhodochrous*" complex and its relationships with allied taxa. In: Goodfellow M, Brownell GH, Serrano JA, eds. The Biology of the Nocardiae, pp. 39–65. London: Academic Press.
2. Lechevalier MP (1976) The taxonomy of the genus *Nocardia:* Some light at the end of the tunnel? In: Goodfellow M, Brownell GH, Serrano JA, eds. The Biology of the Nocardiae, pp. 1–38. London: Academic Press.
3. Barksdale L, Kim KS (1977) *Mycobacterium.* Bacteriol Rev 41:217–312.
4. Goodfellow M, Minnikin DE (1977) Nocardioform bacteria. Annu Rev Microbiol 31:159–180.
5. Goodfellow M, Minnikin DE (1981) Introduction to the coryneform bacteria. In: Starr MP, Stolp H, Trüper HG, Balows A, Schlegel HG, eds. The Prokaryotes, Volume II, pp. 1811–1826. New York: Springer-Verlag.
6. Goodfellow M, Minnikin DE (1981) The genera *Nocardia* and *Rhodococcus.* In: Starr MP, Stolp H, Trüper HG, Balows A, Schlegel HG, eds. The Prokaryotes, Volume II, pp. 2016–2027. New York: Springer-Verlag.
7. Goodfellow M, Minnikin DE (1984) Circumscription of the genus. In: Kubica GP, Wayne LG, eds. The Mycobacteria. A Sourcebook, pp. 1–24. New York: Marcel Dekker.
8. Koch R (1882) Die Aetiologie der Tuberculose. Berliner Klin Wochenschr 19:221–238.
9. Timpe A, Runyon EH (1954) The relationship of "atypical" acid-fast mycobacteria to human disease. J Lab Clin Med 44:202–209.
10. Runyon EH (1958) Mycobacteria encountered in clinical laboratories. Leprosy Briefs 9:21.
11. Runyon EH (1959) Anonymous mycobacteria in pulmonary disease. Med Clin North Am 43:273–290.
12. Lehmann KB, Neumann R (1896) Atlas und Grundis der Bakteriologie und Lehrbuch der speciellen bakteriologischen Diagnostik, 1st ed. Munchen: J.F. Lehmann.
13. Hansen GA (1880) *Bacillus leprae.* Virchow's Arch 79:32–42.
14. Zopf W (1883) Die Spaltpilze. Breslau: Edward Trewendt.
15. Skerman VBD, McGowan V, Sneath PHA (1980) Approved lists of bacterial names. Int J System Bacteriol 30:225–420.
16. Goodfellow M, Wayne LG (1982) Taxonomy and nomenclature. In: Ratledge C, Stanford JL, eds. The Biology of the Mycobacteria, pp. 471–521. London: Academic Press.

17. Wayne LG, Kubica GP (1986) Genus *Mycobacterium.* In: Sneath PHA, Mair NS, Sharpe ME, Holt JG, eds. Bergey's Manual of Systematic Bacteriology, pp. 1436–1457. Baltimore, MD: Williams & Wilkins.

18. Lévy-Frébault VV, Portaels F (1992) Proposal for recommended minimal standards for the genus *Mycobacterium* and for newly described slowly growing *Mycobacterium* species. Int J System Bacteriol 42:315–323.

19. Lechevalier MP, and Lechevalier HA (1970) Chemical composition as a criterion in the classification of aerobic actinomycetes. Int J System Bacteriol 20:435–443.

20. Schleifer KH, Kandler O (1972) Peptidoglycan types of bacterial cell walls and their taxonomic implications. Bacteriol Rev 36:407–477.

21. Suzuki K, Goodfellow M, O'Donnell AG (1993) Cell envelopes and classification. In: Goodfellow M, O'Donnell AG, eds. Handbook of New Bacterial Systematics, pp. 195–250. London: Academic Press.

22. Cummins CS, Harris H (1958) Studies on the cell wall composition and taxonomy of *Actinomycetales* and related groups. J Gen Microbiol 18:173–189.

23. Brennan PJ (1988) *Mycobacterium* and other actinomycetes. In: Ratledge C, Wilkinson SG, eds. Microbial Lipids, Volume 1, pp. 203–298. London: Academic Press.

24. Minnikin DE, Goodfellow M (1980) Lipid composition in the classification and identification of acid fast bacteria. In: Goodfellow M, Board RG, eds. Microbiological Classification and Identification, pp. 189–256. London: Academic Press.

25. Minnikin DE (1988) Isolation and purification of mycobacterial wall lipids. In: Hancock IC, Poxton IC, eds. Bacterial Cell Surface Techniques, pp. 125–135. Chichester: John Wiley & Sons.

26. Minnikin DE (1993) Mycolic acids. In: Kumar D, Weber N, eds. CRC Handbook of Chromatography: Analysis of Lipids, pp. 339–348. Cleveland, OH: CRC Press.

27. Rainey FA, Klatte S, Kroppenstedt RM, Stackebrandt E (1995) *Dietzia,* a new genus including *Dietzia maris* comb. nov., formerly *Rhodococcus maris.* Int J System Bacteriol 45:101–103.

28. Embley TM, Wait R (1994) Structural lipids of eubacteria. In: Goodfellow M, O'Donnell AG, eds. Chemical Methods in Procaryotic Systematics, pp. 121–161. Chichester: John Wiley & Sons.

29. Magee JG (1995) Clinically significant mycobacteria: classification and identification. Ph.D. thesis, University of Newcastle upon Tyne, UK.

30. Lechevalier MP, DeBiévre C, Lechevalier HA (1977) Chemotaxonomy of aerobic actinomycetes: Phospholipid composition. Biochem System Ecol 5:249–260.

31. Lechevalier MP, Stern AE, Lechevalier HA (1981) Phospholipids in the taxonomy of actinomycetes. Zentbl Bakt ParasitKde Abt I Suppl 11, 111–116.

32. Collins MD (1994) Isoprenoid quinones. In: Goodfellow M, O'Donnell AG, eds.

Chemical Methods in Prokaryotic Systematics, pp. 265–309. Chichester: John Wiley & Sons.

33. Goodfellow M, Orlean PAB, Collins MD, Alshamaony L, Minnikin DE (1978) Chemical and numerical taxonomy of strains received as *Gordona aurantiaca*. J Gen Microbiol 109:57–68.
34. Howarth OW, Grund E, Kroppenstedt RM (1986) Structural determination of a new naturally occurring cyclic vitamin K. Biochem Biophys Res Commun 140:916–923.
35. Collins MD, Howarth OW, Grund E, Kroppenstedt RM (1987) Isolation and structural determination of new members of the vitamin K series in *Nocardia brasiliensis*. FEMS Microbiol Lett 41:35–39.
36. Goren MB, Cernich M, Brokl O (1978) Some observations on mycobacterial acid-fastness. Am Rev Respir Dis 118:151–154.
37. Draper P (1982) The anatomy of mycobacteria. In: Ratledge C, Stanford JL, eds. The Biology of the Mycobacteria, pp. 9–52. London: Academic Press.
38. Sneath PHA (1992) International Code of Nomenclature of Bacteria. Washington DC: American Society for Microbiology.
39. Ludwig W, Neumaier J, Klugbauer N, Brockmann E, Roller C, Jilg S, Reetz K, Schachter I, Ludvigsen A, Bachleitner M, Fischer U, Schleifer KH (1993). Phylogenetic relationships of bacteria based on comparative sequence analysis of elongation factor Tu and ATP-synthase $\beta$-subunit genes. Antonie van Leeuwenhoek 64:285–305.
40. Woese C (1987) Bacterial evolution. Microbiol Rev 51:221–271.
41. Embley TM, Stackebrandt E (1994) The molecular phylogeny and systematics of the actinomycetes. Annu Rev Microbiol 48:257–289.
42. Chun J (1995) Computer assisted classification and identification of actinomycetes. Ph.D. thesis, University of Newcastle upon Tyne, UK.
43. Rainey FA, Burghardt J, Kroppenstedt RM, Klatte S, Stackebrandt E (1995) Phylogenetic analysis of the genera *Rhodococcus* and *Nocardia* and evidence for the evolutionary origin of the genus *Nocardia* from within the radiation of *Rhodococcus* species. Microbiology 141:523–528.
44. Pascual C, Lawson PA, Farrow JAE, Gimenez MN, Collins MD (1995) Phylogenetic analysis of the genus *Corynebacterium* based on 16S rRNA gene sequences. Int J System Bacteriol 45:724–728.
45. Ruimy R, Riegel P, Boiron P, Monteil H, Christen R (1995) Phylogeny of the genus *Corynebacterium* deduced from analyses of small-subunit ribosomal DNA sequences. Int J System Bacteriol 45:740–746.
46. Lehmann KB, Neumann R (1907) Lehmann's Medizin, Handatlanten. X. Atlas und Grundriss der Bakteriologie und Lehrbuch der speciellen bacteriologischen Diagnostik, 4 auflage Teil 2. Munich: J.F. Lehmann.
47. Chester FD (1897) Report of the mycologist: Bacteriological work. Delaware Agric Exp Station Bull 9:38–145.

48. Castellani A, Chalmers AJ (1919) Manual of Tropical Medicine, 3rd. edn. New York: William, Wood and Co.

49. Jukes TH, Cantor CR (1969) Evolution of protein molecules. In: Munro HN, ed. Mammalian Protein Metabolism, pp. 21–132. New York: Academic Press.

50. Saitou N, Nei M (1987) The neighbor-joining method: A new method for reconstructing phylogenetic trees. Molec Biol Evol 4:406–425.

51. Fitch WM, Margoliash E (1967) Construction of phylogenetic trees. Science 155:279–284.

52. Kluge A, Farris FS (1969) Quantitative phyletics and the evolution of anurans. System Zool 18:1–32.

53. Goodfellow M (1992) The family *Nocardiaceae.* In: Starr MP, Stolp H, Trüper HG, Balows A, Schlegel HG, eds. The Prokaryotes, Volume II, pp. 1188–1213. New York: Springer-Verlag.

54. Boiron P, Provost F, Dupont B (1993) Laboratory Methods for the Diagnosis of Nocardiosis. Paris: Institut Pasteur.

55. Holt JG, Krieg NR, Sneath PHA, Staley JT, Williams ST (1994) Bergey's Manual of Determinative Bacteriology, 9th ed. Baltimore, MD: Williams and Wilkins.

56. Uchida K, Aida K (1979) Taxonomic significance of cell wall acyl type in *Corynebacterium–Mycobacterium–Nocardia* group by a glycolate test. J Gen Appl Microbiol 25:169–183.

57. Minnikin DE, Dobson G, Goodfellow M, Draper P, Magnusson M (1985) Quantitative comparison of the mycolic and fatty acid composition of *Mycobacterium leprae* and *Mycobacterium gordonae.* J Gen Microbiol 131:2013–2021.

58. Minnikin DE, Minnikin SM, Parlett JH, Goodfellow M, Magnusson M (1984) Mycolic acid patterns of some species of *Mycobacterium.* Arch Microbiol 139:225–231.

59. Minnikin DE, Minnikin SM, Hutchinson IG, Goodfellow M, Grange JM (1984) Mycolic acid patterns of representative strains of *Mycobacterium fortuitum, "Mycobacterium peregrinum"* and *Mycobacterium smegmatis.* J Gen Microbiol 130:363–367.

60. Dobson G, Minnikin DE, Minnikin SM, Parlett JH, Goodfellow M, Ridell M, Magnusson M (1985) Systematic analysis of complex mycobacterial lipids. In: Goodfellow M, Minnikin DE, eds. Chemical Methods in Bacterial Systematics, pp. 237–265. London: Academic Press.

61. Chun J, Kang S-O, Hah Y-C, Goodfellow M (1996) Phylogeny of mycolic acid-containing actinomycetes. J Ind Microbiol 17:205–213.

62. Collins MD, Burton RA, Jones D (1988) *Corynebacterium amycolatum* sp. nov. a new mycolic acid-less *Corynebacterium* species from human skin. FEMS Microbiol Lett 49:349–352.

63. Funke G, Stubbs S, Altwegg M, Carlotte A, Collins MD (1994) *Turicella otitidis* gen. nov., sp. nov., a coryneform bacterium isolated from patients with otitis media. Int J System Bacteriol 44:270–273.

64. Fox GE, Stackebrandt E (1987) The application of 16S rRNA sequencing to bacterial systematics. Methods Microbiol 19:406–458.

65. Rogall T, Wolters J, Flohr T, Böttger EC (1990) Towards a phylogeny and definition of species at the molecular level within the genus *Mycobacterium.* Int J System Bacteriol 40:323–330.

66. Stahl DA, Urbance JW (1990) The division between fast- and slow-growing species corresponds to natural relationships among the mycobacteria. J Bacteriol 172:116–124.

67. Pitulle C, Dorsch M, Kazda J, Wolters J, Stackebrandt E (1992) Phylogeny of rapidly growing members of the genus *Mycobacterium.* Int J System Bacteriol 42:337–343.

68. Stackebrandt E, Smida J (1988) The phylogeny of the genus *Mycobacterium* as determined by 16S rRNA sequences and development of DNA probes In: Okami Y, Beppu T, Ogawara H, eds. Biology of Actinomycetes ’88, pp. 244–250. Tokyo: Japan Scientific Societies Press.

69. Böttger EC, Hirschel B, Coyle MB (1993) *Mycobacterium genavense* sp. nov. Int J System Bacteriol 43:841–843.

70. Meier A, Kirschner P, Schröder K-H, Wolters J, Kroppenstedt RM, Böttger EC (1993) *Mycobacterium intermedium* sp. nov. Int J System Bacteriol 43:204–209.

71. Springer B, Kirschner P, Rost-Meyer G, Schröder K-H, Kroppenstedt RM, Böttger EC (1993) *Mycobacterium interjectum,* a new species isolated from a patient with chronic lymphadenitis. J Clin Microbiol 31:3083–3089.

72. Wayne LG (1959) Quantitative aspects of neutral red reactions of typical and “atypical” mycobacteria. Am Rev Tuber Pulm Dis 79:526.

73. Wayne LG (1967) Selection of characters for an Adansonian analysis of mycobacterial taxonomy. J Bacteriol 93:1382–1391.

74. Kubica GP, Silcox VA, Kilburn JO, Smithwick RW, Beam RE, Jones WD, Stottmeier KD (1970) Differential identification of mycobacteria. VI. *Mycobacterium triviale* Kubica sp. nov. Int J System Bacteriol 20:161–174.

75. Bremer H, Dennis PP, (1987) Modulation of chemical composition and other parameters of the cell growth rate. In: Neidhardt FC et al., eds. *Escherichia coli* and *Salmonella typhimurium:* Cellular and Molecular Biology, Washington, DC. American Society for Microbiology.

76. Bercovier H, Kafri O, Sela S (1986) Mycobacteria possess a surprisingly small number of ribosomal RNA genes in relation to the size of their genome. Biochem Res Commun 136:1136–1141.

77. Ji YE, Colston MJ, Cox RA (1994) Nucleotide sequence and secondary structures of precursor 16S rRNA of slow-growing mycobacteria. Microbiology 140:123–132.

78. Ji YE, Colston MJ, Cox RA (1994) The ribosomal RNA (rrn) operons of fast-growing mycobacteria: Primary and secondary structures and their relation to rrn operons of pathogenic slow-growers. Microbiology 140:2829–2840.

79. Tsukamura M (1967) Identification of mycobacteria. Tubercle, London 48:311–318.

80. Wayne LG, Dietz TM, Gernez-Rieux C, Jenkins PA, Käppler W, Kubica GP,

Kwapinski JBG, Meissner G, Pattyn SR, Runyon EH, Schröder KH, Silcox VA, Tacquet A, Tsukamura M, Wolinsky E (1971) A co-operative numerical analysis of scotochromogenic slowly growing mycobacteria. J Gen Microbiol 66:255–271.

81. Wayne LG, Andrade L, Froman S, Käppler W, Kubala E, Meissner G, Tsukamura M (1978) A co-operative numerical analysis of *Mycobacterium gastri, Mycobacterium kansasii* and *Mycobacterium marinum.* J Gen Microbiol 109:319–327.

82. Wayne LG, Good RC, Krichevsky MI, Beam RE, Blacklock Z, Chaparas SD, Dawson D, Froman S, Gross W, Hawkins J, Jenkins PA, Juhlin I, Käppler W, Kleeberg HH, Krasnow I, Lefford MJ, Mankiewicz E, McDurmont C, Meissner G, Morgan P, Nel EE, Pattyn SR, Portaels F, Richards PA, Rüsch S, Schröder KH, Silcox VA, Szabo I, Tsukamura M, Vergmann B (1981) First report of the co-operative open-ended study of slowly growing mycobacteria by the International Working Group on Mycobacterial Taxonomy. Int J System Bacteriol 31:1–20.

83. Wayne LG, Good RC, Krichevsky MI, Beam RE, Blacklock Z, David HL, Dawson D, Gross W, Hawkins J, Jenkins PA, Juhlin I, Käppler W, Kleeberg HH, Krasnow I, Lefford MJ, Mankiewicz E, McDurmont C, Nel EE, Portaels F, Richards PA, Rüsch S, Schröder KH, Silcox VA, Szabo I, Tsukamura M, Vandenbzeen I, Vergmann B (1983) Second report of the co-operative open-ended study of slowly growing mycobacteria by the International Working Group on Mycobacterial Taxonomy. Int J System Bacteriol 33:265–274.

84. Wayne LG, Good RC, Krichevsky MI, Blacklock Z, David HL, Dawson D, Gross W, Hawkins J, Jenkins PA, Juhlin I, Käppler W, Kleeberg HH, Levy-Frébault V, McDurmont C, Nel EE, Portaels F, Rüsch-Gerdes S, Schröder KH, Silcox VA, Szabo I, Tsukamura M, Van Den Breen L, Vergmann B, Yakrus MA (1989) Third report of the co-operative, open-ended study of slowly growing mycobacteria by the International Working Group on Mycobacterial Taxonomy. Int J System Bacteriol 39:267–278.

85. Wayne LG, Good RC, Krichevsky MI, Blacklock Z, David HL, Dawson D, Gross W, Hawkins J, Vincent-Lévy-Frébault V, McManus C, Portaels F, Rusch-Gerdes S, Schröder KH, Silcox VA, Tsukamura M, Van Der Breen L, Yakrus MA (1991) Fourth report of the co-operative, open-ended study of slowly growing mycobacteria by the International Working Group on Mycobacterial Taxonomy. Int J System Bacteriol 41:463–472.

86. Wayne LG, Good RC, Tsang A, Butler R, Dawson D, Groothuis D, Gross W, Hawkins J, Kilburn J, Kubin M, Schröder KH, Silcox VA, Smith C, Thorel MF, Woodley C, Yakrus MA (1993) Serovar determination and molecular taxonomic correlation in *Mycobacterium avium, Mycobacterium intracellulare* and *Mycobacterium scrofulaceum:* A co-operative study of the International Working Group on Mycobacterial Taxonomy. Int J System Bacteriol 43:482–489.

87. Wayne LG, Good RC, Böttger EC, Butler R, Dorsch M, Ezaki T, Gross W, Jonas V, Kilburn J, Kirschner P, Krichevsky MI, Ridell M, Shinnick TM, Springer B, Stackebrandt E, Tarnok I, Tarnok Z, Tasaka H, Vincent V, Warren NG, Knott CA, Johnson R (1996) Semantide- and chemotaxonomy-based analyses of some

problematic phenotypic clusters of slowly growing mycobacteria, a comparative study of the International Working Group on Mycobacterial Taxonomy. Int J System Bacteriol 46:280–297.

88. Goodfellow M, Lind A, Mordarska H, Pattyn S, Tsukamura M (1974) A co-operative numerical analysis of cultures considered to belong to the "rhodochrous" taxon. J Gen Microbiol 85:291–302.
89. Kubica GP, Baess I, Gordon RE, Jenkins PA, Kwapinski JBG, McDurmont C, Pattyn SR, Saito H, Silcox V, Stanford JL, Takeya K, Tsukamura M (1972) A co-operative numerical analysis of rapidly growing mycobacteria. J Gen Microbiol 73:55–70.
90. Meissner G, Schröder KH, Amadio GE, Anz W, Chaparas S, Engel HWB, Jenkins PA, Käppler W, Kleeberg HH, Kubala E, Kubin M, Lauterbach D, Lind A, Magnusson M, Mikova ZD, Pattyn SR, Schaefer WB, Stanford JL, Tsukamura M, Wayne LG, Willers I, Wolinsky E (1974) A co-operative numerical analysis of nonscoto- and nonphoto-chromogenic slowly growing mycobacteria. J Gen Microbiol 83:207–235.
91. Saito H, Gordon RE, Juhlin I, Käppler W, Kwapinski JBG, McDurmont C, Pattyn SR, Runyon EH, Stanford JL, Tárnok I, Tasaka H, Tsukamura M, Weiszfeiler J (1977) Co-operative numerical analysis of rapidly growing mycobacteria. Int J System Bacteriol 27:75–85.
92. Minnikin DE (1982) Lipids: Complex lipids, their chemistry, biosynthesis and role. In: Ratledge C, Stanford JL, eds. The Biology of the Mycobacteria, Volume 1, pp. 95–184. London: Academic Press.
93. Stanford JL, Grange JM (1974) The meaning and structure of species as applied to mycobacteria. Tubercle 55:143–52.
94. Lind A, Ridell M (1976) Serological relationships between *Nocardia, Mycobacterium, Corynebacterium* and the "*rhodochrous*" complex. In Goodfellow M, Brownell GH, Serrano JA, eds. The Biology of the Nocardiae, pp. 220–235. London: Academic Press.
95. Chaparas SD, Brown TM, Hyman IS (1978) Antigenic relationships among species of *Mycobacterium* studied by fused rocket immunoelectrophoresis. Int J System Bacteriol 28:547–560.
96. Daffé M, Lanéelle MA, Asselineau C, Lévy-Frébault V, David H (1983) Intérêt taxonomique des acides gras des mycobactéries: Proposition d'une méthode d'analyse. Ann Microbiol (Institut Pasteur) 134B:241–256.
97. Minnikin DE, Hutchinson IG, Caldicott AB, Goodfellow M (1980) Thin-layer chromatography of methanolysates of mycolic acid-containing bacteria. J Chromatogr 188:221–233.
98. Ausina V, Luquin M, Garcia-Barcelo M, Lanéelle MA, Lévy-Frébault V, Belda F, Prats G (1992) *Mycobacterium alvei* sp. nov. Int J System Bacteriol 42:529–535.
99. Butler WR, O'Connor SP, Yakrus MA, Smithwick RW, Plikaytis BB, Moss CW, Floyd MM, Woodley CW, Kilburn JO, Vadney FS, Gross WM (1993) *Mycobacterium celatum* sp. nov. Int J System Bacteriol 43:539–548.

100. Häggblom MM, Nohynek LJ, Palleroni NJ, Kronqvist K, Nurmiaho-Lassila EL, Salkinoja-Salonen MS, Klatte S, Kroppenstedt RM (1994) Transfer of polychlorophenol-degrading *Rhodococcus chlorophenolicus* (Apajalahti *et al.,* 1986) to the genus *Mycobacterium* as *Mycobacterium chlorophenolicum* comb. nov. Int J System Bacteriol 44:485–493.

101. Kazda J, Stackebrandt E, Smida J, Minnikin DE, Daffé M, Parlett JH, Pitulle C (1990) *Mycobacterium cookii* sp. nov. Int J System Bacteriol 40:217–223.

102. Kazda J, Múller HG, Stackebrandt E, Daffé M, Müller K, Pitulle C (1992) *Mycobacterium madagascariense* sp. nov. Int J System Bacteriol 42:524–528.

103. Kazda J, Cooney R, Monaghan M, Quinn PJ, Stackebrandt E, Dorsch M, Daffé M, Müller K, Cook BR, Tárnok ZS (1993) *Mycobacterium hiberniae* sp. nov. Int J System Bacteriol 43:352–357.

104. Kirschner P, Teske A, Schröder KH, Kroppenstedt RM, Wolters J, Böttger EC (1992) *Mycobacterium confluentis* sp. nov. Int J System Bacteriol 42:257–262.

105. Koukila-Kähkölä P, Springer B, Böttger ER, Paulin L, Jantzen E, Katila M-L (1995) *Mycobacterium branderi* sp. nov., a new potential human pathogen. Int J System Bacteriol 45:549–553.

106. Luquin M, Ausina V, Vincent-Lévy-Frébault V, Lanéelle MA, Belda F, Garcia-Barcelo M, Prats G, Daffé M (1993) *Mycobacterium brumae* sp. nov., a rapidly growing, nonphotochromogenic *Mycobacterium.* Int J System Bacteriol 43:405–413.

107. Lévy-Frébault V, Rafidinarivo E, Prome J-C, Grandry J, Boisvert H, David HL (1983) *Mycobacterium fallax* sp. nov. Int J System Bacteriol 33:336–343.

108. Padgitt PJ, Moshier SE (1987) *Mycobacterium poriferae* sp. nov., a scotochromogenic, rapidly growing species isolated from a marine sponge. Int J System Bacteriol 37:186–191.

109. Tsukamura M, Yano I, Imaeda T (1986) *Mycobacterium moriokaense* sp. nov., a rapidly growing, nonphotochromogenic *Mycobacterium.* Int J System Bacteriol 36:333–338.

110. Valero-Guillén PL, Martin-Luengo F, Larsson L, Jiménez J, Juhlin I, Portaels F (1988) Fatty and mycolic acids of *Mycobacterium malmoense.* J Clin Microbiol 26:153–154.

111. Yassin AF, Binder C, Schaal KP (1993) Identification of mycobacterial isolates by thin-layer and capillary gas-chromatography under diagnostic routine conditions. Zentbl Bakteriol 278:34–48.

112. Ratledge C (1982) Nutrition, growth and metabolism. In: Ratledge C, Stanford JL, eds. The Biology of the Mycobacteria, Volume 1. Physiology, Identification and Classification, pp. 185–271. London: Academic Press.

113. Hall RM, Ratledge C (1984) Mycobactins as chemotaxonomic characters for rapidly growing mycobacteria. J Gen Microbiol 130:1883–1892.

114. Wayne LG, Sramek HA (1992) Agents of newly recognised or infrequently encountered mycobacterial diseases. Clin Microbiol Rev 5:1–25.

115. Falkinham JO, III (1996) Epidemiology of infection by non-tuberculous mycobacteria. Clin Microbiol Rev 9:177–215.

116. Gordon RE, Smith MM (1953) Rapidly-growing acid-fast bacteria. I. Species descriptions of *Mycobacterium phlei* Lehmann and Neumann and *Mycobacterium smegmatis* (Trevisan) Lehmann and Neumann. J Bacteriol 66:41–48.

117. Gordon RE, Smith MM (1955) Rapidly-growing acid-fast bacteria. II. Species descriptions of *Mycobacterium fortuitum* Cruz. J Bacteriol 69:502–507.

118. Gordon RE, Mimh JM (1959) A comparison of species of mycobacteria. J Gen Microbiol 21:736–748.

119. Bojalil LF, Cérbon J, Trujillo A (1962) Adansonian classification of mycobacteria. J Gen Microbiol 28:333–346.

120. Wayne LG (1964) The mycobacterial mystique: Deterrent to taxonomy. Am Rev Respir Dis 90:255–257.

121. Wayne LG (1967) Selection of characters for an Adansonian analysis of mycobacterial taxonomy. J Bacteriol 93:1382–1391.

122. Wayne LG (1981) Numerical taxonomy and co-operative studies: Roles and limits. Rev Infect Dis 3:822–828.

123. Wayne LG (1984) Mycobacterial speciation. In: Kubica GP, Wayne LG, eds. The Mycobacteria: A Sourcebook, pp. 25–65. New York: Marcel Dekker.

124. Wayne LG (1985) Computer-assisted analysis of data from co-operative studies on mycobacteria. In: Goodfellow M, Jones D, Priest FG, eds. Computer-Assisted Bacterial Systematics, pp. 91–105. London: Academic Press.

125. Wayne LG, Engbaek HC, Engel HWB, Froman S, Gross W, Hawkins J, Käppler W, Karlson AG, Kleeberg HH, Krasnow I, Kubica GP, McDurmont C, Nel EE, Pattyn SR, Schröder KH, Showalter S, Tárnok I, Tsukamura M, Vergmann B, Wolinsky E (1974) Highly reproducible techniques for use in systematic bacteriology in the genus *Mycobacterium:* Tests for pigment, urease, resistance to sodium chloride, hydrolysis of Tween 80 and $\beta$-galactosidase. Int J System Bacteriol 24:412–419.

126. Wayne LG, Engel HWB, Grassi C, Gross W, Hawkins J, Jenkins PA, Käppler W, Kleeberg HH, Krasnow I, Nel EE, Pattyn SR, Richards PA, Showalter S, Slosárek M, Szabo I, Tárnok I, Tsukamura M, Vergmann B, Wolinsky E (1976) Highly reproducible techniques for use in systematic bacteriology in the genus *Mycobacterium:* Tests for niacin and catalase and for resistance to isoniazid, thiophene-2-carboxylic acid hydrazide, hydroxylamine and $p$-nitrobenzoate. Int J System Bacteriol 26:311–318.

127. Wayne LG, Krichevsky EJ, Love LI, Johnson R, Krichevsky MI (1980) Taxonomic probability matrix for use with slowly growing mycobacteria. Int J System Bacteriol 30:528–538.

128. Wayne LG, Krichevsky MI, Portyrata D, Jackson CK (1984) Diagnostic probability matrix for the identification of slowly growing mycobacteria in clinical laboratories. J Clin Microbiol 20:722–729.

129. Goodfellow M, Manfio GP, Chun J (1996) Towards a practical species concept for

cultivable bacteria. In: Claridge MF, Dawah HA, Wilson MR, eds. Species: The Units of Biodiversity, pp. 25–59. Oxford: Chapman & Hall.

130. Stackebrandt, E., and M. Goodfellow, eds. 1991. Nucleic Acid Techniques in Bacterial Systematics. John Wiley & Sons, Chichester.
131. Goodfellow M, O'Donnell AG (1993) Roots of bacterial systematics. In: Goodfellow M, O'Donnell AG, eds. Handbook of New Bacterial Systematics, pp. 3–54. London Academic Press.
132. Goodfellow M, O'Donnell AG (1994) Chemosystematics: Current state and future prospects. In: Goodfellow M, O'Donnell AG, eds. Chemical Methods in Procaryotic Systematics, pp. 1–20. Chichester: John Wiley & Sons.
133. Ravin AW (1960) The origin of bacterial species: Genetic recombination and factors limiting it between bacterial populations. Bacteriol Rev 24:201–220.
134. Ravin AW (1963) Experimental approaches to the study of bacterial phylogeny. Am Naturalist 97:307–318.
135. Colwell RR (1970) Polyphasic taxonomy of bacteria. In: Iizuka H, Hasegawa T, eds. Culture Collections of Microorganisms, pp. 421–436. Tokyo: University of Tokyo Press.
136. Wayne LG, Brenner DJ, Colwell RR, Grimont PAD, Kandler P, Krichevsky MI, Moore LH, Moore WEC, Murray RGE, Stackebrandt E, Starr MP, Trüper HG (1987) Report of the *ad hoc* committee on reconciliation of approaches to bacterial systematics. Int J System Bacteriol 37:463–464.
137. Britten RJ, Kohne DE (1968) Repeated sequences in DNA. Science 161:529–540.
138. Stackebrandt E, Goebel BM (1994) Taxonomic note; A place for DNA:DNA reassociation and 16S rRNA sequence analysis in the present species definition in bacteriology. Int J System Bacteriol 44:846–849.
139. Kusunoki S, Ezaki T (1992) Proposal of *Mycobacterium peregrinum* sp. nov., nom. rev., and elevation of *Mycobacterium abscessus* (Kubica *et al.*) to species status: *Mycobacterium abscessus* comb. nov. Int J System Bacteriol 42:240–245.
140. Chester FD (1901) A Manual of Determinative Bacteriology. New York: The Macmillan Co.
141. Bergey DH, Harrison FC, Breed RS, Hammer BW, Huntoon FM, eds. (1923) Bergey's Manual of Determinative Bacteriology, 1st ed. Baltimore, MD: The Williams and Wilkins Co.
142. Hurley SS, Splitter GA, Welch RA (1988) Deoxyribonucleic acid relatedness of *Mycobacterium paratuberculosis* and other members of the family *Mycobacteriaceae.* Int J System Bacteriol 38:143–146.
143. Saxegaard F, Baess I, Jantzen E (1988) Characterisation of clinical isolates of *Mycobacterium paratuberculosis* by DNA: DNA hybridisation and cellular fatty acid analysis. Acta Pathol Microbiol Scand 96:497–502.
144. Yoshimura HH, Graham DY, Estes MK, Merkel RS (1987) Investigation of association of mycobacteria with inflammatory bowel disease by nucleic acid hybridisation. Clin Microbiol 25:45–51.

145. Springer B, Böttger E, Kirschner P, Wallace RJ Jr (1995) Phylogeny of the *Mycobacterium chelonae*-like organisms based on partial sequencing of the 16S rRNA gene and the proposal of *Mycobacterium mucogenicum* sp. nov. Int J System Bacteriol 45:262–67.

146. Wayne LG (1985) The "atypical" mycobacteria; recognition and disease association. Crit Rev Microbiol 12:185–222.

147. Salfinger M, Pfyffer GE (1994) The new diagnostic mycobacteriology laboratory. Eur J Clin Microbiol Infect Dis 13:961–979.

148. Drake TA, Hindler JA, Berlin OGW, Bruckner DA (1987) Rapid identification of *Mycobacterium avium* complex in culture using DNA probes. J Clin Microbiol 25:1442–1445.

149. Musial CE, Tice LS, Stockman L, Roberts GD (1988) Identification of mycobacteria from culture by using Gen-Probe Rapid Diagnostic System for *Mycobacterium avium* complex and *Mycobacterium tuberculosis* complex. J Clin Microbiol 26:2120–2123.

150. Saito H, Tomioka H, Sato K, Tasaka H, Dawson DJ (1990) Identification of various serovar strains of *Mycobacterium avium* complex by using DNA probes specific for *Mycobacterium avium* and *Mycobacterium intracellulare*. J Clin Microbiol 28:1694–1697.

151. Tasaka H, Kiyotani K, Matsuo Y (1983) Purification and antigenic specificity of alpha protein (Yoneda and Fukui) from *Mycobacterium tuberculosis* and *Mycobacterium intracellulare*. Hiroshima J Med Sci b32:1–8.

152. Tasaka H, Nomura T, Matsuo Y (1985) Specificity and distribution of alpha antigens of *Mycobacterium avium-intracellulare, Mycobacterium scrofulaceum*, and related species of mycobacteria. Am Rev Respir Dis 132:173–174.

153. Wayne LG, Diaz GA (1987) Intrinsic catalase dot blot immunoassay for identification of *Mycobacterium tuberculosis, Mycobacterium avium*, and *Mycobacterium intracellulare*. J Clin Microbiol 25:1687–1690.

154. Luquin M, Ausina V, López Calahorra F, Belda F, Garcia Barceló M, Celma C, Prats G (1991) Evaluation of practical chromatographic procedures for identification of clinical isolates of mycobacteria. J Clin Microbiol 29:120–130.

155. Butler WR, Thibert L, Kilburn JO (1992) Identification of *Mycobacterium avium* complex strains and some similar species by high performance liquid chromatography. J Clin Microbiol 30:2698–2704.

156. Murray RGE, Brenner DJ, Colwell RR, de Vos P, Goodfellow M, Grimont PAD, Pfennig, NP, Stackebrandt E, and Zavarzin GA (1990) Report of the *ad hoc* committee on approaches to taxonomy within the *Proteobacteria.* International J System Bacteriol 40:213–215.

157. Castets, M, Rist N, Boisvert H (1969) La variété africaine du bacille tuberculeux humain. Méd Afrique Noire 16:321–322.

158. Karlsen AG, Lessel EF (1970) *Mycobacterium bovis* nom. nov. Int J System Bacteriol 20:273–282.

159. Aronson JD (1926) Spontaneous tuberculosis in salt water fish. J Infect Dis 39:315–320.

160. Breed RS, Murray EGD, Smith NR, eds. (1957) Bergey's Manual of Determinative Bacteriology, 7th ed. Baltimore, MD: The Williams and Wilkins Co.

161. Fenner F (1950) The significance of the incubation period in infectious diseases. Med J Australia 2:813–818.

162. Grange JM (1981) Koch's tubercle bacillus—a century reappraisal. Zentralblatt Bakteriol Originale A251:297–307.

163. Calmette A, Guérin C (1908) Sur quelques propriétés du bacille tuberculeux culture sur la bile. C R Acad Séances 147:1456–1459.

164. Minnikin DE, Minnikin SM, Dobson G, Goodfellow M, Portaels F, Van Den Breen L, Sesardic D (1983) Mycolic acid patterns of four vaccine strains of *Mycobacterium bovis* BCG. J Gen Microbiol 129:889–891.

165. Sisson PR, Freeman R, Magee JG, Lightfoot NF (1991) Differentiation between mycobacteria of the *Mycobacterium tuberculosis* complex by pyrolysis mass spectrometry. Tubercle 72:206–209.

166. David HL, Jahan M-T, Jumin A, Grandry J, Lehman EH (1978) Numerical taxonomy analysis of *Mycobacterium africanum.* Int J System Bacteriol 28:467–472.

167. Wells AQ (1937) Tuberculosis in wild voles. Lancet 232:1221.

168. Smith N (1960) The "Dassie" bacillus. Tubercle 41:203–212.

169. Magnusson M (1980) Classification and identification of mycobacteria on the basis of sensitin specificity. In: Meissner G, Scheidel A, Nelles A, Pfaffenberg R, eds. Mykobakterien und mykobakterielle Krankheiten. Teil 1. Systematik det Mykobakterien, pp. 319–348. Jena: Gustav Fischer Verlag.

170. Wayne LG, Diaz GA (1979) Reciprocal immunologic distances of catalase derived from *Mycobacterium avium, M. tuberculosis* and closely related species. Int J System Bacteriol 29:19–24.

171. Baess I (1979) Deoxyribonucleic acid relatedness among species of slowly-growing mycobacteria. Acta Pathol Microbiol Scand B87:221–226.

172. Magnusson M (1967) Identification of species of *Mycobacterium* on the basis of the specificity of the delayed type reaction in guinea pigs. Zeitsch Tuberkulose Erkrankungen Thoraxorgane 127:55–56.

173. Wayne LG, Diaz GA (1982) Serological, taxonomic and kinetic studies of the T and M classes of mycobacterial catalase. Int J System Bacteriol 32:296–304.

174. Thorel M-F, Krichevsky M, Lévy-Frébault VV (1990) Numerical taxonomy of mycobactin dependent mycobacteria, emended description of *Mycobacterium avium,* and description of *Mycobacterium avium* subsp. *avium* subsp. nov., *Mycobacterium avium* subsp. *paratuberculosis* subsp. nov., and *Mycobacterium avium* subsp. *silvaticum* subsp. nov. Int J System Bacteriol 40:254–260.

175. Cuttino JT, McCabe AM (1949) Pure granulomatous nocardiosis, a new fungus disease distinguished by intracellular parasitism. Am J Clin Pathol 25:1–34.

176. Runyon EH (1959) Anonymous mycobacteria in pulmonary disease. Med Clin North Am 43:273–290.

177. Marchaux E, Sorel F (1912) Recherches sur la leprae. Ann Institut Pasteur 26:675–700.

178. Gross WM, Wayne LG (1970) Nucleic acid homology in the genus *Mycobacterium.* J Bacteriol 104:630–634.

179. Hawkins JE (1977) Scotochromogenic mycobacteria which appear intermediate between *M. avium/intracellulare* and *M. scrofulaceum.* Am Rev Respir Dis 116:963–964.

180. Wayne LG, Diaz GA (1988) Detection of a novel catalase in extracts of *Mycobacterium avium* and *Mycobacterium intracellulare.* Infect Immun 56:936–941.

181. Wayne LG, Diaz GA (1986) A double staining method for differentiation between two classes of mycobacterial catalase in polyacrylamide electrophoresis gels. Anal Biochem 157:89–92.

182. Baess I (1983) Deoxyribonucleic acid relationships between different serovars of *Mycobacterium avium, Mycobacterium intracellulare* and *Mycobacterium scrofulaceum.* Acta Pathol Microbiol Scand B91:201–203.

183. McFadden JJ, Butcher PD, Thompson J, Chiodini R, Hermon-Taylor J (1987) The use of DNA probes identifying restriction-fragment-length polymorphisms to examine the *Mycobacterium avium* complex. Molec Microbiol 1:283–291.

184. Whipple DL, LeFebre B, Andrews RE, Thiermann AB (1987) Isolation and analysis of restriction endonuclease digestive patterns of chromosomal DNA from *Mycobacterium paratuberculosis* and from other *Mycobacterium* species. J Clin Microbiol 25:1511–1515.

185. Crawford JT, Bates JH (1986) Analysis of plasmids in *M. avium-intracellulare* isolates from persons with acquired immunodeficiency syndrome. Am Rev Respir Dis 134:659–661.

186. Meissner PS, Falkinham JO (1986) Plasmid DNA profiles of epidemiologic markers for clinical and environmental isolates of *Mycobacterium avium, Mycobacterium intracellulare* and *Mycobacterium scrofulaceum.* J Infect Dis 153:325–331.

187. Brennan PJ, Mayer H, Aspinall GO, Nam Shin JE (1981) Structures of the glycopeptidolipid antigens from serovars in the *Mycobacterium avium/ Mycobacterium intracellulare/Mycobacterium scrofulaceum* serocomplex. Eur J Biochem 115:7–15.

188. Wayne LG, Good RC, Tsang A, Butler R, Dawson D, Groothuis D, Gross W, Hawkins J, Kilburn J, Kubin M, Schröder KH, Silcox VA, Smith C, Thorel MF, Woodley C, Yakrus MA (1993) Serovar determination and molecular taxonomic correlation in *Mycobacterium avium, Mycobacterium intracellulare* and *Mycobacterium scrofulaceum:* A cooperative study of the International Working Group on Mycobacterial Taxonomy. Int J System Bacteriol 43:482–489.

189. Wards BJ, Collins DM, DeLisle GW (1987) Restriction endonuclease analysis of members of the *Mycobacterium avium-M. intracellulare-M. scrofulaceum* serocomplex. J Clin Microbiol 25:2309–2313.

190. Johne HA, Frothingham L (1895) Ein eigenthümlicher fall von Tuberculose beim Rind. Deutsche Zeitschr Tiermedizine 21:438–454.

191. McFadden JJ, Collins J, Beaman B, Arthur M, Gitnick G (1992) Mycobacteria in Crohn's disease: DNA probes identify the wood pigeon strains of *Mycobacterium avium* and *Mycobacterium paratuberculosis* from human tissue. J Clin Microbiol 30:3070–3073.

192. Green EP, Tizard MLV, Moss MT, Thomson J, Winterbourne DW, McFadden JJ, Hermon-Taylor J (1989) Sequence and characteristics of IS900, an insertion element identified in a human Crohn's disease isolate of *Mycobacterium paratuberculosis.* Nucl Acids Res 17:9063–9073.

193. Weiszfeiler G, Karasseva V, Karzag E (1971) A new *Mycobacterium* species: *Mycobacterium asiaticum* n. sp. Acta Microbiol Acad Sci Hung 18:247–252.

194. Karasseva V, Weiszfeiler J, Krasznay E (1965) Occurrence of atypical mycobacteria in *Macacus rhesus.* Acta Microbiol Acad Sci Hung 12:275–282.

195. Baess I, Magnusson M (1982) Classification of *Mycobacterium simiae* by means of comparative reciprocal intra-dermal sensitin testing on guinea pigs and deoxyribonucleic acid hybridisation. Acta Pathol Microbiol Immunol Scand 90:101–107.

196. Wayne LG, Diaz GA (1985) Identification of mycobacteria by specific precipitation of catalase with absorbed sera. J Clin Microbiol 21:721–725.

197. Wayne LG (1966) Classification and identification of mycobacteria. III. Species within group III. Am Rev Respir Dis 93:919–928.

198. Hauduroy P (1955) Derniers Aspects du Monde des Mycobacteries. Paris: Masson et Cie.

199. Magnusson M (1971) A comparative study of *Mycobacterium gastri* and *Mycobacterium kansasii* by delayed type skin reactions in guinea pigs. Am Rev Respir Dis 104:377–384.

200. Huang ZH, Ross BC, Dwyer R (1991) Identification of *Mycobacterium kansasii* by DNA hybridisation. J Clin Microbiol 29:2125–2129.

201. Ross BC, Jackson K, Yang M, Sievers A, Dwyer B. Identification of a genetically distinct subspecies of *Mycobacterium kansasii.* J Clin Microbiol 30:2930–2933.

202. Ross BC, Raios K, Jackson K, Dwyer B (1992) Molecular cloning of a highly repeated DNA element from *Mycobacterium tuberculosis* and it's use as an epidemiological tool. J Clin Microbiol 30:942–946.

203. Anz W, Schröder KH (1970) Photochromogene Stämme von *Mycobacterium gastri* Zentralblatt Bakteriol A214:553–554.

204. Norlin M, Lind A, Ouchterlony Ö (1969) A serologically based taxonomic study of *Mycobacterium gastri.* Zeitschr Immunitatsforsch Allergie Klinische Immunol 137:241–248.

205. Tsukamura M (1965) A group of mycobacteria from soil sources resembling nonphotochromogens (group III), a description of *Mycobacterium nonchromogenicum.* Med Biol 71:110–113.

206. Imaeda T, Broslawski G, Imaeda S (1988) Genomic relatedness among mycobacterial species by nonisotopic blot hybridisation. Int J System Bacteriol 38:151–156.

207. Valdivia Alvarez JA, Ferra Salazar C, Echemendia Font M, Dumas Valdivisieso S (1975) Estudio de 16 cepas de microbacterias cromogenicas y de crecimiento rapido. Rev Cubana Med Trop 27:213–233.

208. Meissner G, Schröder KH (1975) Relationship between *Mycobacterium simiae* and *Mycobacterium habana.* Am Rev Respir Dis III:196–200.

209. Thorel MF (1976) Utilisation d'une methode d'immunoelectrophorese bidimensionalle dans l-etude des antigenes de *Mycobacterium simiae* et *Mycobacterium habana.* Ann Institut Pasteur Microbiol 127B:41–55.

210. Weiszfeiler JG, Karczag E (1976) Synonymy of *Mycobacterium simiae* Karasseva *et al.* 1965 and *Mycobacterium habana* Valdivia *et al.* 1971. Int J System Bacteriol 26:474–477.

211. Coyle MB, Carlson LDC, Wallis CK, Leonard RB, Raisys VA, Kilburn JO, Sanadpour M, Böttger EC (1992) Laboratory aspects of "*Mycobacterium genavense*", a proposed species isolated from AIDS patients. J Clin Microbiol 30:3206–3212.

212. Schröder KH, Juhlin I (1977) *Mycobacterium malmoense* sp. nov. Int J System Bacteriol 27:241–246.

213. Marks J, Jenkins PA, Tsukamura M (1972) *Mycobacterium szulgai*—A new pathogen. Tubercle 53:210–214.

214. Sompolinsky D, Lagziel A, Naveh D, Yankilevitz T (1978) *Mycobacterium haemophilum* sp. nov., a new pathogen of humans. Int J System Bacteriol 28:67–75.

215. Sompolinsky D, Lagziel A, Rosenberg I (1979) Further studies of a new pathogenic mycobacterium (*M. haemophilum* sp. nov.). Can J Microbiol 25:217–26.

216. Dawson DJ, Jennis F (1980) Mycobacteria with a growth requirement for ferric ammonium citrate, identified as *Mycobacterium haemophilum.* J Clin Microbiol 11:190–192.

217. Damato JJ, Collins MT (1984) Radiometric studies with gas-liquid and thin-layer chromatography for rapid demonstration of haem dependence and characterisation of *Mycobacterium haemophilum.* J Clin Microbiol 20:515–518.

218. Portaels F, Dawson DJ, Larsson L, Rigouts L (1993) Biochemical properties and fatty acid composition of *Mycobacterium haemophilum:* Study of 16 isolates from Australian patients. J Clin Microbiol 31:26–30.

219. Besra GS, Minnikin DE, Rigouts L, Portaels F, Ridell M (1990) A characteristic phenolic glycolipid antigen from *Mycobacterium haemophilum.* Lett Appl Microbiol 11:202–204.

220. Kauppinen J, Pelkonen J, Katila M-L (1994) RFLP analysis of *Mycobacterium malmoense* strains using ribosomal RNA gene probes: An additional tool to examine intraspecies variation. J Microbiol Methods 19:261–267.

221. Springer B, Tortoli E, Richter I, Grünewald R, Rüsch-Gerdes S, Uschmann K, Suter F, Collins MD, Kroppenstedt R, Böttger EC (1995) *Mycobacterium conspicuum* sp. nov., a new species isolated from patients with disseminated infections. J Clin Microbiol 33:2805–2811.

222. Castelnuovo G, Morellini M (1962) Gli antigeni di alcuni dei cosidetti "Micobatten atipici" O "anonimi". Ann Institute Carlo Forlanini 22:1–20.

223. Schaefer WB (1965) Serologic identification and classification of atypical mycobacteria by their agglutination. Am Rev Respir Dis 92(Suppl):85–93.

224. Schaefer WB (1968) Incidence of the serotypes of *Mycobacterium avium* and atypical mycobacteria in human and animal diseases. Am Rev Respir Dis 97:18–23.

225. Jenkins PA, Marks J, Schaefer WB (1972) Thin layer chromatography of mycobacterial lipids as an aid to classification. The scotochromogenic mycobacteria, including *Mycobacterium scrofulaceum, M. xenopi, M. aquae, M. gordonae* and *M. flavescens.* Tubercle 53:118–127.

226. Brennan PJ (1981) Structures of the typing antigens of atypical mycobacteria: A brief review of present knowledge. Rev Infect Dis 3:905–913.

227. Goslee S, Rynearson JK, Wolinsky E (1976) Additional serotypes of *M. scrofulaceum, M. gordonae, M. marinum* and *M. xenopi* determined by agglutination. Int J System Bacteriol 26:136–142.

228. Prissick FA, Masson AM (1956) Cervical lymphadenitis in children caused by chromogenic mycobacteria. J Can Med Assoc 75:798–803.

229. Judicial Commission (1978) Opinion 53. Rejection of the species name *Mycobacterium marianum* Penso 1953. Int J System Bacteriol 28:334.

230. Tsukamura M (1982) *Mycobacterium shimoidei* sp. nov., nom. rev., a lung pathogen. Int J System Bacteriol 32:67–69.

231. Schwabacher H (1959) A strain of mycobacterium isolated from skin lesions of a cold blooded animal, *Xenopus laevis,* and it's relation to atypical acid fast bacilli occurring in man. J Hyg Cambridge 57:57–67.

232. Hansen GA (1874) Undersøgelser angående spedalskhedens årsager. Norsk Magazin Laegevidenskaben (Oslo, Norge) 3:88, LIII.

233. Shepard CC (1960) The experimental disease that follows the injection of human leprosy bacilli into foot-pads of mice. J Exp Med 112:445–454.

234. Kirchheimer WF, Storrs EE (1971) Attempts to establish the armadillo (*Dasypus novemcinctus* Linn.) as a model for the study of leprosy. (1). Report of lepromatoid leprosy in an experimentally infected armadillo. Int J Leprosy 39:693–702.

235. Portaels F, Francken A, Pattyn SR (1982) Bacteriological studies of armadillo livers infected with *Mycobacterium leprae.* Ann Soc Belge méd trop 62:233–245.

236. Portaels F, De Ridder K, Pattyn SR (1985) Cultivable mycobacteria isolated from organs of armadillos uninoculated and inoculated with *Mycobacterium leprae.* Ann Inst Pasteur/Microbiol 136A:181–190.

237. Portaels F, Asselineau C, Baess I, Daffé M, Dobson G, Draper P, Gregory D, Hall RM, Imaeda T, Jenkins PA, Lanéelle MA, Larsson L, Magnusson M, Minnikin DE, Pattyn SR, Wieten G, Wheeler PR (1986) A comparative taxonomic study of mycobacteria isolated from armadillos with *Mycobacterium leprae.* J Gen Microbiol 132:2693–2707.

238. De Kesel M, Coene M, Portaels F, Cocito C (1987) Analysis of deoxyribonucleic acids from armadillo-derived mycobacteria. Int J System Bacteriol 37:317–322.

239. Wallace RJ Jr (1994) Recent changes in taxonomy and disease manifestations of the rapidly growing mycobacteria. Eur J Clin Microbiol Infect Dis 13:953–960.

240. Tsukamura M, Nemoto H, Yugi H (1983) *Mycobacterium porcinum* sp. nov., a porcine pathogen. Int J System Bacteriol 33:162–165.

241. Gordon RE (1937) The classification of acid-fast bacteria. J Bacteriol 34:617–630.

242. Gordon RE, Hagan WA (1938) The classification of acid-fast bacteria. J Bacteriol 36:39–46.

243. Lind A, Ridell M (1984) Mycobacterial species: Immunological classification. In: Kubica GP, Wayne LG, eds. The Mycobacteria: A Sourcebook, Part A, pp. 67–82. New York: Marcel Dekker.

244. Ridell M, Baker R, Lind A, Ouchterlony Ö (1979) Immunodiffusion studies of ribosomes in classification of mycobacteria and related taxa. Int Arch Allergy Appl Immunol 59:162–172.

245. Takeya K, Tokiwa H (1972) Mycobacteriocin classification of rapidly growing mycobacteria. Int J System Bacteriol 22:178–180.

246. Baess I, Bentzon W (1969) Rapidly growing mycobacteria. Susceptibility to bacteriophages, reactions in the amidase test, production of acids from carbohydrates and growth at various temperatures. Acta Pathol Microbiol Scand B75:331–347.

247. Butler WR, Jost KC Jr, Kilburn JO (1991) Identification of mycobacteria by high-performance liquid chromatography 29:2468–2472.

248. Yassin AF, Brzezinka H, Schaal KP (1993) Cellular fatty acid methyl ester profiles as a trial in the differentiation of members of the genus *Mycobacterium.* Zentralblatt Bakteriol 279:316–329.

249. Lévy-Frébault F, Grimont F, Grimont PAD, David HL (1986) Deoxyribonucleic acid relatedness study of the *Mycobacterium fortuitum–Mycobacterium chelonae* complex. Int J System Bacteriol 36:458–460.

250. Domenech P, Menendez MC, Garcia MJ (1994) Restriction fragment length polymorphisms of 16S rRNA genes in the differentiation of fast-growing mycobacterial species. FEMS Microbiol Lett 116:19–24.

251. Springer B, Böttger EC, Kirschner P, Wallace RJ Jr (1995) Phylogeny of *Mycobacterium chelonae*-like organism based on partial sequencing of the 16S rRNA gene and proposal of *Mycobacterium mucogenicum* sp. nov. Int J System Bacteriol 45:262–267.

252. Moore M, Frerichs B (1953) An unusual acid-fast infection of the knee with subcutaneous, abscess-like lesions of the gluteal region; Report of a case with a study of the organism, *Mycobacterium abscessus,* n. sp. J Invest Dermatol 20:133–169.

253. Hill LR, Skerman VBD, Sneath PHA (1984) Corrigenda to the Approved Lists of Bacterial Names. Int J System Bacteriol 34:508–511.

254. Redpath F, Seabury JH, Sanders CV, Domer J (1976) Prosthetic valve endocarditis due to *Mycobacterium chelonae.* South Med J 69:1244–1246.

255. Bönicke R, Stottmeier JL (1965) Erkennung und Identifizierung von Stämmen der Species *Mycobacterium borstelense.* Beitr Klinck Erforsch Tuberkulose Lungenkrankheiten 130:210–222.

256. Stanford JL, Beck A (1969) Bacteriological and serological studies of fast growing mycobacteria identified as *Mycobacterium friedmannii.* J Gen Microbiol 58:99–106.

257. Tsukamura M, Mizuno S, Tsukamura S (1968) Classification of rapidly growing mycobacteria. Japan J Microbiol 12:151–166.

258. Tsukamura M (1970) Differentiation between *Mycobacterium abscessus* and *Mycobacterium borstelense.* Am Rev Respir Dis 101:426–428.

259. Weiszfeiler JG, Karasseva V, Karczag E (1969) Comparative studies on the taxonomic relationship between *Mycobacterium abscessus* and *Mycobacterium borstelense.* Acta Microbiol Acad Sci Hung 16:317–379.

260. Jenkins PA, Marks J, Schaefer WB (1971) Lipid chromatography and seroagglutination in the classification of rapidly growing mycobacteria. Am Rev Respir Dis 103:179–187.

261. Silcox VA, Good RC, Floyd MM (1981) Identification of clinically significant *Mycobacterium fortuitum* complex isolates. J Clin Microbiol 14:686–691.

262. Baess, I (1982) Deoxyribonucleic acid relatedness among species of rapidly growing mycobacteria. Acta Path Microbiol Scand B90:371–375.

263. Häggblom M, Nohynek LJ, Palleroni NJ, Kronqvist K, Nurmiaho-Lassila E-L, Salkinoja-Salonen MS, Klatte S, Kroppenstedt RM (1994) Transfer of polychlorophenol-degrading *Rhodococcus chlorophenolicus* (Apajalahti *et al.* 1986) to the genus *Mycobacterium* as *Mycobacterium chlorophenolicum* comb. nov. Int J System Bacteriol 44:485–493.

264. Tsukamura M, Mizuno S, Tsukamura S (1981) Numerical analysis of rapidly growing, scotochromogenic mycobacteria, including *Mycobacterium obuense* sp. nov., comb. nov., *Mycobacterium rhodesiae* sp. nov., nom. rev., and *Mycobacterium tokaiense* sp. nov., nom. rev. Int J System Bacteriol 31:263–275.

265. Chamoiseau G (1973) *Mycobacterium farcinogenes* agent causal du farcin du boeuf en Afrique. Ann Microbiol Inst Pasteur (Paris) 124:215–222.

266. da Costa Cruz JC (1938) *Mycobacterium fortuitum* um novo bacillo acidoresistance pathogenico para o homen. Acta Med Rio de Janiero 1:297–301.

267. Chamoiseau G (1979) Etiology of farcy in African bovines: Nomenclature of causal organisms *Mycobacterium farcinogenes* Chamoiseau and *Mycobacterium senegalense* (Chamoiseau) comb. nov. Int J System Bacteriol 29:407–410.

268. Stanford JL, Gunthorpe WJ (1969) Serological and bacteriological investigation of *Mycobacterium ranae (fortuitum).* J Bacteriol 98:375–383.

269. Runyon EH (1972) Conservation of the specific epithet *fortuitum* in the name of the organism known as *Mycobacterium fortuitum* da Costa Cruz—Request for an opinion. Int J System Bacteriol 22:50–51.

270. Judicial Commission of the International Committee on Systematic Bacteriology

(1974) Opinion 51. Conservation of the epithet fortuitum in the combination *Mycobacterium fortuitum* da Costa Cruz. Int J System Bacteriol 24:552.

271. Haas H, Michel J, Sachs T (1974) Identification of *Mycobacterium fortuitum, Mycobacterium abscessus,* and *Mycobacterium borstelense* by polyacrylamide gel electrophoresis of their cell proteins. Int J System Bacteriol 24:366–369.

272. Nakayama Y (1967) The electrophoretical analysis of esterase and catalase and its use in taxonomical studies of mycobacteria. Japan J Microbiol 11:95–101.

273. Van den Berghe DA, Pattyn SR (1979) Comparison of proteins from *Mycobacterium fortuitum, Mycobacterium nonchromogenicum* and *Mycobacterium terrae* using flat bed electrophoresis. J Gen Microbiol 111:283–291.

274. Tsukamura M, Mizuno S, Tsukamura S, Tsukamura J (1979) Comprehensive numerical classification of 369 strains of *Mycobacterium, Rhodococcus* and *Nocardia.* Int J System Bacteriol 29:110–129.

275. Castelnuovo G, Guadiano A, Morellini M, Penso G, Rossi C (1960) Gli antigeni der micobatteri. Rend Inst Superiore de Sanita, Roma 23:1222–1233.

276. Gimpl F, Lanyi M (1965) Use of the gel-precipitation method for determining the type of mycobacteria and the clinical diagnosis. Bull Int Union Tuberculosis 36:22–25.

277. Drzinš E (1960) Tuberculose das gias (*Leptodactylus pentadactylus*). Arch Inst Brasil Invest Tuberculose, Bahia 9:23–37.

278. Penso G, Castelnuovo G, Guadiana A, Prinivalle M, Vella L, Zampiere A (1952) Studi e recerche sui micobatteri. VIII. Un nuovo bacillo tuberculare; il *Mycobacterium minettii* n. sp.-studio microbiologico e patogenetico. Rend Inst Superiore de Sanita, Roma 15:491–548.

279. Wallace RJ Jr, Brown BA, Silcox VA, Tsukamura M, Nash DR, Steele LC, Steingrube A, Smith J, Sumter G, Zhang Y, Blacklock Z (1991) Clinical disease, drug susceptibility, and biochemical patterns of the unnamed third biovar complex of *Mycobacterium fortuitum.* J Infect Dis 163:598–603.

280. Kirschner P, Kiekenbeck M, Meissner D, Wolters J, Böttger EC (1992) Genotypic heterogeneity within *Mycobacterium fortuitum* complex species: Genotypic criteria for identification. J Clin Microbiol 30:2772–2775.

281. Wallace RJ Jr, Swenson JM, Silcox VA, Good RC, Tschen JA, Stone MS (1983) Spectrum of disease due to rapidly growing mycobacteria. Rev Infect Dis 5:657–679.

282. Wolinsky E (1979) Non-tuberculous mycobacteria and associated diseases. Am Rev Respir Dis 119:107–159.

283. Ingram CW, Tanner DC, Durack DT, Kernodie GW Jr, Corey GR (1993) Disseminated infection with rapidly growing mycobacteria. Clin Infect Dis 16:463–471.

284. Wallace RJ Jr (1994) Recent changes in taxonomy and disease manifestations of the rapidly growing mycobacteria. Eur J Clin Microbiol Infect Dis 13:953–960.

285. Wallace RJ Jr, Musser JM, Hull SI, Silcox VA, Steele LC, Forester GD, Labidi A,

Selander RK (1989) Diversity and sources of rapidly growing mycobacteria associated with infections following cardiac surgery. J Infect Dis 159:708–716.

286. Wallace RJ Jr, Steele LC, Labidi A, Silcox VA (1989) Heterogeneity among isolates of rapidly growing mycobacteria responsible for infections following augmentation mammoplasty despite case clustering in Texas and other southern coastal states. J Infect Dis 160:281–288.

287. Wallace RJ Jr, O'Brien R, Glassroth J, Raleigh J, Dutt A (1990) Diagnosis and treatment of disease caused by non-tuberculous mycobacteria. Am Rev Respir Dis 142:940–953.

288. Wallace RJ Jr, Hull SL, Bobey DG, Price KE, Swensen JH, Steele LC, Christensen L (1985) Mutational resistance as the mechanism of acquired drug resistance to aminoglycosides and antibacterial agents in *Mycobacterium fortuitum* and *Mycobacterium chelonei.* Am Rev Respir Dis 132:409–416.

289. Pattyn SR, Magnusson M, Stanford JL, Grange JM (1974) A study of *Mycobacterium fortuitum (ranae).* J Med Microbiol 7:67–76.

290. Minnikin DE, Minnikin SM, Hutchinson IG, Goodfellow M, Grange JM (1984) Mycolic acid pattern of *Mycobacterium fortuitum, "Mycobacterium peregrinum",* and *Mycobacterium smegmatis.* J Gen Microbiol 130:363–367.

291. Tsang AY, Barr VL, McClatchy JK, Goldberg M, Drupa I, Brennan PJ (1984) Antigenic relationship of the *Mycobacterium fortuitum–M. chelonae* complex. Int J System Bacteriol 34:35–44.

292. Asselineau J, Lanéelle MA, Chamoiseau G (1969) De l'etiologie du farcin de zebus tchadiens: Nocardiose ou mycobactériose? II. Composition lipidique. Rev Elevage Méd Véterin Pays Trop 22:205–209.

293. Chamoiseau G (1969) De l'etiologie du farcin de zebus tchadiens. Nocardiose ou mycobactériose? I. Étude bacteriologique et biochemique. Rev Elevage Méd Véterin Pays Trop 22:195–204.

294. El-Sanousi SM, Tag El-Din MH, Abdel Wahab SM (1977) Classification of bovine farcy organism. Trop Animal Health Product 9:124.

295. Lanélle G, Asselineau J, Chamoiseau G (1971) Présence de mycosides C′ (formes simplifié de mycoside C) dans les bactéries isolées de bovins atteint du farcin. FEBS Lett 19:109–111.

296. Baess I, Weiss Bentzon M (1978) Deoxyribonucleic acid hybridisation between different species of mycobacteria. Acta Pathol Microbiol Scand B86:71–76.

297. Ridell M, Goodfellow M, Minnikin DE, Minnikin SM, Hutchinson IG (1982) Classification of *Mycobacterium farcinogenes* and *Mycobacterium senegalense* by immunodiffusion and thin-layer chromatography of long-chain components. J Gen Microbiol 128:1299–1307.

298. Hall R, Ratledge C (1985) Equivalence of mycobactins from *Mycobacterium senegalense, Mycobacterium farcinogenes* and *Mycobacterium fortuitum.* J Gen Microbiol 131:1691–1696.

299. Hamid ME, Minnikin DE, Goodfellow M, Ridell M (1993) Thin layer

chromatographic analysis of glycolipids and mycolic acids from *Mycobacterium farcinogenes, Mycobacterium senegalense* and related taxa. Zentralblatt Bakteriol 279:354–367.

300. Besra GS, Gurcha SS, Khoo K-H, Morris HR, Dell A, Hamid ME, Minnikin DE, Brennan PJ (1994) Characterisation of the specific antigenicity of representatives of *M. senegalense* and related bacteria. Zentralblatt Bakteriol 281:415–432.

301. Ridell M, Goodfellow M (1983) Numerical classification of *Mycobacterium farcinogenes, Mycobacterium senegalense* and related taxa. J Gen Microbiol 129:599–611.

302. Ridell M (1981) Immunodiffusion analysis of *Mycobacterium farcinogenes, Mycobacterium senegalense* and some other mycobacteria. Zentbl Bakteriol ParasitenKde Abt I Suppl II:235–241.

303. Ridell M (1983) Immunodiffusion analysis of *Mycobacterium farcinogenes, Mycobacterium senegalense* and some other mycobacteria. J Gen Microbiol 129:613–619.

304. Tsukamura M, van der Meulen HJ, Grabow WOK (1983) Numerical taxonomy of rapidly growing, scotochromogenic mycobacteria of the *Mycobacterium parafortuitum* complex: *Mycobacterium austroafricanum* sp. nov. and *Mycobacterium diernhoferi* sp. nov., nom. rev. Int J System Bacteriol 33:460–469.

305. Tsukamura M (1972) A new species of rapidly growing, scotochromogenic mycobacteria, *Mycobacterium neoaurum* Tsukamura n. sp. Med Biol 85:229–233.

306. Tsukamura M, Mizuno S (1977) Numerical analysis of relationships among rapidly growing, scotochromogenic mycobacteria. J Gen Microbiol 98:511–517.

307. Davison MB, McCormack JG, Blacklock ZM, Dawson DJ, Tilse MH, Cummins FB (1988) Bacteremia caused by *Mycobacterium neoaurum.* J Clin Microbiol 26:762–764.

308. Bönicke R, Juhasz SE (1965) *Mycobacterium diernhoferi* nom. sp., eine in der Umgebung des Rindes häufig vorkommende neue Mycobacterium-Species. Zentbl Bakteriol ParasitenKde Abt I Orig 197:292–294.

309. Stanford JL, Wang JKC (1974) A study of the relationship between *Nocardia* and *Mycobacterium diernhoferi*—A typical fast growing *Mycobacterium.* Br J Exp Pathol 55:291–295.

310. Tsukamura M (1966) Adansonian classification of mycobacteria. J Gen Microbiol 45:253–273.

311. Bönicke R, Juhasz SE (1964) Beschreibung der neuen species *Mycobacterium vaccae* n. sp. Zentbl Bakteriol ParasitenKde Abt I Orig 197:292–294.

312. Pattyn SR (1970) Agglutination with rapidly growing (Runyon's Group IV) mycobacteria. Zentbl Bakteriol ParasitenKde Abt I Orig A 215:99–105.

313. Tsukamura M (1967) *Mycobacterium chitae:* A new species. Japan J Microbiol 11:43–47.

314. Trevisan V (1889) I Generi e le Specie delle Bacteriacea. Milano: Zanaboni and Gabuzzi.

315. Alvarez E, Tavel E (1885) Rechérches sur le bacillé de lustgarten. Arch Physiol Normal Pathol 6:303–321.

316. Lehmann KB, Neumann R (1899) Lehmann's Medizin Handatlanten. X. Atlas und Grundriss der Bakteriologie und Lehrbuch der speziellen bakteriologischen Diagnostik. 2 Auflage. Munchen, Germany.

317. Lind A (1960) Serological studies of mycobacteria by means of the diffusion-in-gel techniques. IV. The precipitinogenic relationships between different species of mycobacteria with special reference to *M. tuberculosis, M. phlei, M. smegmatis* and *M. avium.* Int Arch Allergy Appl Immunol 17:300–322.

318. Norlin M (1965) Unclassified mycobacteria, a comparison between a serological and biochemical classification method. Bull Int Union Tuberculosis 36:25–32.

319. Magnusson M (1962) Specificity of sensitins. III. Further studies in guinea pigs with sensitins of various species of *Mycobacterium* and *Nocardia.* Am Rev Respir Dis 86:395–404.

320. Wallace RJ Jr, Nash DR, Tsukamura M, Blacklock ZM, Silcox VA (1988) Human disease due to *Mycobacterium smegmatis.* J Infect Dis 158:52–59.

321. Casal M, Rey Calero J (1974) *Mycobacterium gadium* sp. nov. a new species of rapidly growing scotochromogenic mycobacteria. Tubercle 55:299–308.

322. Stanford JL, Gunthorpe WJ (1971) A study of some fast-growing scotochromogenic mycobacteria including species descriptions of *Mycobacterium gilvum* (new species) and *Mycobacterium duvalii* (new species). Br J Exp Pathol 52:627–637.

323. Kazda J, Müller K (1979) *Mycobacterium komossense* sp. nov. Int J System Bacteriol 29:361–365.

324. Kazda J (1980) *Mycobacterium sphagni* sp. nov. Int J System Bacteriol 30:77–81.

325. Möeller A (1898) Microorganismen, die den Tuberkelbacillen verwandt sind und bei Thieren eine miliare Tuberkelkrankheit verursachen. Deutsche Med Wochenschr 24:376–379.

326. Möeller A (1898) Ueber den Tuberkelbacillus verwandte Microorganismen. Therapeut Monalshefte 12:607–613.

327. Jones WD Jr, Kubica GP (1968) Fluorescent antibody techniques with mycobacteria. III. Investigation of five serologically homogeneous groups of mycobacteria. Zentbl Bakteriol ParasitKde Abt I Orig A207:58–62.

328. Kubica GP, Silcox VA, Hall E (1973) Numerical taxonomy of selected slow growing mycobacteria. J Gen Microbiol 74:159–167.

329. Tsukamura M, Mizuno S (1975) Differentiation among mycobacterial species by thin-layer chromatography. Int J System Bacteriol 25:271–280.

330. Tsukamura M, Mizuno S, Tsukamura S (1967) Numerical classification of atypical mycobacteria. Japan J Microbiol 11:233–241.

331. Tsukamura M (1968) Classification of scotochromogenic mycobacteria. Japan J Microbiol 12:63–75.

332. Tsukamura M, Mizuno S (1978) A further study on the method of identification of mycobacteria by thin-layer chromatography after incubation with $^{35}$S-methionine. Kekkaku 54:15–27.

333. Tsukamura M (1972) *Mycobacterium agri* Tsukamura sp. nov. A relatively thermophilic *Mycobacterium.* Med Biol Tokyo 85:153–156.

334. Tsukamura M (1981) Numerical analysis of rapidly growing, nonphotochromogenic mycobacteria, including *Mycobacterium agri* (Tsukamura 1972) Tsukamura sp. nov., nom. rev. Int J System Bacteriol 31:247–258.

335. Tsukamura M, Mizuno S, Toyama H (1983) *Mycobacterium pulveris* sp. nov., a nonchromogenic *Mycobacterium* with an intermediate growth rate. Int J System Bacteriol 33:811–815.

336. Band JD, Ward JI, Frazer DW, Peterson NJ, Silcox VA, Good RC, Ostroy PR, Kennedy J (1982) Peritonitis due to a *Mycobacterium chelonae*-like organism associated with intermittent chronic peritoneal dialysis. J Infect Dis 145:9–17.

337. Silcox VA, Good RC, Floyd MM (1981) Identification of clinically significant *Mycobacterium fortuitum* isolates. J Clin Microbiol 14:686–691.

338. Duvall CW, Wellman C (1912) A critical study of the organisms cultivated from the lesions of human leprosy, with consideration of their etiological significance. J Infect Dis 11:116–139.

339. Tsukamura M (1966) *Mycobacterium parafortuitum:* A new species. J Gen Microbiol 42:7–12.

340. Canhos VP, Manfio GP, Blaine LD (1993) Software tools and databases for bacterial systematics and their dissemination via global networks. Antonie van Leeuwenhoek 64:205–229.

341. O'Donnell AG, Embley TM, Goodfellow M (1993) Future of bacterial systematics. In: Goodfellow M, O'Donnell AG, eds. Handbook of New Bacterial Systematics, pp. 513–524. London: Academic Press.

# 2

# Drug Resistance in Tubercle Bacilli

*Pattisapu R.J. Gangadharam*

From a microbiological point of view, every culture of tubercle bacilli would contain some proportion of mutant organisms which are "resistant" to the drug. Thus, resistance and susceptibility are relative terms. What is generally accepted and what is important for our discussion in this chapter is the proportion of resistant mutants in the culture which will entail clinical failure. Patients who have organisms with a high proportion of resistant bacilli in their bacterial population are, in general, nonresponsive or, at the most, poorly responsive to the treatment with the same drugs (1). From a laboratory standpoint, optimal concentrations of the drugs to be used and the critical proportions of the resistant bacilli in the cultures which carry therapeutic penalties are arrived at by statistical exercises. Those conditions vary from drug to drug (2). In general, most cultures that carry more than 1% of the organisms resistant to the concentration of the drug which inhibits susceptible bacilli have been shown to be indicative of clinical failure of the patient when treated with that drug alone (3).

On epidemiological grounds, drug resistance has been divided into three broad categories discussed below. These definitions are used in mass surveys to assess the prevalence of drug resistance in the community or the country. Such surveys conducted at frequent intervals could be used to monitor the success of treatment programs, particularly if a national treatment policy is implemented, and thus serve as epidemiological "yardsticks" (4). The categories are as follows:

(i) **Primary drug resistance,** where drug-resistant bacilli are isolated from previously untreated patients.

(ii) **Acquired drug resistance,** where resistant bacilli are isolated from patients who had originally susceptible bacilli.

(iii) **Initial drug resistance** denotes drug resistance in patients who deny a history of previous chemotherapy. In actual terms, it consists of true primary resistance with a mixture of an unknown amount of undisclosed acquired drug resistance. Because it takes elaborate efforts on the part of the health authorities to obtain accurate information on the possibility of previous treatment which is unreported, initial drug resistance has been gaining practical importance.

Besides these three types, some countries assessed the prevalence of drug resistance in patients as they first report to chest clinics (5). The available information on drug resistance in the world is briefly discussed later in this chapter. In all these discussions, the term multidrug resistance (MDR) is used most frequently; such bacilli (most commonly) are resistant to rifampin and isoniazid, and often to other drugs as well.

## 1. Historical Aspects of Drug Resistance in Tuberculosis

Drug resistance as a limiting factor for success of chemotherapy of tuberculosis was recognized almost immediately following the introduction of streptomycin, the first effective antituberculosis drug. Youmans and associates (6) found in 1946 that when streptomycin was given alone, there was, at first, a striking improvement in the patient's symptoms, followed by a rapid decrease in the number of bacilli in the sputum. The number of bacilli, however, soon increased and the condition of the patient deteriorated. The bacilli isolated at that time, instead of being killed by the drug, continued to grow in vitro in the presence of high concentrations of the drug. This classical observation was soon followed by another important finding by Pyle (7) who showed that during treatment with streptomycin alone, the proportion of drug-resistant bacilli increased progressively from about 1 in 88,750 organisms before the therapy was started, to about 1 in 367 after 15 weeks of treatment. These observations were the pioneering studies on drug resistance in mycobacteria. Subsequent studies proved that the increase in drug resistance following monotherapy was due to the selection of drug-resistant mutants. Other significant landmarks which soon followed were the findings by Temple and associates (8) in the United States and by the British Thoracic Society (9), which showed that multiple-drug therapy would prevent development of drug resistance. These valuable findings have been confirmed by several studies carried out worldwide and have led to the clinical maxim that tuberculosis patients should be treated with multiple drugs. Parallel to these clinical observations, enormous literature on the techniques of detection of drug resistance have been developed and have been accumulating (1,10,11). The currently accepted methods are reviewed in Chapters 11 and 12.

## 2. Theoretical Aspects of Drug Resistance

### *2.1. Origin of Drug Resistance*

Investigations on the origin of drug resistance started almost immediately after the discovery of drug resistance in mycobacteria and have been advanced significantly by the parallel growth of knowledge in bacterial genetics. Early investigators in this area concerned with two broadly divergent aspects dealing with either the mutation or genetic origin, and the adaptation or training mechanisms. These two pathways differed significantly in many aspects, although there was an overlap or a sort of compromise between these two theories as advanced by some authorities. In the first one, dealing with genetic transfer, the main emphasis is on the fundamental operation of mutation, followed by selection. In the second one, on adaptation theory, the fundamental basis is on the "training" of the organisms following exposure to the toxic environment (i.e., the drug), and no operation of random genetic accidents is considered important. For nearly three decades around the 1950s, great controversy existed between the proponents of both the theories, with most authorities accepting the mutation theory of drug resistance, although some (12–14) argued in favor of the adaptation theory for a long time. Working as a part of arbitration or compromise between the two theories, Szybalski and Bryson (15) indicated that "conceptually, the two theories are opposed, but operationally they may co-exist. At present, the most ardent extremist would hardly dare to claim that all resistance comes about by way of either method. Yet, in fairness, we must admit that the rigorous demonstration of drug-induced resistance is exceptional, notwithstanding the ingenious efforts of a minority of microbiologists to prove otherwise." Elaborating on their discussion, these authors further stressed that the selective process operated by the drug is the more important component in the origin of drug resistance, with mutation or adaptation playing only minor roles. They stated that

> unless *all* cells respond *homogeneously* to a toxic agent (drug) thus eliminating selection, drug resistance originates more correctly by (i) genetic adaptation (e.g. mutation) + selection or by (ii) phenotypic adaptation + selection. Genetic adaptation which is actually mutation, is more frequent and may occur in the absence of drugs; whereas phenotypic adaptation usually results from the action of an inducing drug directly on the cells. Both processes therefore could occur simultaneously or in sequence in the origin of resistance.

By giving such statements, these authors (15) ingeniously formed a bridge between the two basically controversial viewpoints by terming mutation as "genetic adaptation" and adaptation–training as "phenotypic adaptation." Another similar approach, which can also be considered as an arbitration, was advanced by Mitchison (16), who expressed the view that the mutation theory can explain most, if not all, aspects of drug resistance observed in mycobacteria, and the role of other mechanisms, principally adaptation, should not be ruled out.

Earlier, using the replica plating technique, Schaefer (17) calculated the mutation rates of *Mycobacterium tuberculosis* resistant to streptomycin and isoniazid. Likewise, Grosset and Canetti (18) have calculated the frequency of occurrence of spontaneous resistant mutants in H37Rv strain and several wild strains of *M. tuberculosis* for the various drugs using several inoculum sizes. By far, the most convincing evidence in favor of the mutation theory for the origin of drug resistance in tuberculosis came from the work of David and Newman (19), who used fluctuation analysis and direct titration. Their results showed frequencies of resistance of 1–5 in $10^8$ organisms for rifamycin, 1 in $10^5$–$10^6$ organisms for isoniazid and streptomycin, and 1 in $10^3$ organisms for ethionamide, capreiomycin, viomycin, ethionamide, cycloserine, and thiacetazone.

In the clinical perspective, a single drug will be able to kill all susceptible bacilli and the few resistant bacilli present therein should be handled by the host immune mechanisms. In cases where such mechanisms do not operate, the few resistant bacilli will be able to multiply and eventually make the whole population resistant. This aspect is exemplified by the pioneering studies of Youmans et al. (6) and Pyle (7), discussed earlier.

## *2.2. Mechanism of drug resistance*

Several years ago, Pollock (20) listed three broad categories of mechanisms of drug resistance to antimicrobial agents. These include (a) susceptible metabolic pathways, (b) interference to uptake or penetration of drugs into the cells, and (c) destruction of the drug. Review of the available literature indicates some information on only few antimycobacterial drugs: isoniazid, para amino salicylic acid (PAS), streptomycin and other aminoglycosides, cycloserine, rifampin, and pyrazinamide. Recent developments in genetics of mycobacteria provided considerable knowledge on the molecular mechanisms of drug resistance development with some drugs (e.g., isoniazid, ethionamide, rifampin, and streptomycin).

### MECHANISM OF DEVELOPMENT OF RESISTANCE TO ISONIAZID

Most of the earlier literature on this aspect dealt with the uptake or permeability of the drug into the cell and very little dealt with the differences in the metabolic pathways between susceptible and resistant organisms, and still much less with the destruction of the drug. Recent evidence, however, dealt with the molecular aspects of the mechanism of drug resistance (discussed later). Barclay and associates (21,22) showed that the isoniazid-resistant strains exposed in the growth period bound markedly less of the $^{14}C$-labeled isoniazid than the susceptible strains. The uptake by the susceptible growing bacilli reached a peak after 1 day, then fell and only slowly increased thereafter to a plateau at 7 days, when the bactericidal action was essentially complete. Their results suggested that the development of resistance was due to a failure to accumulate the drug. Furthermore, because resistant strains are usually characterized by a loss of some enzymes, particularly peroxidase and catalase, there is a strong suggestion that the loss of these enzyme activities might be responsible for the failure of the resistant strains to take up the drug.

Barclay et al. (22) also suggested a two-phase action of isoniazid. The first phase of a reversible uptake takes place in only susceptible cells, whether growing or not, and which probably depends on the peroxidase or catalase enzymes. This will be followed by a second phase, which takes place only when the susceptible cells are growing, in which the drug which is already taken up is bound irreversibly. These interesting findings were later extended by Youatt (23,24), who used dense suspensions and high concentrations of isoniazid suspended with the bacteria and saline in subgrowth medium. Youatt might have studied only the reversible phase of isoniazid uptake and, therefore, found that the bound isoniazid could be removed in larger part by overnight washing. She found that the uptake required oxygen and is inhibited by cyanide or azide but not by pH or chelating agents. In her studies, resistant strains showed a greatly reduced uptake of isoniazid, unless carbon dioxide was allowed to accumulate. On the whole, Youatt's studies were in conformity with the hypothesis that peroxidase or catalase is involved in the isoniazid uptake (23–25). Several other hydrazides, which are less active than isoniazid and which interfere with the bactericidal action of isoniazid, also inhibited isoniazid uptake, presumably by competing for the same uptake mechanism. This finding and that the fact that mutants resistant to these hydrazides also lack peroxidase and catalase activity suggest that they, too, are taken up through the action of peroxidase and catalase. The involvement of catalase on the mode of action and in relation to recent molecular biological studies is discussed later in this section.

One significant difference between the findings of Youatt and of Barclay was that the former found that isonicotinic acid also is apparently taken by susceptible bacilli, whereas Barclay and associates could not show this uptake. Likewise, nontuberculosis mycobacteria, which are naturally resistant to isoniazid, have also been shown to take up small amounts of isoniazid. There is some controversy about the uptake of isoniazid by saprophytic mycobacteria, which are also naturally resistant to isoniazid; some have reported these bacilli to have a slow initial uptake of isoniazid, whereas others have reported much higher uptakes (25,26). This point has some relevance to the recent studies with molecular mechanisms (discussed later), most of which involved *M. smegmatis.*

Another interesting finding is the independence of the uptake from the metabolism of the organism. For instance, pyridoxal stimulates the uptake of isoniazid, particularly by the dead bacilli. This suggests that the hydrazone is taken up by a route different from that of isoniazid and not dependent on metabolism. This might also explain why pyridoxal increases the activity of isoniazid against resistant organisms, even though it decreases the activity against susceptible strains.

Overall, the evidence suggests that the primary factor in the mechanism of development of resistance to isoniazid by tubercle bacilli seems to be lack of permeability. The accumulation process appears to be an active transport mechanism, because heated cells show very little radioactivity and the uptake is also inhibited by anaerobic culture conditions. Permeability factors might also explain the difference in the absorption of isoniazid compared with the inactive metabolites.

Specific studies using strains which are resistant to 100 $\mu$g/ml of isoniazid have also shown that the resistant bacilli possess a cell wall or membrane-impermeable factor to the drug. Another evidence in support of the difference in biochemical mechanisms is by Quemard et al. (27), who showed that the mycolic acid synthesis in cell-free extracts from isonicotic acid hydrazide (INH)-resistant strains of *M. aurum* was not inhibited by the drug, whereas that with susceptible organisms was inhibited. Because mycolic acid synthesis has been shown to be the primary target for the action of isoniazid, this finding also shows a strong biochemical mechanism on the differences between the susceptibility and resistance to this drug. Besides this evidence in favor of permeability factors and biochemical factors, only one finding (28) is available on the presence of the isoniazid-destroying substances in tubercle bacilli, on the model of penicillinase activity in penicillin-resistant bacteria. However, no further confirmation or extension of this finding has been published.

## MOLECULAR ASPECTS ON THE MECHANISM OF DRUG RESISTANCE TO ISONIAZID

Some of the recent studies which involved molecular genetics in the understanding of the drug resistance of isoniazid have created a wave of excitement in many who consider them as very important starting points for understanding the basic mechanisms of drug resistance in mycobacteria in general. Soon after the discovery of strains of tubercle bacilli which are resistant to isoniazid, it was found that most of them had diminished levels of catalase and peroxidase activities. There was also a wealth of information to show that the loss of catalase activity is related to the loss of virulence of such strains in guinea pigs and possibly in humans. Working on the premise that the enzyme catalase was the key to the activity and the resistance development to isoniazid, Young and associates showed that the strains of *M. smegmatis,* which were resistant to high concentrations of isoniazid, could be made susceptible to low concentrations of the drug by adding to its genome, a stretch of *M. tuberculosis* DNA containing the *katG* gene (29). Probing isoniazid-resistant mutants of *M. tuberculosis* showed that in two out of the three catalase-negative strains resistant to >50 $\mu$g/ml of INH, the *katG* gene was deleted, leading to the hypothesis that isoniazid resistance in tubercle bacilli might be attributed to the deletion of the *katG* gene. They also found that this gene was missing in two of the eight drug-resistant *M. tuberculosis* strains. In subsequent studies (30), they isolated several resistant mutants of *M. smegmatis* and *M. aurum* on media containing isoniazid, and among the survivors, they examined the catalase-negative clones using the catalase–peroxidase gene (*katG*) of *M. tuberculosis* as a probe. The probe hybridized with the DNA extracted from the INH-resistant mutants of both species of mycobacteria (31).

Several other studies by Young and Zhang using point mutations, insertions, and deletions of the gene showed that the wild-type *katG* gene can be transformed into isoniazid-resistant mycobacteria, restoring susceptibility to this drug. The wild-type gene can also confer isoniazid susceptibility to catalase-negative *E. coli*. The *E. coli* version of the *katG* gene, when transformed into drug-resistant *M. tuberculosis* mutants partially restored their susceptibility to isoniazid. However, the authors suggest that once isoniazid is inside the bacteria, its *katG* gene product converts it into an activated intermediate. This step might not be specific for the drug because the *E. coli* enzyme also restores drug susceptibility to the active intermediate, which, in turn, can act as a second target which is yet unknown. Because the catalase and peroxidase enzymes are involved in the uptake of the drug by the susceptible bacilli as proposed earlier by Gayathri Devi et al. (32), this observation can support, at least in part, the involvement of these enzymes in the development of drug resistance to isoniazid. On the other hand, Rastogi and David (33) argued that the evidence with the *katG* gene might be inconclusive for two reasons: (i) In actual clinical practice, many of the tubercle bacilli strains which are resistant to isoniazid are still catalase positive and are susceptible to 50 $\mu$g/ml of the drug. (ii) The mutants examined by Young and Zhang (29) were isolated using two selective steps; first with regard to resistance to isoniazid and then among those mutants without catalase activity.

Available evidence on the mechanism of action of isoniazid relies heavily on the observation that the drug inhibits mycolic acid synthesis, although involvement in the nicotinamide adenine dinucleotide (NAD) and pyridoxal phosphate metabolisms are also considered by many. Based on this hypothesis, Banerjee and associates (34) identified another gene from an isoniazid-resistant mycobacterial strain that may represent the drug's cellular target. This gene, called the *inhA* gene, which is isolated from the cloned fragment from a genomic library of a resistant mutant of tubercle bacilli, conferred resistance to isoniazid-susceptible tubercle bacilli. The resistance phenotype was later located in an open reading frame containing the *inhA* gene. The mycobacterial *inhA* gene shows a strong homology to genes in *E. coli* and *Salmonella typhimuriun* that are known to be involved in fatty acid biosynthesis. According to Banerjee et al. (34), isoniazid or its derivative, which is produced by the action of the mycobacterial catalase or peroxidase, forms the target for the inhibition of the mycolic acid synthesis, thereby disrupting the mycobacterial cell wall. Mutations in the wild-type *inhA* gene, which is conserved among many mycobacteria, would then be responsible for the development of resistance to the drug. More information on this aspect dealing with several clinical isolates is needed before this concept is established. On the other hand, this discovery along with the *katG* gene constitutes an important landmark in mycobacterial drug resistance studies.

## MECHANISM OF RESISTANCE TO PAS

Several years ago Barclay (35) showed that the $^{14}C$-labeled PAS is taken up by *M. tuberculosis,* the maximum level reached in about 3–4 days. The drug is not simply adsorbed, but it is taken up metabolically. With repeated washings, there was a 50% loss of activity from the labeled cells for 7 days. In contrast to the susceptible strains, there was more rapid and higher uptake by the resistant strains, and none of the label was lost on repeated washings. This suggested that this increase is not due to a decrease in permeability but is more likely to be due to alteration in the affinity of the enzyme for the drug. However, the actual increase in uptake in the resistant strain could not be explained easily by this assumption alone and it is suggested that more PAS is metabolized by the resistant strains to some product which is firmly bound to the cell. Mycobacteria which are comparatively less susceptible to PAS have been shown to convert it to colored products, in some cases by decarboxylation, followed by oxidation, and in other cases by direct oxidation. Some of these colored compounds remain bound to the cells and, therefore, can still keep the radioactivity intact and to a higher degree. Thus, the development of resistance to PAS is probably due to an altered enzyme system which does not recognize PAS.

## MECHANISM OF RESISTANCE TO CYCLOSERINE

Cycloserine is an inhibitor of two important mycobacterial enzymes: alanine racemase and D-alanyl-D-alanine synthetase, which are involved in the conversion of L-alanine to D-alanine and D-alanine to D-alanyl-D-alanine, respectively (36). The dipeptide D-alanyl-D-alanine is essential for the biosynthesis of mycobacterial cell wall and is transferred to a UDP-acetyl-muramyl-tripeptide precursor to form a pentapeptide precursor of mucopeptides in the mycobacterial cell walls. Cycloserine is actively transported intracellularly by *M. tuberculosis* and inhibits the synthetase enzyme which is present in small amounts inside the bacilli. A single transport system accumulates L- and D-alanyl glycine and D-serine and the antimicrobials, D-cycloserine and O-carbamyl D-serine, are transported with the same system (37). A cycloserine-resistant mutant which is permeate deficient but still susceptible to the activity of D-cycloserine was isolated. These findings indicated that enough of the drug is accumulated by simple diffusion to inhibit the enzyme and, subsequently, bacterial growth. Based on these studies, it is concluded that resistance to D-cycloserine on tubercle bacilli is due primarily to mutations in the gene(s) which control D-alanyl-D-alanine synthetase.

## MECHANISM OF RESISTANCE TO PYRAZINAMIDE

Susceptible strains of *M. tuberculosis* can metabolize pyrazinamide to pyrazinoic acid, whereas pyrazinamide-resistant strains of *M. tuberculosis* are not capable to doing so (38). It has also been stated that pyrazinamide, after penetrating the monocytes, would be dominated by the amidase of susceptible tubercle bacilli to deliver pyrazinoic acid directly to them. Pyrazonoic acid is effective on both susceptible and resistant strains of *M. tuberculosis* and *M. bovis*. Thus, these studies indicate that pyrazonoic acid is the active antimicrobial moiety of pyrazinamide, although it is not directly effective in vivo. This acid is not absorbed by animals in the gastrointestinal tract, and monocytes are more permeable to the pyrazinamide than pyrazonoic acid. In other words, in vivo, the amidase of the susceptible strains of tubercle bacilli is used to convert the pyrazinamide to the active metabolite, which is the responsible factor. The amidase is absent from the resistant bacilli.

## MECHANISM OF RESISTANCE DEVELOPMENT TO ETHAMBUTOL

Ethambutol is equally active on *M. smegmatis* and *M. tuberculosis*. Based on this fact, most of the earlier studies dealt with *M. smegmatis* because it is a saprophyte and because of the ease with which the organisms can be grown. Subsequent studies with this drug used the *M. avium* complex. Earlier work by Takayama (39) showed that ethambutol inhibits the transfer of mycolic acid to cell wall of *M. smegmatis*, althrough the mycolyte acetyl transhalose enzyme leads to the accumulation of trehalose mono and di-mycolates. Subsequently, Takayama and Kilburn (40) showed that ethambutol inhibited the transfer of $^{14}$C-labeled glucose into the D-arabinose residue of arabinogalactan in the drug-susceptible strain of *M. smegmatis* but not in the drug-resistant mutant. In cell-free enzyme systems, it has been shown that the drug inhibits the synthesis of arabinose containing oligo saccharides. Takayama and Kilburn (40) also suggested that ethambutol triggers a series of changes in the lipid metabolism of mycobacteria. These studies were confirmed by Hoffner et al. (41) and by Rastogi et al. (42). Based on this fact, it has been shown that ethambutol enables the transfer of drugs into drug-resistant mutants, thereby making those drugs more active on bacilli to which they were originally resistant. Using in vitro, macrophage and animal models, Jagannath and associates have shown that ethambutol enhances the activity of isoniazid, streptomycin, or rifampin on multiple-drug resistant tubercle bacilli (43).

## MECHANISM OF RESISTANCE DEVELOPMENT TO STREPTOMYCIN AND OTHER AMINOGLYCOSIDE ANTIBIOTICS

The mechanism of development of resistance to these antibiotics was originally elucidated with non mycobacterial organisms e.g. *E. coli* and subsequently with mycobacteria. In the absence of any evidence to the contrary, most of the conclusions are expected to be valid with both systems.

*Data with Nonmycobacterial Systems.* Streptomycin is reversibly adsorbed on the surface of *E. coli* whether or not the particular strain of the organism is susceptible, resistant, or dependent (44). (Contrary to many other drugs, dependency of the microorganisms to streptomycin has been frequently observed.) With resistant organisms, however, little further action takes place. In susceptible strains, the drug probably damages the membrane, perhaps by interfering with the synthesis of the membrane protein in the same way as it interferes with the cytoplasmic protein synthesis. This damage enables streptomycin to penetrate freely into the cells. Within the cell, it influences the ribosomal stage of protein synthesis, probably by interfering with the function of natural polyamines therein, by being adsorbed on ribosomes along with or in place of them, thus probably preventing the normal attachment to the messenger RNA. The ribosomes from the resistant mutants, however, are much less affected by streptomycin. In the streptomycin-dependent strains, the ribosomes require the drug for normal functioning, and some messenger RNAs probably do not become attached to the ribosomes.

*Data with Mycobacterial Systems.* Several of these dealt with some saprophytic organisms like *M. friburgensis* and *M. smegmatis* (45). Both the whole cells of the susceptible and resistant strains of these organisms as well as the cell-free systems were used to study the incorporation of $^{14}C$ tyrosine into the ribosomal protein. In the wild-type drug-susceptible strain of *M. friburgensis,* incorporation of $^{14}C$ tyrosine into the protein was inhibited by streptomycin; with the streptomycin-resistant strain under similar conditions, it was not affected. However, this cell-free system was devoid of ribosomes.

With whole cells of *M. friburgensis,* it was noted that streptomycin caused only a slight inhibition of the tyrosine incorporation; however, it caused almost complete inhibition of protein synthesis in the whole cells of the streptomycin-susceptible *M. tuberculosis* H37Rv strain but not in the streptomycin-resistant mutant

(46). The inhibitory activity of several antituberculosis drugs was studied with both *M. smegmatis* and *M. tuberculosis* using the ribosomal amino acid incorporation techniques. Among them, streptomycin, kanamycin, and capreomycin* were found to be highly effective at low concentrations, maximal inhibition of the order of 80–85% being obtained with streptomycin. The fact that some drugs like chloramphenicol and tetracycline do not inhibit the growth of *M. tuberculosis* but can inhibit the protein synthesis in vitro indicates that the intact cells are impermeable to the drugs. Low concentrations of streptomycin ($2.5 \times 10^{-9}$ *M*) could inhibit the protein synthesis in vitro, and apparently only one molecule of the drug is necessary per ribosome. In strains which are resistant to high concentrations of streptomycin, a 10,000-fold increased concentration of the drug is needed to produce a 50% inhibition. Susceptibility and high-level resistance appear to reside in the ribosomes and not in the supernatant fluid; in this respect, the mycobacterial system is quite similar to that of *E. coli.* Low-level streptomycin-resistant mutants show the same degree of inhibition of in vitro protein synthesis as the wild type, but the whole cells are not susceptible. This suggests that in these resistant mutants, the mechanism of resistance does not reside in the ribosomes.

Sahila et al. (46) have shown two major phases of the action of streptomycin on *M. tuberculosis:* (a) initiation which involve protein synthesis and which probably represents an effect of streptomycin on the membrane, resulting in the entry of the drug into the cells and (b) the lethal phase, which involves a further attack by the drug, possibly on a process other than protein synthesis. The change in permeability of the membrane must be selective, as the rate of escape of several components of the soluble pool is not affected. The fact that streptomycin can inhibit bacteriophage multiplication in streptomycin-resistant cells, if added when the phage nucleic acid is still near the cells, suggests that streptomycin can have an effect on nonribosomal nucleic acid in the cell. Other literature on viomycin, neomycin, and kanamycin has also indicated that they have similar modes of action as that of streptomycin in the nonmycobacterial systems (47).

Other evidence concerning the mechanism of resistance development to aminoglycosides was concerned with the permeability studies. Taber and Halfenger (48) showed that tubercle bacilli resistant to kanamycin failed to accumulate the drug. Similarly, Nishida (49) and Tsukamura and associates (50) independently studied streptomycin resistance of the *M. avium* complex in relation to the permeability to the drug. Nishida correlated the streptomycin resistance of *M. avium*

---

*In the broad classification of the aminoglycoside antibiotics, drugs like viomycin and capreomycin are also included. Viomycin, a basic peptide antibiotic, is included in the aminoglycosed family, because its mode of action is rather similar to streptomycin and it acts on ribosomes and inhibits protein synthesis. It should be stressed, however, that viomycin does not cause miscoding, as do the regular aminoglycoside antibiotics. Similarly, capreomycin, which is a cyclic peptide antibiotic, is also included in the aminoglycoside family because of its similarity in action to streptomycin.

with changes in the permeability but could not correlate this with any development of resistance in susceptible strains. At least, reduced binding was not observed to be a characteristic of drug resistance in all instances. More recently, Jagannath and associates (43) have investigated the action of streptomycin on streptomycin-resistant bacilli in conjunction with ethambutol and dimethyl sulfoxide (DMSO), which alter the cell permeability. At sub-inhibitory concentrations, both these agents enhanced the activity of streptomycin, isoniazid, or rifampin on bacilli which are resistant to them initially.

In contrast to these findings, alterations of the cellular permeability by other agents like sodium dodecyl sulfate (SDS), which allowed streptomycin to enter the cells and inhibit the growth of the ribosomal functions, showed that it only did so in susceptible cells and not in resistant strains (51). Thus, even if the cellular permeability has been altered by other agents, in this case SDS, the drug did not seem to exercise the action on the ribosome in resistant bacilli. In summary, however, most of the evidence shows that permeability factors operate in the uptake of streptomycin by susceptible bacilli and the mechanism of resistance to this drug in mycobacteria is similar to those found with *E. coli* and other non-mycobacterial organisms.

## MOLECULAR BIOLOGICAL BASIS OF DEVELOPMENT OF RESISTANCE TO STREPTOMYCIN

Heym and associates (31) have studied extensively the molecular basis of development of resistance of several drugs using genetic analysis. In these pursuits, they used the procedures involving the amplification of a segment of the gene encoding the drug target by means of the polymerase chain reaction (PCR), and the comparison of the products obtained from susceptible and resistant strains to streptomycin by means of a single-strand confirmation polymorphism (SSCP) analysis. The combined procedure, called PCR–SSCP, was used to identify usually the altered SSCP profiles of mutational elements mainly leading to drug resistance. Cole and associates used the SSCP–PCR analysis of several strains of *M. tuberculosis* whose drug-susceptibility patterns have been established previously by standard microbiological methods. In this study, the authors used the PCR procedure specific for the various genes: *katG* (for INH), *inhA* (for INH and ETH), *rpoB* (for rifampin), and *rpsL* and *rrs* (for streptomycin). Concerning streptomycin, they have shown that the mutation takes place at the two genes, *rpsL* and *rrs*, and the SSCP–PCR analysis revealed alterations in the *rpsL* and *rrs* genes of $^{13}/_{25}$ and $^{2}/_{25}$ resistant isolates, respectively. The same mutation in the *rpsL* gene (k42R) was found in all cases, whereas two different mutations were detected in the *rrs* gene of two streptomycin-resistant strains. These authors found an excellent correlation between high-level resistance to streptomycin [minimal inhibitory concentration (MIC) $>1280$ $\mu$g/ml] in the presence of these mutations. Among the few remaining streptomycin-resistant isolates, where no mutants were detectable with the PCR–SSCP technique or by sequencing the key regions of the target gene, their resistance was low (MIC $< 90\ \mu$g/ml). However, no alterations in either the *rpsL* or *rrs* genes were detected in the streptomycin-susceptible strains.

## MECHANISMS OF THE DEVELOPMENT OF RESISTANCE TO RIFAMPIN

Because rifampin also is active equally well on mycobacterial and nonmycobacterial organisms, the mechanisms of development of resistance to it have been studied extensively with nonmycobacterial systems and subsequently with mycobacterial systems. Furthermore, because of the development of molecular biology of mycobacteria, genetic aspects of the development of resistance to rifampin, just like with isoniazid and streptomycin, have also been studied. For these reasons, the mechanism of resistance development to rifampin will be discussed under three categories: (a) with nonmycobacterial systems, (b) with mycobacterial systems, and (c) recent genetic studies.

*Data with Nonmycobacterial Systems.* Soon after the introduction of rifampin as a broad-spectrum antibiotic, it has been shown that it is a potent inhibitor of DNA-dependent RNA polymerase enzymatic activity in many bacteria, the inhibition being brought about by the formation of a stable drug–enzyme complex. Earlier studies on the mechanism of inhibition of *E. coli* RNA polymerase by this drug indicated that it acted at some step of transcription prior to the formation of first phosphodiester bond (52). Maximum inhibition of RNA synthesis occurs when rifampin and the enzyme are incubated, prior to the addition of other components of the assay. It has also been observed subsequently that the DNA-dependent RNA polymerase from *E. coli* was able to catalyze the formation of the dinucleotide tetraphosphate bond (53).

*Data with Mycobacterial Systems.* Similar to non-acid-fast organisms, the action of the drug in mycobacteria also seems to be related to the inhibition of the DNA-dependent RNA polymerase, although, interestingly, the sensitivity of this enzyme to the drug is different between mycobacterial and nonmycobacterial sources (59). For instance, with cell-free extracts, Harshey and Ramakrishnan (54) showed that the DNA-dependent RNA polymerase enzyme from *M. tuberculosis* was 1000 times more sensitive than that of *E. coli.* Similarly, the enzyme from *M. smegmatis,* which is naturally resistant to this drug, is also susceptible to the same concentration of the drug as with *M. tubercolosis.* However, it was shown that the drug concentration needed to inhibit the enzyme from the resistant mutants was much higher; in fact, in most cases the enzyme was not at all inhibited. Based on these findings, it was concluded that rifampin acts on mycobacteria to inhibit the DNA-dependent RNA polymerase activity in the same manner as it does in gram-negative bacteria. The uptake of $^{14}$C-uridine by susceptible but not from rifampin-resistant strains of *M. tuberculosis* is inhibited by rifampin.

In contrast to those studies with *M. tuberculosis* and *M. smegmatis,* the studies with *M. intracellulare,* an organism which is naturally resistant to rifampin, gave some interesting results. The DNA-dependent RNA polymerase from *M. intracellulare* has also been shown to be sensitive to rifampin (55). Apparently, in intact *M. intracellulare,* the enzyme is inaccessible to rifampin. With the addition of Tween-80 to growing cultures of *M. intracellulare,* there by altering the permeability, marked increase in susceptibility has been demonstrated. Based on these studies, it was concluded that *M. intracellulare* are resistant to rifampin by virtue of a permeability barrier (55). Similar results were seen with *M. bovis.* In contrast, both rifampin-susceptible and rifampin-resistance strains of *M. phlei* were found to incorporate similar amounts of $^{14}$C-rifampin, suggesting that resistance to rifampin in this organism depends on conformational changes in the polymerase enzyme but not on permeability barrier (56).

*Molecular Basis of the Resistance Development to Rifampin.* Genetic studies on the mechanism of drug-resistance development to rifampin used a similar background of the DNA-dependent RNA polymerase enzyme from *E. coli* as was used in the earlier biochemical studies. The enzyme in *E. coli* has been shown to be complex oligomer containing four different subunits ($\alpha$, $\beta$, $\beta^1$, and $\delta$) and assembled in two major forms: the core enzyme ($\alpha$, $\beta$, $\beta^1$) and the holoenzyme ($\alpha 2$, $\beta$, $\beta^1$) plus a $\delta$ subunit (57). The core enzyme has been shown to perform the RNA polymerization but requires a $\delta$ subunit to initiate the site-specific transcription at the promoter sites (64). The genes encoding the subunits $\alpha$, $\beta$, $\beta^1$, and $\delta$ have been designated as *rpoA, rpoB, rpoC,* and *rpoD,* respectively (58). Earlier studies using rifampin-susceptible and rifampin-resistant strains of *E. coli* have demonstrated that the RNA polymerase containing the $\alpha$ and $\beta^1$ subunits from a resistant strain and the $\beta$ subunit from a susceptible strain are sensitive to rifampin, whereas the RNA polymerase containing the $\alpha$ and $\beta^1$ subunits from a susceptible strain and the $\beta$ subunit from a resistant strain are resistant to rifampin (58). It was shown that in strains of *E. coli* which are susceptible to rifampin, the drug binds to the $\beta$ subunit of the RNA polymerase and leads to abhortive or premature initiation of transcription (59). Rifampin has a similar mechanism of action on RNA polymerase from *M. smegmatis* (60). More recently, the mapping of the *rpoB* gene in *M. tuberculosis* and its relationship to rifampin resistance have also been established following the leads from the work of Ovchinnikov et al. (61) and of Jin and Gross (62) on the mapping and sequencing of the mutations of the *rpoB* gene in *E. coli* and established its relation to rifampin. Miller et al. (63), who have established the entire sequence of the *rpoB* gene in *M. tuberculosis,* have drawn their guidance from these observations that portions of the amino acid sequences of RNA polymerase $\beta$ subunits from different bacteria, plants, and some eukaryotes are highly conserved. They also established using PCR to amplify a portion of the *rpoA* gene from *M. tuberculosis* chromosomal DNA that the oligo nucleotide primers corresponded to conserved domains of the $\beta$ subunits. Thus, the work of Miller et al. (63) has established that the 69 base-pair region and the presence of the histidine and serine codons at positions 5, 2, 6 and 5, 3, 1, respectively are similar to the *E. coli rpoB* sequence alignment. More recently, polymorphism in the *rpoB* gene has been described in the rifampin-resistant isolates of *M. tuberculosis* from New York City and Texas (64); with few exceptions, rifampin resistance is associated with the 69 base-pair coregion of the *rpoB* gene. Other studies by Honore and Cole (65) have established the importance of the *rpoB* gene in the mechanism of resistance of rifampin in *M. leprae.* Similarly, very recent studies by Whelen et al. (66) have identified a small region in the *rpoB* gene of the RNA polymerase in *M. tuberculosis* to be harboring the mutation potency of rifampin in *M. tuberculosis.* To sum up, the rapidly developing molecular biology studies have pinpointed the *rpoB* gene of the DNA-dependent RNA polymerase of the organism to be instrumental in the development of resistance to rifampin.

## EMERGENCE OF DRUG RESISTANCE

Although the earlier discussions dealt with the theoretical aspects of the origin and mechanism of development of drug resistance in individual cultures, mostly with a laboratory viewpoint, the actual situation of the emergence of drug-resistant bacilli with respect to clinical significance is of paramount importance. As can be judged by the earlier discussion, emergence of drug resistance is primarily due to the selection of preexisting resistant mutants in the original bacterial population. Because the proportion of drug-resistant mutants in the original population should be very low, large bacterial populations, such as those occurring in large lung cavities, are needed for the appearance of the drug-resistant mutants in the lesion. In such situations, susceptible organisms will be replaced by the resistant mutants in the presence of the selective drug, especially if single drug is used. Mutation to resistance thus emerging can be of either high or low degree. Twenty seven years ago, David (67) has shown that the organisms which are resistant to ethambutol are usually of the low degree, whereas those resistant to rifampin or pyrazinamide will be of high degree. Similarly, Mitchison (68) has shown that with streptomycin, three types of mutants with different levels of resistance could be demonstrated.

In clinical situations, the levels of resistance developed will be of great importance, as high doses of certain drugs (e.g., isoniazid) can be useful in eliminating the organisms with lower levels of drug resistance. However, because combination chemotherapy with at least two drugs is universally used for the treatment of tuberculosis, all the mutants resistant to the individual drugs will be eliminated, except those which are resistant to both drugs, which are very few in number ($1 \times 10^{10}$ or $1 \times 10^{11}$). As such, emergence of drug resistance should profitably be discussed separately for single- and multiple-drug regimens. Due to the widespread use of multiple-drug treatment worldwide, information on the emergence of resistance with single-drug treatment is limited. However, a few studies done in Madras thirty-three years ago afforded some opportunity to investigate the emergence of resistance with isoniazid alone (69).

*Emergence of Resistance Under Monotherapy.* Using the degree of resistance in the first-month and six-month cultures, the rate of inactivation, and the attainable peak serum concentration of isoniazid as the parameters, it was shown that isoniazid resistance emerges in two stages. In the first stage, which occurred early in treatment, resistant bacilli grew at any isoniazid dose, but low-resistance mutants failed to grow when the peak serum concentration of isoniazid was sufficiently high. This probably explains the lower proportion of patients with resistant organisms as the dosage of isoniazid was increased. In the second stage, low-resistance mutants continued to multiply, although still partially inhibited by isoniazid, and became more resistant; this is seen more in slow inactivators of isoniazid. It is to be stressed that only the first stage determines the emergence of resistance, because once resistance emerged, its existence was unrelated to the patient's eventual progress.

With streptomycin, it was shown that resistance appeared readily, and about 85% of the patients had streptomycin-resistant organisms at the end of the first month (70). The period by which complete resistance to streptomycin emerges has been calculated to be 45 days. Similar data on other antituberculosis drugs when used alone are not available, as monotherapy in the initial treatment of tuberculosis is not given.

*Emergence of Resistance Under Multiple Drug Therapy.* Very little emergence of resistance is seen in patients on daily isoniazid and streptomycin therapy with initially susceptible bacilli to both drugs, as most (95%) of the patients would have a bacteriologically favorable response. In contrast, treatment with daily isoniazid and PAS, or daily streptomycin plus PAS, resulted in the emergence of resistance to isoniazid or streptomycin, (but not to PAS) in about 10% of the patients, thus showing the inability to PAS to prevent the growth of isoniazid- or streptomycin-resistant mutants (71). Failure of PAS to inhibit streptomycin-resistant mutants, which was seen more frequently when the bacterial population was large, was also shown to be related to the dose of PAS (72). Thus, resistance to streptomycin emerged in 21% of the patients treated with streptomycin and 5–10 g of PAS daily, and in only 4% of the patients when the dose of PAS was raised to 20 g. The smaller dose, however, was much better in preventing emergence of resistance to streptomycin than when streptomycin was used alone.

*Emergence of Resistance with Intermittent Chemotherapy.* With the twice-weekly intermittent regimen of streptomycin and isoniazid, the outcome is the same as with daily regimens, but the once-weekly regimens of streptomycin and isoniazid resulted in the emergence of resistance to either streptomycin or isoniazid, even at the end of a 1-year treatment; 15–20% of the patients had drug susceptible organisms at 1 year, a finding at variance with the experience with daily regimens (73). More detailed analysis has shown that in these once-weekly regimens, the low level of the companion drug, which is due either to the frequency of administration or inactivation, is responsible for the emergence of resistance to the companion drug (74).

*Emergence of Resistance Under Short-Course Chemotherapy.* With short-course chemotherapy, there is no, or only minimal, emergence of drug resistance, because almost all positive cultures isolated at 6 or 9 months of chemotherapy are still drug susceptible and, therefore, could be retreated with the same drug combinations (75).

## 3. Factors Responsible for the Development of Resistance

### *3.1. Primary Drug Resistance*

Primary drug resistance which is due to the infection of patients with drug-resistant bacilli generated somewhere else could be a serious epidemiological problem. The increase in such resistance is due to the increase in resistant cases in the community and the chances of exogenous infections. The larger numbers of resistant cases and the greater frequency of infections in the country are reflected in the increase in primary drug resistance. The recent outbreaks of resistance in closed environments (e.g., prisons, schools, churches, etc.), as reviewed elsewhere in this book, have great relevance to the spread of drug resistance.

### *3.2. Acquired Drug Resistance*

Acquired drug resistance, which develops in patients who had originally drug-susceptible organisms but whose bacilli become resistant subsequently, can be due to several reasons:

1. Biological
   a. Initial bacterial population
   b. Local factors inside the host which are favorable for the multiplication of drug-resistant bacilli

   c. Presence of drug in insufficient concentrations
   d. Patients' drug inactivation status

2. Clinical
   a. Treatment with single drugs
   b. Inadequate dosage of the drugs
   c. Insufficient duration of treatment
   d. Adding a single drug to a failing regimen
   e. Interference by occult or quack medicine
   f. Interference by other indigenous systems of medicine

3. Pharmaceutical and Pharmacological
   a. Insufficient concentration of the pure drug
   b. Inadequate standardization of the bioavailability of the drugs
   c. Improper storage conditions of the preparations
   d. Improper bioavailability of combined tablet preparations
   e. Confusion created by the trade names of the various preparations of combination tablets
   f. Improper or incorrect dispensing of the drugs

4. Administrative
   a. Insufficient supplies of the drugs
   b. Bureaucratic influence in the ordering and supply of the drugs
   c. Substandard drugs purchased because of cost considerations and government regulations
   d. Administrative delays in the release of the drugs from the ports of entry
   e. Administrative controls on the drug dispensing

5. Sociological (Patient's Cooperation)
   a. Noncompliance
   b. Irregularity in drug intake
   c. Premature discontinuation of drug intake
   d. Avoidance of other exogenous infections with drug-resistant bacilli

## 4. Clinical Significance of Drug Resistance

The critical question in the study of drug resistance is whether resistance, as identified by the in vitro tests, carries any therapeutic penalties, and if so, does it vary under different treatment protocols with the available drugs? Based on this, proper approaches to manage such patients are to be decided; this latter aspect is discussed by Shimao in the chapter "Chemotherapy of drug-Resistant Tuberculosis in the Context of Developed and Developing Countries." (see Volume II, Chapter 7)

Because of the considerable differences of acquisition, clinical significance should be discussed separately for primary and acquired drug resistance. In the former case, resistance to the organisms is developed in another individual, and the fresh contact might still have a good chance of cure if an in vivo concentration of the concerned drug is greater than the minimal inhibitory concentration for the resistant bacilli could be obtained. Some clinical studies have indeed proved these suggestions (76). With acquired drug resistance, on the other hand, because resistance develops in the same host, its level would normally be determined by the attainable in vivo drug concentration. Such a situation obviously offers little benefit by further administration of the same drug. Because monotherapy, in contrast to multiple-drug treatment, favors development of drug resistance, this aspect should also be considered with that perspective, although only scant information is available for single-drug use.

The limited evidence obtained in Madras with the use of isoniazid alone has shown that only 29% of the patients with primary isoniazid resistance had a favorable response at the end of 1 year compared to 61% of 360 patients with isoniazid-susceptible bacilli (77). Even with the double-drug therapy containing isoniazid plus PAS, only a 29% response was seen with patients with initial primary drug resistance compared to 86% with susceptible organisms (78). Treatment with isoniazid plus thiacetazone, or isoniazid plus PAS also showed 77% response of patients with drug-susceptible organisms, and only 36–53% response in those having resistant organisms (79). Similar findings were obtained with the streptomycin and isoniazid combination. Thus, it is clear that the treatment of patients with primary drug resistance with isoniazid alone or its combination with PAS, thiacetazone, or streptomycin results in a poor outcome. With acquired drug resistance, a similar poor response is shown with monotherapy or combination chemotherapy; only 16% of 57 patients with acquired isoniazid resistance who were treated with isoniazid alone or with isoniazid and PAS had bacteriological conversion compared to 80% of patients with susceptible organisms (80).

A thorough review, involving the results of 12 controlled clinical trials carried out in Africa, Hong Kong, and Singapore, of the influence of initial drug resistance on the response of short-course chemotherapy of pulmonary tuberculosis, the now common method of treatment of tuberculosis, has been presented by Mitchison and Nunn (81). Failures during chemotherapy were encountered in 17% of 23 patients with initial resistance to isoniazid and/or streptomycin and given a 6-month regimen of isoniazid and rifampin, and in 12% of 264 patients given rifampin only in an initial 2-month intensive phase of their regimen. The proportion of failures decreased as the number of drugs in the regimen and the duration of treatment with rifampin were increased, to reach 2% of 246 patients receiving four or five drugs, including rifampin in the 6-month regimen. The sterilizing activity of the regimen, whether it included rifampin or pyrazinamide, was little influenced by initial resistance; the sputum conversion rate at 2 months was similar to that in patients with initially susceptible or resistant bacilli, and the relapse rates after the chemotherapy were only a little higher. The response in the 11 patients with initial rifampin resistance was, however, much poorer, failure during chemotherapy occurring in five and relapse afterward in an additional three patients. This review demonstrated the value of rifampin in preventing failures caused by the emergence of resistance during treatment and the greater sterilizing activity of rifampin and pyrazinamide, compared with that of isoniazid and streptomycin.

Another important study on the clinical significance of drug resistance in short-course chemotherapy regimens has been presented by Shimao (82). Using a monthly scoring system for the intensity of chemotherapeutic regimens for the bactericidal drugs (1.0 for isoniazid, pyrazinamide, and streptomycin; 1.5 for rifampin; and 0 for the resistant drugs), response to the several short-course chemotherapy regimens were calculated for both susceptible and resistant cases for either isoniazid, streptomycin, or both drugs. As can be expected, failure rates of patients with drug-susceptible organisms was minimal, with a score >9 in all cases. The failure rates of patients with isoniazid- and streptomycin-resistant organisms was higher but the rate was low when the score was ≥9. In patients with resistant organisms to both the drugs, the failure rate was high, even in cases with a score of 9–12. The relapse rates of patients with drug-susceptible organisms were low when the score was ≥17, and high when the score was <13. Overall, the failure and relapse rates were low with scores >13. In fact, the pretreatment drug susceptibility of the organism seems to have a minimal influence if the score was 13 or above.

Several studies (83–85), which were undertaken to guide "the choice of the first course of chemotherapy to be given to the patient," expressed the belief that chemotherapy failure due to initial drug resistance had been greatly exaggerated. Among these, the one from Hong Kong (85) drew the greatest attention, some (86,87) supporting it and several others (88–90) questioning the significance of its recommendations. If nothing else, these studies have temporarily, at least, reduced the interest of many people in continuing drug-susceptibility testing. This aspect, along with the overall reduction of interest in tuberculosis worldwide, has perhaps contributed to the tragic upsurge of tuberculosis and, more unfortunately, the increase of multiple-drug-resistant strains worldwide. An exhaustive discussion of this topic is given elsewhere (1). Opinions, however, have now changed diagonally, forging several attempts, including molecular biological procedures to develop tests for rapid detection of drug resistance.

## 5. Epidemiological Aspects of Drug Resistance

Data on the prevalence of drug resistance are being collected from many countries from time to time. Earlier, such information was obtained from planned studies, but, more recently, it involved only local areas and some retrospective analyses. Both developing and developed countries have been active in such endeavors. In developing countries, such information has helped in formulating national programs of chemotherapy and served as epidemiological tools or "yardsticks" for measuring the success of chemotherapy programs when the surveys are repeated at scheduled intervals. In developed countries, such data could assist in planning proper approaches, especially in communities dealing with large numbers of patients expected to harbor drug-resistant bacilli (e.g., immigrants from high-incidence areas).

As Canetti (4) suggested, the prevalence of drug resistance must be discussed separately for developed and developing countries. Recent studies are centering on initial drug resistance. On the other hand, some workers, including this author (1), have questioned the logic of this kind of dichotomy, especially when it is realized that the increase in drug resistance in developed countries has been shown to be due, to a large extent, to immigrants from countries with high prevalence rates (Tables 2.1 and 2.2). The exhaustive data on primary, initial, and acquired drug-resistance prevalence rates in several parts of the world, as well as the

Table 2.1 Immigrants and drug resistance

| | | Never treated | | Previously treated | |
|---|---|---|---|---|---|
| | Race | No. | % | No. | % |
| U.S. born | White | 1452 | 6.0 | 125 | 15.2 |
| | Black | 1011 | 7.4 | 92 | 23.9 |
| | American Indian | 236 | 7.2 | 59 | 6.8 |
| Hispanic (any area) | | 209 | 11.0 | 34 | 15.2 |
| Foreign born | Europe | 53 | 11.3 | — | — |
| | Central/S. America | 342 | 11.4 | 53 | 37.7 |
| | Asia | 410 | 16.1 | 86 | 27.9 |
| Overall | U.S. born | 2922 | 6.9 | 312 | 17.0 |
| | Foreign born | 837 | 13.9 | 152 | 32.9 |

*Source:* Data from CDC, Ref. 91.

Table 2.2 Incidence of drug-resistance according to country of origin

| | United States, Canada | Asia, Africa Caribbean, Latin America |
|---|---|---|
| Isoniazid | 7.0 | 39 |
| Streptomycin | 5.0 | 29 |
| Rifampin | 7.0 | 19 |

*Source:* Data from Ref. 92.

changes over a period of years in certain countries, have been discussed extensively elsewhere (1,93). It should be stressed, however, that most of the data are obtained from unplanned, retrospective, or sporadic studies and should not be taken at face value. Differences in techniques, intensity of questioning, and other important relevant parameters will also limit their usefulness. Assuming the same laboratory and interrogation parameters are used, the rates of primary drug resistance do not seem to change, except in some Asian countries, where rifampin resistance increased at a greater rate than in South American and African countries. It is gratifying to note that the World Health Organization is now coordi-

Table 2.3 Resistance to antituberculosis drugs[a] in culture-positive TB cases, United States, First Quarter, 1991 versus 1992

| Drug | Resistance 1991 first quarter Percentage (Resistant/Tested) | Resistance 1992 first quarter Percentage (Resistant/Tested) |
|---|---|---|
| **Any drug** | | |
| All cases[b] | 14.2 (472/3,313) | 13.1 (410/3,141) |
| New cases[b] | 13.4 (399/2,980) | 12.8 (371/2,903) |
| Recurrent cases[b] | 26.6 (51/192) | 19.4 (32/165) |
| **Isoniazid** | | |
| All cases[b] | 9.1 (301/3,303) | 9.4 (295/3,137) |
| New cases[b] | 8.2 (243/2,971) | 9.2 (266/2,899) |
| Recurrent cases[b] | 21.5 (41/191) | 15.8 (26/165) |
| **Rifampin** | | |
| All cases[b] | 3.9 (128/3,260) | 3.8 (119/3,109) |
| New cases[b] | 3.5 (103/2,931) | 3.8 (110/2,873) |
| Recurrent cases[b] | 9.0 (17/188) | 4.3 (7/163) |
| **Isoniazid and rifampin (MDR TB)** | | |
| All cases[b] | 3.5 (114/3,256) | 3.3 (104/3,106) |
| New cases[b] | 3.2 (94/2,927) | 3.4 (97/2,870) |
| Recurrent cases[b] | 6.9 (13/188) | 3.7 (6/163) |
| **Isoniazid and/or rifampin** | | |
| All cases | 9.5 (315/3.307) | 9.9 (310/3,138) |
| **Ethambutol** | | |
| All cases | 2.4 (77/3,175) | 2.5 (76/3,030) |
| **Streptomycin** | | |
| All cases | 5.7 (172/3,035) | 5.9 (174/2,925) |

[a]Results for pyrazinamide are excluded because of the law percentage (<30%) of culture-positive cases that were tested against this drug.

[b]Cases with "unknown new versus recurrent" disease status are excluded from rows of "new cases" and "recurrent cases" but are included in the rows of "all cases."

nating a multinational, multicenter study to collect meaningful information, using identical procedures in the laboratories as well as in eliciting history information on prior treatment and residence (94). It is hoped that such data will surface in the literature very soon. Considering the rates in the United States (Table 2.3), the overall percentage seems to be same or increasing slightly, although some states have shown an upward trend and others, the reverse.

The significant aspect in all these studies is the occurrence of drug resistance in previously treated patients who contribute the bulk of drug-resistant cases, not only in their countries of origin but, more importantly, in the countries of their immigration. Those patients, many of whom might not have been identified as having drug-resistant tubercle bacilli at the time of immigration, happen to be the potential source of resistant bacilli to new contacts in the developed countries.

### 5.1. Drug Resistance and HIV Infection

The poor prognostic influence of HIV infection on tuberculosis is discussed by Shaw and Coker in Chapter 8 of Volume II. The serious nature of HIV infection on the influence of drug resistance in tuberculosis, which has been shown to cause rapid progression, and death, has also been recognized. Although it is not possible to identify the specific reasons for this unfortunate outcome or to see whether and how the immunodeficiency could foster the spread of drug-resistant bacilli, the dismal outcome is obvious. The studies performed in this country in the "mini-epidemics" as discussed by Jarvis in Chapter 7 of this volume have brought this distressing situation to the forefront. The dangers of dual infections (*M. tuberculosis* and HIV) in African countries also are great and the growing drug resistance is adding further weight to this human tragedy.

### 5.2. A Hypothetical View of Spread of Drug Resistance

In developing and underdeveloped countries, *acquired* drug resistance will emerge because of (i) noncompliance, (ii) substandard drugs, (iii) monotherapy, and (iv) indiscriminate use of drugs (especially rifampin) for nonmycobacterial diseases. HIV infections can also have a great influence; however, the magnitude of this effect is not clear. All these reasons will contribute to *primary* drug resistance in these countries because of enormous opportunities of exposure to tuberculosis and poor isolation practices. In developed countries, *acquired* drug resistance emerges because of (i) noncompliance and associated factors like homelessness and frequent dislocation, (ii) alcohol or drug abuse (which increases noncompliance), and (iii) HIV disease. Although these factors contribute mostly to the spread of *primary* drug resistance (except in some special cases like the mini-epidemics where gross failures of isolation practice are involved), the bulk of primary drug resistance is shown to be due to immigrants from high-prevalence countries. Some of these immigrants who had undiagnosed tuberculosis or drug resistance (a consequence of the enormous delay in laboratory identification) would soon develop multiple-drug resistance. Thus, in the absence of facilities to obtain rapid information on drug resistance, the patient receiving both isoniazid and rifampin (the major drugs in short-course chemotherapy) will, in essence, receive a single drug if he has undetected drug resistance.

To minimize this problem, careful isolation and observation of patients plus the use of rapid methods for the identification of drug resistance are of great importance. Parallel to these approaches, discovery and development of new powerful drugs which are bactericidal in nature, of the caliber equal to or greater than that of isoniazid and rifampin, are urgently needed. Whereas this is the recommendation for the advanced countries such as the United States, it is not an approach realizable in the foreseeable future in many developing countries, where, unfortunately, the bulk of tuberculosis patients with drug resistance live.

**Acknowledgments**

The author is grateful to Dr. Cauthen, Dr. Bach, and others from CDC, Dr. Laszlo from Canada, Dr. De Kantor from Brazil, Dr. O'Brien, Dr. Spinaci, and associates from the WHO for unpublished information of the prevalence rates in the United States, Canada, and several other countries.

**References**

1. Gangadharam PRJ (1984) Drug Resistance in Mycobacteria. Boca Raton, FL: CRC Press.
2. Canetti G, Fox W, Khomenko A, Mahler HT, Mennon NK, Mitchison D, Rist N, Smelev NA (1969) Advances in techniques of testing mycobacterial drug sensitivity and the use of sensitivity tests in tuberculosis testing programs. Bull WHO 41:21.
3. Middlebrook G, Cohn ML (1958) Bacteriology of tuberculosis. Laboratory methods. Am J Pub Hlth 48:844.
4. Canetti G (1962) The eradication of tuberculosis: Theoretical problems and practical solutions. Tubercle 43:301.
5. Indian Council of Medical Research (1968) Prevalence of drug resistance in patients with pulmonary tuberculosis presenting for the first time with symptoms at chest clinics in India. Part I, Findings in urban clinics among patients giving no history of previous chemotherapy. Ind J Med Res 56:1617.
6. Youmans GP, Williston EH, Feldman WH, Hinshaw CH (1946) Increase in resistance of tubercle bacilli to streptomycin. A preliminary report. Proc Mayo Clin 21:126.
7. Pyle MM (1947) Relative numbers of resistant tubercle bacilli in sputum of patients before and during treatment with streptomycin. Proc Mayo Clin 22:465.

8. Temple CW, Hughs EJ, Mardis RE, Towbin MN, Dye WE (1951) Combined intermittent regimens employing streptomycin and para-amino salicylic acid in the treatment of pulmonary tuberculosis. Am Rev Tuberc 63:295.
9. Medical Research Council (1950) A Medical Research Council investigation. Treatment of pulmonary tuberculosis with streptomycin and para-amino salicylic acid. Br Med J ii:1073.
10. Heifets LB (1991) Drug Susceptability in the Chemotherapy of Mycobacterial Infections. Boca Raton, FL: CRC Press.
11. Vestal AL (1973) Procedures for the Isolation and Identification of Mycobacteria. DHEW Publication No. HSM 73-8230. Atlanta, GA: U.S. Department of Health, Education and Welfare, Centers for Disease Control.
12. Dean, ACR, Hinshelwood CN (1953) Observations on bacterial adaptation. In: Gale EF, Davies R, eds. Adaptation in Microorganisms, 3rd Symp. Soc. General Microbiology, p. 21. Cambridge: Cambridge University Press.
13. Pollock MR (1953) Stages in enzyme adaptation. In: Gale EF, Davies R, eds. Adaptation in Microorganisms, 3rd Symp. Soc. General Microbiology, p. 150. Cambridge: Cambridge University Press.
14. Mcdermott W (1970) Microbial drug resistance. The John Barnwell Lecture. Am Rev Respir Dis 102:857.
15. Szybalski W, Bryson V (1958) Origin of drug resistance in microorganisms. In: Sevez MG, Reid RD, Reynolds OE, eds. Origins of Resistance to Toxic Agents, pp. 22–39. New York: Academic Press.
16. Mitchison DA (1962) Microbial genetics and chemotherapy. Br Med Bull 18:74.
17. Schaefer WB (1961), quoted by Russel WF, Middlebrook G. In Chemotherapy of Tuberculosis, p. 21. Springfield, IL: Charles C Thomas.
18. Grosset J, Canetti G (1962) Incidence in wild strains of *Mycobacterium tuberculosis* of variants resistant to minor antibiotics (*p*-amino salicylic acid, ethionamide, cycloserine, viomycin, kanamycin). Ann Inst Pasteur 103:163.
19. David HL, Newman CM (1971) Some observations on the genetics of isoniazid resistance in tubercle bacilli. Am Rev Respir Dis 104:508.
20. Pollock MR (1960) Drug resistance and mechanisms for its development. Br Med Bull 16:16.
21. Barclay WR, Ebert RH, Koch-Weser D (1953) Mode of action of isoniazid, Part I. Am Rev Tuberc 67:490.
22. Barclay WR, Koch-Weser D, Ebert RH (1954) Mode of action of isoniazid, Part II. Am Rev Tuberc 70:784.
23. Youatt J (1958) The uptake of isoniazid by washed cell suspensions of mycobacteria and other organisms. Aust J Exp Biol Med Sci 36:223.
24. Youatt J (1960) The uptake of isoniazid and related compounds by mycobacteria. Aust J Exp Biol Med Sci 38:331.
25. Youatt J (1969) A review of the action of isoniazid. Am Rev Respir Dis 99:729.

26. Koch-Weser D, Palmer RP (1960) Correlation of drug susceptibility and metabolic activity of various strains of mycobacteria. Am Rev Respir Dis 81:949.

27. Quemard A, Lacave C, Laneelle G (1991) Isoniazid inhibition of mycolic acid synthesis by cell free extracts of sensitive and resistant strains of *Mycobacterium aurum.* Antimicrob Agents Chemther 35:1035.

28. Toida I (1962) Isoniazid-hydrolising enzyme of mycobacteria. Am Rev Respir Dis 85:720.

29. Zhang Y, Heym B, Allen B, Young D, Cole S (1992) The catalase-peroxidase gene and isoniazid resistance of *Mycobacterium tuberculosis.* Nature (London) 358:591.

30. Zhang Y, Garbe T, Young D (1993) Transformation with katG restores isoniazid-sensitivity in *Mycobacterium tuberculosis* isolates resistant to a range of drug concentrations. Molec Microbiol 8, 521–524.

31. Heym B, Cole ST (1992) Isolation and characterisation of isoniazid-resistant mutants of *Mycobacterium smegmatis* and *M. aurum.* Res Microbiol 143:721.

32. Gayathri Devi B, Shaila MS, Ramakrishnan T, Gopinathanan KP (1975) The purification and properties of peroxidase in *Mycobacterium tuberculosis* and its possible role in the mechanism of action of isonicotinic acid hydrazide. Biochem J 149:187.

33. Rastogi N, David HL (1993) Mode of action of antituberculosis drugs and mechanisms of drug resistance in *Mycobacterium tuberculosis.* Res Microbiol 144:133.

34. Banerjee A, Dubnau E, Quemard A, Balasubramanian V, Um KS, Wilson T, de Lisle G, Jacobs WR Jr (1994) *inha,* a gene encoding a target for isoniazid and ethionamide in *Mycobacterium tuberculosis.* Science 263:227.

35. Barclay WR (1955). In vitro studies with 14C PAS. Trans Veterans Administration–Armed Forces Conf. Chemother Tuberculosis 14:222.

36. David HL, Takayama K, Goldman DS (1969) Susceptibility of mycobacterial D-analyl-D-alanine synthetase to D-cycloserine. Am Rev Respir Dis 100:579.

37. David HL (1971) Resistance to D-cycloserine in the tubercle bacilli: mutation rate and transport of alanine in parental cells and drug resistant mutants. Appl Microbiol 21:888.

38. Konno K, Feldmann FM, McDermott W (1967) Pyrazinamide susceptibility and amidase activity of tubercle bacilli. Am Rev Respir Dis 95:461.

39. Takayama K, Armstrong EL, Kuugi KA, Kilburn JO (1979) Inhibition by ethambutol of mycolic acid transfer into the cell wall of *Mycobacterium smegmatis.* Antimicrob Agents Chemother 16:240.

40. Takayama K, Kilburn JO (1989) Inhibition of synthesis of arabinogalactan by ethambutol in *Mycobacterium smegmatis.* Antimicrob Agents Chemother 33:1493.

41. Hoffner SE, Svenson SB, Kallenius G (1987) Synergistic effects of antimicobacterial drug combinations on *Mycobacterium avium* complex determined radiometrically in liquid medium. Eur J Clin Microbiol 6:530.

42. Rastogi N, Goh KS, David HL (1990) Enhancement of drug susceptibility of

*Mycobacterium avium* by inhibitors of cell envelope synthesis. Antimicrob Agents Chemother. 34:759.

43. Jagannath C, Reddy MV, Gangadharam PRJ (1995) Enhancement of drug susceptibility of multi-drug resistant strains of *Mycobacterium tuberculosis* by ethambutol and dimethyl sulphoxide. J Antimicrob Chemother 35:381.
44. Winder F (1964) Antibacterial action of streptomycin, isoniazid and PAS. In: Barry VC, ed. Chemotherapy of Tuberculosis, p. 111. London: Butterworths.
45. Erdos T, Ullman A (1959) Effect of streptomycin on the incorporation of amino-acids labelled with carbon-14 into ribonucleic acid and protein in a cell-free system of a mycobacterium. Nature 183:618.
46. Sahila MS, Gopinathanan KP, Ramakrishnan T (1971) Protein synthesis in *Mycobacterium tuberculosis* H37Rv and the effect of streptomycin. In: Proceedings of the Biochemical Society Agenda Papers. Golden Jubilee Meetings of the Department of Biochemistry. p. 59. Bangalore: Indian Institute of Science.
47. Feingold DS, Davis BD (1962) Paradoxical effect of streptomycin, kanamycin and neomycin on *Escherichia coli* ribonucleic acid. Biochim Biophys Acta 55:787.
48. Taber H, Halfenger GM (1976) Multiple-aminoglycoside-resistant mutants of *Bacillus subtilis* deficient in accumulation of kanamycin. Antimicrobial Agents Chemother 9:251.
49. Nichida S (1959) Studies on resistance of microorganisms to various chemicals. 8. Mechanism of acquisition and loss of drug resistance by bacteria. 1. Experiments with anion type of tubercle bacilli. Ann Rept Res Inst Tuberc Kanazawa Univ 15:243; quoted from Biol. Abstr. 33:3847.
50. Tsukamura M, Tsukamura S, Mizuno S, Toyama H (1967) Bacteriological studies on atypical mycobacteria isolated in Japan, Report III. A comparison between pathogenic scotochromogens and soil scotochromogens. Origin of pathogenic scotochromogen. Kekkaku 42:15.
51. McClatchy JK (1980) Antituberculosis drugs: Mechanisms of action, Drug resistance, susceptibility testing and assays of activity in biological fluids. In: Lorian V, ed. Antibiotics in Laboratory Medicine, p. 156. Baltimore MD: Williams & Wilkins.
52. Sippel A, Hartmann GR (1968) Mode of action of rifampicin on the RNA polymerase reaction. Biochim Biophys Acta 157:218.
53. Kessler O, Hartmann GR (1977) The two effects of rifampicin on the RNA polymerase reaction. Biochim Biophys Res Commun 74:50.
54. Harshey RM, Ramakrishnan T (1976) Purification of DNA-dependent RNA-polymerase from *Mycobacterium tuberculosis* H37Rv. Biochim Biophys Acta (Amst) 432:49.
55. Hui J, Gordon N, Kajioka R (1977) Permeability barrier to rifampicin in mycobacteria. Antimicrob Agents Chemother 11:773.
56. Konno K, Oizumi K, Oka S (1973) Mode of action of rifampin on mycobacteria. II Biosynthetic studies on the inhibition of ribonucleic acid polymerase of

*Mycobacterium bovis* BCG by rifampin and uptake of rifampin–14 C by *Mycobacterium phlei.* Am Rev Respir Dis 107:1006.

57. Ishihama A (1988) Promoter selectivity of prokaryotic RNA polymerases. Trends Genet 4:282.
58. Burgess RR, Erickson B, Gentry D, Gribstov M, Hager D, Lesley S, Stricland M, Thompson N (1987) Bacterial RNA polmerase subunits and genes. In: Reznikoff WS, Gross CA, Burgess RR, Record MT Jr, Dahlberg JE, and Wickens, eds. RNA Polymerase and the Regulation of Transcription, p. 3. New York: Elsevier.
59. Johnston DE, McClure WR (1976) Abortive initiation of in vitro RNA synthesis on bacteriophage lambda DNA polymerase. In: Losick R, Chamberlain M, eds. RNA Polymerase, p. 413. Cold Spring Harbor, NY: Cold Spring Harbor Laboratory.
60. Levin ME, Hatfull GF (1993) *Mycobacterium smegmatis* RNA polymerase: DNA supercoiling, action of rifampicin and mechanism of rifampicin resistance. Molec Microbiol 8:277.
61. Ovchinnikov YA, Monastyrskaya GS, Gubanov VV, Guryev SO, Chertov OU, Modyanov NM, Grinkevich VA, Makarova IA, Marchenko TV, Polovnikova IN, Lipkin VM, Sverdlov ED (1981) Primary structure of *Escherichia coli* RNA polymerase: Nucleotide sequence of the beta-subunit. Eur J Biochem 116:621.
62. Jin DJ, Gross CA (1988) Mapping and sequencing of mutations in the *Escherichia coli* rpoB gene that lead to rifampicin resistance. J Molec Biol 202:45.
63. Miller LP, Crawford JT, Shinnick TM (1994) The rpoB gene of *Mycobacterium tuberculosis.* Antimicrob Agents Chemother 38:805.
64. Kapur V, Li LL, Bordanescu S, Hamrick MR, Wanger A, Kreiswirth BN, Musser JM (1994) Characterisation by automated DNA sequencing of mutations in the gene (rpoB) encoding the RNA polymerase *Beta* subunit in rifampin-resistant *Mycobacterium tuberculosis* strains from New York City and Texas. J Clin Microbiol 32:1095.
65. Honore N, Cole ST (1993) Molecular basis of rifampin resistance in *Mycobacterium leprae.* Antimicrob Agents Chemother 37:414.
66. Whelen AC, Felmlee TA, Hunt JM, Williams DL, Roberts GD, Stockman L, Persing DH (1995) Direct genotypic detection of *Mycobacterium tuberculosis* rifampin resistance in clinical specimens by using single-tube heminested PCR. J Clin Microbiol 33:556.
67. David HL (1970) Probability distribution of drug resistant mutants in unselected populations of *Mycobacterium tuberculosis.* Appl Microbiol 20:810.
68. Mitchison DA (1951) The segregation of streptomycin-resistant variants of *Mycobacterium tuberculosis* into groups with characteristic levels of resistance. J Gen Microbiol 5:596.
69. Selkon JB, Devadatta S, Kulkarni KG, Mitchison DA, Narayana ASL, Nair CN, Ramachandran K (1964) The emergence of isoniazid-resistant cultures in patients with pulmonary tuberculosis during treatment with isoniazid alone or isoniazid plus PAS. Bull WHO 31:273.

70. Medical Research Council (1948) Streptomycin treatment of pulmonary tuberculosis. Br Med J 2:769.

71. Tuberculosis Chemotherapy Center, Madras (1959) A concurrent comparison of home and sanatorium treatment of pulmonary tuberculosis in South India. Bull WHO 21:51.

72. Daniels M, Hill AB (1952) The chemotherapy of pulmonary tuberculosis in young adults. Br Med J 1:1162.

73. Tuberculosis Chemotherapy Center, Madras (1970) A controlled comparison of a twice weekly and three once weekly regimens in the initial treatment of pulmonary tuberculosis. Bull WHO 43:143.

74. Tripathy SP (1968) Madras study of supervised once weekly chemotherapy in the treatment of pulmonary tuberculosis: Laboratory aspects. Tubercle 49(Suppl):78.

75. Fox W, Mitchison DA (1975) Short course chemotherapy for pulmonary tuberculosis. Am Rev Respir Dis 111:325.

76. Canetti G, Rist N, Grosset J (1964) Primary drug resistance in tuberculosis. Am Rev Respir Dis 90:792.

77. Ramakrishnan CV, Bhatia AL, Devadatta S, Fox W, Narayana ASL, Selkon JB, Velu S (1962) The course of pulmonary tuberculosis in patients excreting organisms which have acquired resistance to isoniazid. Response to continued treatment for a second year with isoniazid plus PAS. Bull WHO 26:1.

78. Devadatta S, Bhatia AL, Andrews RH, Fox W, Mitchison DA, Radhakrishnan CV, Selkon JB, Velu S (1961) Response of patients infected with isoniazid resistant tubercle bacilli to treatment with isoniazid plus PAS or isoniazid alone. Bull WHO 25:807.

79. East African/British Medical Council Pretreatment Drug Resistance Report (1963) Influence of pretreatment bacterial resistance to isoniazid, thiacetazone or PAS on the response to chemotherapy of African patients with pulmonary tuberculosis. Tubercle 44:393.

80. Medical Research Council (1955). Tuberculosis Trials Committee. Various combinations of isoniazid with streptomycin or with PAS in the treatment of pulmonary tuberculosis. Br Med J 1:435.

81. Mitchison DA, Nunn AJ (1986) Influence of initial drug resistance on the response of short-course chemotherapy of pulmonary tuberculosis. Am Rev Respir Dis 133:423.

82. Shimao T (1987) Drug resistance in tuberculosis control. Tubercle 68(Suppl):5.

83. World Health Organisation (1963) Introduction. Bull WHO 29:559.

84. International Union Against Tuberculosis (1964) An international investigation of the efficacy of chemotherapy in previously untreated patients with pulmonary tuberculosis. Bull Int Union Tuberc 34:79.

85. Hong Kong Tuberculosis Treatment Services/British Medical Council Investigation (1972) A study in Hong Kong to evaluate the role of pretreatment susceptibility tests in the selection of regimens of chemotherapy for pulmonary tuberculosis. Am Rev Respir Dis 106:1.

86. Mitchison DA (1972) Implications of the Hong Kong study of policies of sensitivity testing. Bull Int Union Tuberc 47:9.
87. Fox W (1974) Drug sensitivity tests. Paper presented at the Annual Conference of the American Thoracic Society.
88. Wolinsky E (1974) Discussion of the paper by Fox at the Annual Conference of the American Thoracic Society.
89. Corpe R (1974) Discussion on drug sensitivity tests. Annual Conference of the American Thoracic Society.
90. Gangadharam PRJ (1980) Drug resistance and drug susceptibility tests. Am Rev Respir Dis 122:660.
91. Cauthen G (1992) Personal communication.
92. Hershfield E (1987) Drug resistance—response to Dr. Shimao's paper. Tubercle 68(Suppl): 17.
93. Gangadharam PRJ (1993) Drug resistance in tuberculosis. In: Reichman LB, Herschfield ES eds. Tuberculosis, A Comprehensive International Approach, p. 293. Marcel Dekker Inc.
94. Laszlo A (1996) Personal communication.

# 3

# Molecular Biology of *Mycobacterium Tuberculosis*

*Thomas M. Shinnick*

Molecular biology is simply the study of the biology of organisms at the molecular level. It includes investigations to define cellular architecture, biochemical pathways, macromolecular composition, structure, and function, and gene structure, regulation, and expression. This broad discipline of science encompasses diverse techniques ranging from sophisticated chemistry such as mass spectroscopy and x-ray crystallography to simple biochemistry such as assays for $\beta$-galactosidase activity. This chapter will concentrate on the aspect of molecular biology usually referred to as molecular genetics, which is concerned with defining gene structure, regulation of gene expression, protein products of genes, and functions of genes and gene products. The reader is referred to other chapters in this volume and several recent excellent reviews for discussions of other aspects of the molecular biology of *Mycobacterium tuberculosis,* including cellular architecture, macromolecular composition, and biochemistry (1–3).

Research into the molecular nature of the causative agents of tuberculosis and leprosy began more than century ago with their identification and microscopic description. In 1874, Hansen first described the presence of *Mycobacterium leprae* in a tissue biopsy specimen and suggested that it was the etiologic agent of leprosy (4). Eight years later, Koch identified *M. tuberculosis* as the causative agent of tuberculosis and formulated his now famous postulates for establishing a causal relationship between a microbe and a given disease (5). Early studies of mycobacterial molecular biology used classic biochemical techniques to investigate the physiology and biochemistry of the mycobacteria. For example, Koch produced crude filtrates of *M. tuberculosis* cultures known as "old tuberculin" for use as immunologic and therapeutic agents (6) and Seibert subsequently produced "purified protein derivatives" from the culture filtrates by simple chemical fractionation methods such as precipitation with trichloroacetic acid (7). Culture filtrate proteins were then further defined in a series of cooperative studies sponsored by the U.S.–Japan Medical Sciences Program (reviewed in Ref. 8), and studies are currently under way to define the structure, function, and immunoreactivity of individual culture filtrate antigens (reviewed in Ref. 9). Similarly, studies of mycobacterial lipids began in the 1930s with the description of the lipid composition of *M. tuberculosis* (10) and have proceeded to a precise definition of the chemistry and structure of individual lipid moieties such as lipoarabinomannan (reviewed in Ref. 11). Indeed, these elegant biochemical studies coupled with careful electron microscopic investigations have produced a detailed picture of the structure of the mycobacterial cell wall (reviewed in Refs. 1 and 12).

In contrast to the research into the biochemistry and physiology of mycobacteria during the past century, the study of the genetics of mycobacteria is still in its infancy. Until quite recently, few mycobacterial genes had been identified, few mutant strains had been isolated and characterized, and essentially no reliable genetic techniques had been developed. The lack of progress in mycobacterial genetics was primarily because (i) mycobacteria grow very slowly (*M. tuberculosis* has a generation time of 12–24 h and *M. leprae* has yet to be cultivated in vitro), (ii) mycobacteria are rather hydrophobic and tend to grow in clumps, which makes it quite difficult to purify individual cells for genetic analyses, (iii) there is no known naturally occurring genetic exchange between mycobacteria, and (iv) as a consequence, very few genetic markers have been identified in mycobacteria.

## 1. Analysis of Mycobacterial Genes in *Escherichia coli*

Because of these difficulties, many investigators began using recombinant DNA techniques to identify and characterize mycobacterial genes and gene products in *E. coli.* For example, *M. leprae* genomic DNA was cleaved with the restriction enzyme *Pst*I to generate fragments of 15,000–30,000 base-pairs and ligated to *Pst*I-cleaved pHC79, an *E. coli* cosmid vector carrying a tetracycline-resistance determinant (13,14). (A cosmid is a cloning vehicle that can accommodate large pieces of foreign DNA and can be packaged into phage particles for efficient delivery to recipient bacteria, because the vector contains a *cos* site of phage lambda.) The recombinant cosmids were packaged into phage using an in vitro packaging extract, *E. coli* bacteria were infected with the resulting phage particles, and transformants were selected by growth in the presence of tetracycline. This recombinant DNA library contained approximately 1000 independent recombinant cosmids, each of which carried 40,000–45,000 base-pairs of *M. leprae* genomic DNA. To identify recombinant cosmids in this library that encoded certain enzymes, the cosmids were screened for the ability to complement enzymatic defects in several mutant strains of *E. coli.* To do this, the recombinant library was transformed into *E. coli* strains carrying a mutation in the *proA, trpE, thyA,* *asd, araC,* or *lacZ* gene, and the transformants were plated on the appropriate selective media (e.g., on media lacking proline for the *proA* recipients). However, none of the recovered transformants contained mycobacterial genes that could complement any of these mutations. Because the library contained virtually all of the *M. leprae* genome and because mycobacterial enzymes displaying the required enzymatic activities were known to exist, these studies suggested that the failure to observe complementation of the *E. coli* defects was probably due to failure of the mycobacterial genes to be properly expressed in *E. coli.*

To circumvent this problem of gene expression, *M. leprae* genomic DNA was cloned into the plasmid expression vector pYA626, which directs transcription of the inserted DNA using the *Streptococcus mutans* aspartate $\beta$-semialdehyde dehydrogenase (*asd*) promoter (15). When the complementation experiments were repeated with this expression library, complementation of a defect in citrate synthase (*gltA16*) and a defect in dehydroquinate synthase (*aroB15*) were observed at frequencies of $10^{-4}$–$10^{-7}$. One conclusion from these results and the requisite control experiments was that the failure to express the mycobacterial genes in *E. coli* was due to a failure to transcribe the mycobacterial genes and not to problems with translation of the messenger RNA.

Bacteriophage vectors have also been used to express mycobacterial proteins in *E. coli.* For example, random fragments of the *M. tuberculosis* and *M. leprae* genomes were cloned into the lambda-gt11 vector to generate libraries that should contain recombinants that express each of the more than 3000 mycobacterial genes (16,17). Key features of this expression vector are that transcription of the inserted genes is driven by a promoter present in the vector, and translation of the mycobacterial genes is facilitated by fusion of the inserted open reading frame to that in the $\beta$-galactosidase (18). The lambda-gt11 libraries have been particularly useful in the identification of immunoreactive mycobacterial proteins. In initial studies, phage plaques were tested with a panel of monoclonal antibodies that had been collected and characterized by the World Health Organization, and genes encoding seven mycobacterial proteins were isolated (reviewed in Ref. 19). As is typically done with cloned genes, the nucleotide sequences of the genes were determined, the encoded proteins were expressed, and the immunoreactivity of the proteins characterized in detail. It turns out that many of these initially identified immunoreactive mycobacterial proteins displayed homology with heat shock proteins, which led to a great deal of experimentation into their possible roles in autoimmunity and arthritis (reviewed in Refs. 20–24).

Two of these genes, one encoding a 65-kDa antigen and one encoding a 10-kDa antigen, received a great deal of attention because studies suggested that they were highly immunoreactive B-cell and T-cell antigens (perhaps even immunodominant) and that they were homologous to highly conserved heat shock proteins (reviewed in Refs. 20–22). Molecular analysis of these mycobacterial genes in *E.*

*coli* revealed that the gene (*cpn10*) encoding the 10-kDa antigen was adjacent to a gene (*cpn60-1*) encoding a 60-kDa protein and that the gene (*cpn60-2*) encoding the 65-kDa antigen was located elsewhere in the genome (25,26). Analysis of the predicted amino acid sequences of these genes revealed that the 10-kDa protein was a member of the HSP10 family of heat shock proteins and that the 60-kDa and 65-kDa proteins were members of the HSP60 family of heat shock proteins. The presence of two genes encoding HSP60 homologs in a bacteria genome appears to be unique to species of *Mycobacterium* and *Streptomyces* (25–27) and has led to interesting insights into the evolution of these species (28). Detailed immunologic analyses of these genes included careful measurement of the antibody and T-cell response to these antigens in infected persons and controls, precise mapping of B-cell and T-cell epitopes, and determination of their ability to induce protective immunity, arthritis, and autoimmunity in experimental animal models (reviewed in Refs. 20–24).

The recombinant DNA libraries in *E. coli* were also used to identify mycobacterial genes that were related to already characterized genes such as those encoding ribosomal RNAs (29) or the $\beta$-subunit of RNA polymerase (30). Here, the conservation of the nucleotide sequences of the homologous genes was exploited to isolate the mycobacterial genes by virtue of their ability to cross-hybridize with the gene from other bacteria.

In general, the analysis of mycobacterial genes in *E. coli* can provide much information about some genes and gene products, but the general usefulness of this analysis is limited by the fact that the mycobacterial genes are being studied in a heterologous species (*E. coli*) and many questions can only be answered by studying mycobacterial genes in mycobacteria. For example, when the *recA* gene of *M. tuberculosis* was cloned and analyzed in *E. coli* (reviewed in Ref. 31), it was discovered that it could encode an 85,000-Da protein, although the mature *M. tuberculosis* RecA protein has a molecular mass of only approximately 38,000 Da. The primary translation product of the *M. tuberculosis recA* gene has subsequently been shown to contain an internal amino acid sequence, called an intein, which was removed during maturation by a protein-splicing mechanism to produce an enzymatically active RecA protein. This is an interesting phenomenon because so few examples of protein splicing are known (reviewed in Ref. 32) and because it may be a novel way to regulate protein expression and function. However, because no examples of protein splicing have been described for *E. coli* proteins, the biologic significance or regulation of protein splicing cannot be studied in *E. coli* but must be studied in the native organism. In other words, the biology of the splicing of the mycobacterial RecA precursor protein must be studied in *M. tuberculosis.* (On the other hand, the biochemistry of the protein splicing event can be analyzed in *E. coli* because splicing of the *M. tuberculosis* RecA precursor appears to occur by an autocatalytic process.)

## 2. Analysis of Mycobacterial Genes in Mycobacteria

The analysis of mycobacterial genes in mycobacteria was hampered by the factors listed above for genetic studies as well as by the inability to manipulate genes in mycobacteria, and attempts to develop genetic maps of the mycobacterial genome using classic genetic techniques (e.g., genetic crosses, matings, transformation, transduction) or molecular genetic techniques (e.g., site-directed mutagenesis, gene replacement) met with little success (reviewed in Refs. 33 and 34). For example, one molecular genetic approach to analyzing the structure and function of a gene is to (i) clone the gene into a plasmid vector, (ii) make mutations in it by site-directed mutagenesis techniques, (iii) replace the wild-type gene with the altered gene by homologous recombination, and (iv) characterize any alteration in the phenotype of the bacterium. The first two steps can be done in *E. coli,* but the last two steps must be done in the target organism to obtain meaningful data. However, until recently, techniques were not available to introduce foreign DNA reliably into mycobacteria or to promote homologous recombination.

Early attempts to introduce chromosomal DNA (transformation) or mycobacteriophage DNA (transfection) into mycobacteria were only sporadically successful (35–37). For example in 1954, Katanuma and Nakasato (35) demonstrated the transformation of streptomycin-susceptible mycobacteria to resistant ones by the addition of chromosomal DNA from streptomycin-resistant mycobacteria, but in 1959 Bloch et al. (36) were unable to transform a streptomycin-susceptible *M. tuberculosis* strain to a resistant strain using DNA from a streptomycin-resistant *M. tuberculosis* strain. With respect to transfection, Tokunaga and Sellers (37) demonstrated that mycobacteriophage D29 DNA could transfect *M. smegmatis,* although the efficiency of transfection was quite low.

One contributing factor to the low efficiency of transformation observed in these early studies was the mycobacterial cell wall (reviewed in Refs. 1 and 13), which presents a substantial barrier for any large, highly charged molecule such as DNA to cross. In mycobacteria, this barrier includes (i) an outermost hydrophilic layer with exposed mannose and arabinose units, (ii) a rather formidable hydrophobic layer of mycolic and mycocerosic acids, (iii) a peptidoglycan layer, and (iv) a cell membrane. Similar barriers have plagued the development of transformation systems in many species of bacteria and fungi, and a common strategy to overcome such barriers is to generate protoplasts (cells with an incomplete cell wall) by physical or enzymatic means, and indeed, growth of mycobacteria in glycine, which weakens the cell wall, does improve transfection efficiency (38). An efficient transfection system for *M. smegmatis* was finally developed in 1987 based on a protoplast transformation system for *Streptomyces* which involves growing the mycobacteria in 1% glycine for 16 h, treating with lysozyme (2 mg/ml) for 2 h, and then adding phage DNA and polyethylene glycol (39). This system routinely produces transfection efficiencies of $>10^4$ plaque-forming units per microgram of D29 DNA.

The availability of a reliable transfection system made possible the genetic manipulation of the transfecting DNA and the development of cloning vehicles for mycobacteria. The first cloning vehicles were shuttle phasmids, which are DNA molecules that can replicate in *E. coli* as plasmids and in mycobacteria as phage (39). The usefulness of shuttle phasmids is that they allow mycobacterial genes to be manipulated in *E. coli* and then transfected into *M. smegmatis* to generate infectious phage particles that can infect *M. smegmatis* or *M. bovis* BCG (Bacille–Calmette–Guérin) for the study of gene expression and function. The first shuttle phasmids were constructed using mycobacteriophage TM4 because TM4 was thought to be a temperate phage. (A temperate phage can integrate into the host chromosome and be stably maintained as a prophage.) However, it turns out that TM4 phages do not form stable lysogens, but grow and lyse the infected mycobacteria such that the expression or function of mycobacterial genes in mycobacteria cannot be easily studied using TM4-based shuttle phasmids. So, new shuttle phasmids were made using mycobacteriophage L1, which does form stable lysogens of *M. smegmatis* and *M. bovis* BCG (40). One of the key discoveries from these studies was that kanamycin was a usable, selectable genetic marker in mycobacteria, which led to the development of plasmids that could replicate in mycobacteria and *E. coli.* Such shuttle plasmids (e.g., JC85) typically contain a mycobacterial origin of replication (often derived from the *M. fortuitum* plasmid pAL5000), an *E. coli* origin of replication (usually derived from pUC plasmids), and a gene-conferring kanamycin resistance (derived from Tn5 or Tn903 transposons).

Of course, shuttle plasmids are useful only if they can be introduced into mycobacteria. Unfortunately, early attempts to transform them into *M. smegmatis* protoplasts failed, primarily because of the inability to regenerate viable cells from the protoplasts. (Transfection was successful because it did not require regeneration of the protoplasts, but only that the phage DNA enter the cell and be properly expressed.) Fortunately, at about this time, it was discovered that electroporation could be used to introduce foreign DNA into many bacterial and eukaryotic cells. In electroporation, DNA and cells are mixed together and a strong electrical field is applied to drive the DNA through the cell wall and into the cell. Electroporation works well with mycobacteria and is now the standard method for introducing exogenous DNA into mycobacteria (41). Studies with the electroporation of shuttle plasmids led to the development of *M. smegmatis* as a host cell, including the isolation of strains of *M. smegmatis* that could be transformed with a high efficiency (42).

## 3. Tools for Molecular Genetic and Recombinant DNA Techniques

The discovery of the utility of electroporation led to an explosion in the development of new tools and techniques for the molecular genetic analysis of mycobacteria (reviewed in Refs. 41 and 43–47). Table 3.1 contains a list of representative cloning vectors available for use with mycobacteria. These include shuttle phasmids and plasmids, cosmid cloning vehicles, vectors to express the cloned genes, vectors to secrete the products of the cloned genes, vectors to deliver transposons to mycobacteria, vectors to promote homologous recombination, and vectors to isolate and characterize promoters. Most of the mycobacterial plasmid cloning vehicles use an origin of replication from the *M. fortuitum* plasmid pAL5000, which allows the vectors to be maintained in *M. smegmatis* or *M. bovis* BCG with a copy number of 5–10 plasmids per mycobacterium. Other plasmid vectors use origins of replication derived from mycobacteriophage D29 or from naturally occurring plasmids of *M. scrofulaceum, M. avium, Rhodococcus* spp, *Corynebacterium* spp, and gram-negative bacteria, and some of these plasmids are compatible with the pAL5000-based vectors (e.g., pMSC262 and pMVS301). Another type of vector replicates as a plasmid in *E. coli* but integrates into a specific site in the mycobacterial genome. Such integrating vectors are based on the integrase/attachment sites of mycobacteriophage L5 or of *Streptomyces ambofaciens* plasmid pSAM2.

One limitation of these cloning vehicles is that only a few selectable genetic markers are suitable for use with mycobacteria (Table 3.2; reviewed in Ref. 45). Resistance to kanamycin is the most frequently used selectable marker because *M. smegmatis, M. bovis* BCG, and *M. tuberculosis* are quite susceptible to kanamycin, the mutation rate to kanamycin resistance in these species is low, and well-characterized genes that confer kanamycin resistance are available from the transposons Tn5 and Tn903. Importantly, one mycobacterium containing a plasmid expressing the Tn5 *aph* gene can be easily recovered from a mixture of $10^5$–$10^6$ susceptible mycobacteria by plating on solid medium containing 50 $\mu$g kanamycin/ml. Other selectable markers include genes that confer resistance to chloramphenicol, gentamicin, thiostrepton, hygromycin, streptomycin, and sulfonamide, and superinfection by mycobacteriophage L5 and genes that encode biosynthetic enzymes that allow growth of the recipient strains in the absence of specific nutrients such as uridine or valine. However, several of these selection systems (e.g., chloramphenicol and thiostrepton) are not sufficiently powerful for the direct selection of low-frequency events such as transformation because of a high background rate of mutation to resistance or the limited susceptibility of the mycobacteria to the agent. Also, streptomycin resistance is rarely used as a selection system because of concerns regarding the introduction of genes that confer resistance to a first-line antituberculosis drug into the mycobacteria.

Besides selectable genetic markers, easily measured phenotypic markers can sometimes be used to identify recombinant strains, although phenotypic markers are best suited to investigate gene expression and protein structure and function. Phenotypic markers that have been used with the mycobacteria include $\beta$-galac-

Table 3.1. Cloning vehicles

| Replicon source | Market[a] | Examples | References |
|---|---|---|---|
| **Shuttle phasmids** | | | |
| Mycobacteriophage TM4 | Kan | phAE1 | 39 |
| Mycobacteriophage L1 | Kan | phAE15 | 40 |
| Mycobacteriophage L5 | Kan | phGS1 | 48 |
| Mycobacteriophage Ms6 | Kan | E1 | 49 |
| **Shuttle plasmids** | | | |
| *M. fortuitum* (pAL5000) | Kan | pYUB12, pRR3, pJC85 | 42, 50, 51 |
| *M. fortuitum* (pAL5000) | Hyg | p16R1 | 52 |
| *M. fortuitum* (pAL5000) | Gp71 | pMD132 | 53 |
| *M. scrofulaceum* (pMSC262) | Kan | pYT937 | 54 |
| *M. avium* (pRL7) | Kan | pJC20 | 50 |
| Mycobacteriophage D29 | Kan | pBL415 | 55 |
| *Rhodococcus* (pMVS300) | Tsr | pMVS301 | 56, 57 |
| *Corynebacterium* (pNG2) | Kan | pEP2 | 58 |
| *Corynebacterium* (pNG2) | Hyg | pEP3 | 58 |
| *E. coli* (pACYC184) | Cat, Kan | pIJ666 | 59 |
| *E. coli* (pUC19) | Cat | pSGMU37 | 59 |
| *E. coli* (RSF1010) | Str, Sul | RSF1010 | 60 |
| **Shuttle cosmids** | | | |
| *M. fortuitum* (pAL5000) | Kan | pYUB18, pMSC1 | 41, 61 |
| Mycobacteriophage L5 *int* | Kan | pYUB178 | 62 |
| *E. coli* (RSF1010) | Kan, Str | pJRD215 | 63 |
| **Integrating plasmids** | | | |
| Mycobacteriophage L5 *int* | Kan | pMH94 | 64 |
| *Streptomyces* (pSAM2) *int* | Tsr, Kan | pTSN39 | 65 |
| Mycobacteriophage FRAT1 *int* | Kan | pNIV2173 | 66 |
| **Specialty vectors (feature/purpose)** | | | |
| Expression of cloned genes | | pMV261, pmV361 | 67 |
| Secretion of cloned proteins | | pIJK-1 | 68 |
| Fusion to reporter molecules | | pYUB75, pIPJ66 | 69, 70 |
| Expression of reporter enzymes | | pLUC10, pPA3 | 71, 72 |
| Presentation of epitopes | | pIJK-1 | 68 |
| Isolation of promoters | | pYUB75, pSD7 | 69, 73 |
| Isolation of transposons | | pYUB215, pIPC26 | 74, 75 |
| Temperature-sensitive replication | | pCG59 | 76 |
| Transposon-mediated mutagenesis | | pCG79 | 77 |
| Gene replacement | | pY6001 | 78 |

[a]The markers confer resistance to kanamycin (Kan), hygromycin (Hyg), thiostrepton (Tsr), cloramphenicol (Cat), streptomycin (Str), sulfonamide (Sul), or superinfection by phage L5 (Gp71) (see Table 3.2).

Table 3.2. Selectable genetic markers

| Gene | Product | Resistance to | Plasmid | Ref. |
|---|---|---|---|---|
| *aph* | Aminoglycoside phosphotransferase | Kanamycin | pYUB12 | 42 |
| *aacl* | Aminoglycoside acetyltransferase | Gentamicin | pML122 | 60 |
| *cat* | Chloramphenicol acetyltransferase | Chloramphenicol | pMSGU37 | 59 |
| *hyg* | Hygromycin phosphotransferase | Hygromycin | pEP3 | 58 |
| *sul* | Dihydrofolate reductase | Sulfonamide | RSF1010 | 60 |
| *gp71* | L5 repressor | L5 infection | pMD132 | 53 |
| *strA* | Aminoglycoside phosphotransferase | Streptomycin | pJRD215 | 63 |
| *tsr* | Ribosomal RNA methylase | Thiostrepton | pMVS301 | 56 |
| *pyrF* | Orotidine monophosphate decarboxylase | Lack of uridine | pY6001 | 78 |

Table 3.3. Scorable phenotypic markers

| Gene | Enzymatic activity | Example | Refs. |
|---|---|---|---|
| *lacZ* | β-Galactosidase | pYUB76 | 69 |
| *dagA* | Agarase | | 79 |
| *phoA* | Alkaline phosphatase | pIPJ68 | 70 |
| *xylE* | Catechol 2,3-dioxygenase | pHCX-1 | 80 |
| *fflux* | Firefly luciferase | pLUC10 | 71 |
| *luxA,B* | *Vibrio* luciferase | pBL525, pPA3 | 72, 55 |
| *cat* | Chloramphenicol acetyltransferase | pSD7 | 73 |

tosidase, agarase, firefly and *Vibrio* luciferases, alkaline phosphatase, chloramphenicol acetyl transferase, and catechol 2,3-dioxygenase (Table 3.3).

Some of the molecular genetic techniques that have been developed for use with these tools include vectors and protocols for (i) replacing genes in *M. smegmatis* by homologous recombination, which allows one to make mutations in mycobacterial genes in vitro and then to replace the wild-type gene in *M. smegmatis* with the mutated gene in order to analyze its function in mycobacteria; (ii) mutating genes in mycobacteria by the insertion of foreign DNA (e.g., transposon-mediated mutagenesis), which greatly improves the ability to make mutations and investigate the genetics of mycobacteria; (iii) stably expressing proteins in mycobacteria, which allows trans-complementation and merodiploid analyses of gene function; (iv) strongly expressing reporter enzyme activities, which facilitates measuring the effects of agents (e.g., antimicrobial drugs and heat shock) on mycobacterial gene regulation, expression, secretion, and function; and (v) identifying genes that are differentially expressed, which may define virulence factors of mycobacteria (reviewed in Refs. 41 and 43–46). To illustrate how the recently developed cloning vehicles and molecular techniques will help further the understanding of the biology of mycobacteria, specific applications will be discussed in the remainder of this chapter.

## 4. Recent Progress in the Molecular Genetics of the Mycobacteria

### *4.1. Mycobacteriophage*

Mycobacteriophages continue to play important roles in the development of tools and techniques for the molecular genetic analysis of the mycobacteria, and, in addition, because most phages use the transcriptional and translational machinery of the host cell, the study of mycobacteriophages can provide important insights into molecular details of gene expression in mycobacteria (reviewed in Refs. 43 and 48). The most thoroughly analyzed mycobacteriophage is L5. The nucleotide sequence of this phage has been determined, open reading frames and potential protein products have been identified, and gene products and their functions have begun to be characterized (reviewed in Ref. 48). For example, the genes and gene products required for integration of this temperate mycobacteriophage into the mycobacterial chromosome have been identified and the mechanism of integration elucidated. This information has been exploited to develop recombinant DNA vectors that can integrate foreign DNA into the mycobacterial genome (45). One of the advantages of integration-proficient vectors is that the integrated DNA is stably maintained in the transformants in the absence of continual selective pressure (e.g., the presence of an antimicrobial agent is required to maintain most plasmid vectors).

Analysis of the regulation of the transcription of the L5 genes revealed that the product of gene 71 (Gp71) functioned as a repressor of the transcription of phage genes and was responsible for the immunity of L5 lysogens from superinfection by L5 phages in a manner analogous to the lambda repressor (53). This information was then exploited to develop a novel system to select transformants (53); that is, transformants expressing Gp71 are resistant to infection and lysis by mycobacteriophage L5 and related phages, whereas nontransformed bacteria are susceptible to infection and lysis. This novel selection system circumvents many of the problems concerning the use of antimicrobial resistance genes in recombinant organisms (e.g., *M. bovis* BCG recombinant vaccines) that might be used in humans.

In addition, ongoing studies with L5 are providing important insights into regulation of gene expression, codon usage, and structures of genes, promoters, and transcription terminators in mycobacteria (reviewed in Refs. 43 and 48).

### *4.2. Structure of the Mycobacterial Genome*

Biochemical and biophysical analyses of the genomes of mycobacteria revealed that (i) the genome displays a typical bacterial chromosomal structure (i.e., a single large circular DNA molecule), (ii) the genome contains about $3 \times 10^6$ base-pairs of DNA, (iii) the guanine plus cytosine content of mycobacterial genomic DNA ranges from 60 to 70 mol%, and (iv) many species of *Mycobacterium* contain naturally occurring plasmids (reviewed in Ref. 13). Molecular and recombinant DNA analyses of mycobacterial genomes revealed that, among other things, (i) the genomes of most *Mycobacterium* species contain repetitive DNA elements, (ii) mycobacteria have only one or two copies of the genes encoding the ribosomal RNAs (in contrast, *E. coli* has seven copies) and these genes are arranged in the typical rRNA operon, and (iii) the genomic DNAs of *M. leprae* and *M. tuberculosis* display surprisingly little variation among strains isolated from around the world (reviewed in Ref. 13). Interestingly, each of these observations has led to improvements in our ability to combat mycobacteria-caused diseases.

For example, Southern blot analysis of the hybridization of cloned genes to mycobacterial genomic DNA revealed that although most clones hybridized to a single portion of the genome, several clones hybridized to many (>10) genomic DNA fragments (81,82). Some of the multicopy genes, or repeated DNA elements, were found in only certain species of *Mycobacterium* (e.g., IS6110 in the *M. tuberculosis*-complex or IS900 in *M. paratuberculosis*), whereas others were found in many species (e.g., the major polymorphic tandem repeat) (81–84). In addition, the location of some of the repetitive DNA elements varied dramatically among isolates of a given species. The species-specific elements have been used as hybridization probes and targets of polymerase chain reaction assays to detect and identify medically important mycobacterial pathogens in clinical specimens, and the variable locations of the repeated sequences have been exploited in strain identification and fingerprinting systems (reviewed in Ref. 85).

A detailed analysis of the cloned genes that hybridized to single genomic regions revealed that the *M. leprae* genome displays surprisingly little variation among strains isolated from around the world; that is, from analyzing variation in patterns of the hybridization of 15 genes to genomic DNA cleaved with six restriction enzymes, Clark-Curtiss and Walsh (86) estimated that *M. leprae* isolates contain <0.26% base substitutions. Similar studies revealed that *M. paratuberculosis* isolates also display very little sequence variation, <0.15% base substitutions (83). On the other hand, the genome of *M. tuberculosis* appeared to be much more variable than the genomes of *M. leprae* or *M. paratuberculosis* (81) when analyzed according to variation in restriction fragment hybridization patterns, although much of this variability was due to variability in the locations of IS6110 repetitive DNA elements (85). So, to estimate the genetic variability of *M. tuberculosis* strains, Kapur et al. (87) determined the nucleotide sequences of eight genes in seven isolates and found that changes in the sequences of these *M. tuberculosis* genes were exceedingly rare (>1 synonymous substitution per 10,000 synonymous sites) in the absence of selective pressure. Of, course, in the presence of selective pressure, such as exposure to antimicrobial agents, one can readily find changes in the nucleotide sequences of certain *M. tuberculosis* genes. The biologic implications of such unusually high sequence conservation in these three *Mycobacterium* species that are slowly growing intracellular pathogens is unclear.

Molecular analysis of the 16S rRNAs of mycobacteria has led to important insights into the phylogeny of mycobacteria as well as important new tools for the diagnosis of mycobacteria-caused diseases (reviewed in Refs. 88–90). The features that make ribosomal RNA genes particularly useful as "molecular evolutionary clocks" for phylogenetic analyses are the following: (i) rRNA genes are ubiquitous among the bacteria and eukaryotes; (ii) rRNA functions have been highly conserved over extremely long evolutionary times (even a small change in function can dramatically alter sequence relationships); (iii) rRNA sequences are sufficiently long to contain a large amount of phylogenetic information but not too long for convenient sequence determination; and (iv) rRNA molecules display extensive secondary structure, which helps align the nucleotide sequences and identify differences (reviewed in Ref. 91). Phylogenetic trees of the genus *Mycobacterium* have been constructed by using sequences of 16S rRNA genes (89,92), and as expected, all species of *Mycobacterium* are closely related to each other and distantly related to members of other genera. Interestingly, a major phenotypic division of *Mycobacterium* species also appears as a major branch point in the phylogenetic tree (93); that is, rapidly growing species are more closely related to each other than to any of the slowly growing species, and the slowly growing species are also more closely related to each other than to any of the rapidly growing species.

Analysis of 16S rRNA sequences also revealed that although many regions of the sequence were highly conserved among the species, portions of the sequence were highly variable, and each species appeared to display a unique or "signature" sequence in the variable regions. These signature sequences have been used to develop species-specific hybridization probes, which in the diagnostic laboratory are more accurate and cost-effective than conventional biochemical or serologic tests for identifying certain *Mycobacterium* species including *M. avium* and *M. intracellulare* (94). In addition, because the 16S rRNA variable regions are flanked by highly conserved regions, gene amplification procedures can be combined with hybridization or sequence analyses to identify organisms in clinical specimens (90,95). Thus, the observation of a signature sequence (by sequencing or hybridization assays) in an amplicon indicates the presence of the corresponding species in the specimen. This approach is particularly powerful when dealing with difficult-to-grow organisms (e.g., *M. leprae* or *M. haemophilum*) or with a previously unknown *Mycobacterium* species (e.g., *M. genavense*).

Finally, the ultimate definition of the structure of the mycobacterial genome will come from determinating of the nucleotide sequence of the entire genome, and because so much information can be gained from such a nucleotide sequence, great efforts are being expended on determining the sequences of the genomes of *M. leprae* and *M. tuberculosis* (reviewed in Refs. 96 and 97). These data will be a tremendous resource for mycobacterial molecular geneticists and will allow them to concentrate on addressing the biologically important questions of the regulation of gene expression and the structure and function of the gene products.

### 4.3. Characterization of Gene Products and Functions

The nucleotide sequences of many genes have been determined, and many more will be available soon from the genome sequencing projects. The identification of open reading frames in the DNA sequences will be facilitated by the typical pattern of codon usage in mycobacterial genes, which is reflected in an unusually high G + C content in the third position of the codons (98,99). Computer-aided analysis of the deduced amino acid sequences should identify matches with previously described proteins. Although "genbank lotto" might provide hints as to the function of a protein, the determination of the role(s) of the cloned mycobacterial genes and their products in the biology of mycobacteria must be addressed experimentally. One molecular genetic approach to this question is to construct defined mutations in a cloned gene in vitro using standard *E. coli* site-directed mutagenesis techniques, reintroduce the altered gene into mycobacteria, and then characterize the phenotype of the transformants. Such studies have been referred to as fulfilling Koch's postulates at the molecular level (100).

For example, changes in the amino acid sequence of a small region of the $\beta$-subunit of DNA-directed RNA polymerase (encoded by *rpoB*) have been observed in clinical isolates of rifampin-resistant *M. tuberculosis* (101). [Rifampin resistance in *E. coli* is due to mutations in *rpoB* which abrogate the ability of rifampin to inactivate RNA polymerase (102).] To prove that the observed changes in the *rpoB* gene were responsible for rifampin resistance in *M. tuberculosis,* Miller et al. (30) first cloned and sequenced the *rpoB* gene from a rifampin-susceptible strain of *M. tuberculosis* and then, using site-directed mutagenesis techniques, constructed mutations in the susceptible *rpoB* gene corresponding to the changes observed in the rifampin-resistant clinical isolates. The susceptible gene and the altered gene, whose sequences differ by a single base, were cloned into pJC85 and electroporated into *M. smegmatis.* As expected, *M. smegmatis* transformants receiving the rifampin-susceptible *M. tuberculosis rpoB* gene were susceptible to rifampin, whereas *M. smegmatis* recipients receiving the in vitro-mutagenized *M. tuberculosis rpoB* gene were resistant to rifampin. These results confirm that the changes in the *rpoB* sequence observed in the clinical isolates of rifampin-resistant *M. tuberculosis* could, indeed, confer the rifampin-resistant phenotype.

Miller et al. (30) characterized the phenotype of a strain carrying two copies of the *rpoB* gene: one on the chromosome and one on a plasmid. Whereas such trans-complementation can reveal the functions of some genes, the generality of the approach is limited because of possible confounding interactions between the two copies of the gene. A better experimental system would allow replacement of a chromosomal gene with an in vitro-mutagenized gene, and such gene replacement systems have been of tremendous utility in many microorganisms. Husson et al. (78) described a gene replacement system for *M. smegmatis* in which (i) the target gene is cloned into a vector that can replicate in *E. coli* but not in mycobacteria, (ii) the target gene is disrupted by insertion of an *aph* gene, (iii) the plasmid carrying the mutated gene is electroporated into *M. smegmatis,* and (iv) transformants are selected by kanamycin resistance. Because the plasmid cannot replicate in *M. smegmatis,* the transformants arise from the insertion of the *aph* gene into the chromosome via illegitimate or homologous recombination. Insertion via illegitimate recombination often disrupts other chromosomal genes, produces transformants carrying two copies of the target gene, and is of limited value for analyzing the function of the target gene. On the other hand, illegitimate recombination can be used to create random insertional mutations in the mycobacterial chromosome (103). The insertion by homologous recombination into the target gene leads either to insertion of the entire plasmid into the target chromosomal gene (a single-crossover event) or replacement of the chromosomal gene with the mutated gene (a double-crossover event). If the plasmid-borne copy contains only an internal, disrupted segment of the target gene, either a single or double crossover can generate a transformant with a disrupted and presumably nonfunctional target gene. Although homologous recombination and gene replacement occur at a usable frequency in *M. smegmatis* (78), homologous recombination is rare and gene replacement has not yet been observed in members of the *M. tuberculosis* complex (104). The development of a reliable gene replacement system for *M. tuberculosis* has a high research priority.

*N*OTE ADDED IN PROOF:

Systems for allele replacement in slowly growing species have been recently developed (Pelicic V, Reyrat JM, Gicquel B (1996). Positive selection of allelic exchange mutants in *Mycobacterium bovis.* FEMS Microbiol Letters 144:161–166).

### *4.4. Drug Susceptibility*

The cloning and sequencing of the genes responsible for resistance to rifampin, isoniazid, streptomycin, ethambutol, fluoroquinolones, and ethionimide have led to a better understanding of the mechanisms of action of these antimicrobial agents and rapid tests for the detection of the altered genes (reviewed in Ref. 105). For example, Telenti et al. (101) combined polymerase chain reaction (PCR) amplification of a portion of the *rpoB* gene with a detection system called single-strand conformation–polymorphism electrophoresis to detect changes in the *rpoB* gene and, using this assay, successfully identified changes in 64 of the 66 rifampin-resistant *M. tuberculosis* isolates tested within 2 days of obtaining the sample. This is quite an improvement over the 6–12 weeks required for conventional drug-susceptibility tests. Furthermore, the limited sequence variation observed in *M. tuberculosis* isolates (see above) makes it likely that the observation of a change in the sequence of a target of an antimicrobial agent reflects the development of resistance to that antimicrobial agent.

Two limitations of the genetic approach to determining drug susceptibility are that each possible mechanism of resistance to a particular drug must be defined (there are at least two mechanisms of streptomycin resistance and three of isoniazid resistance) and that a different genetic assay may be needed for each mechanism of resistance. One way to circumvent these limitations is to assess drug susceptibility functionally because a common feature of drug-resistant strains is that they grow in the presence of the corresponding drug. One type of functional assay uses the inhibition of a reporter activity as a surrogate marker for rapidly determining the drug susceptibility of an isolate. The reporter phage system pioneered by Jacobs et al. (106) takes advantage of the exquisite sensitivity of assays to detect the generation of light by the enzyme luciferase and the fact that the production of luciferase activity is sensitive to the metabolic state of the mycobacterium. Luciferase activity is dependent on the transcription and translation of the luciferase gene carried by the reporter phage and is influenced by the ATP levels within a cell. Because antimicrobial agents can decrease gene expression and/or ATP levels, the luciferase–reporter system distinguishes susceptible from resistant strains by virtue of their ability to produce luciferase activity in the presence of the antimicrobial agent. For example, in one possible assay, a specimen would be infected with the luciferase–reporter phage and incubated in media with or without drugs. In the absence of a drug, both resistant and susceptible strains would produce luciferase activity. In the presence of a drug, infection of a drug-resistant strain would lead to the production of reporter activity, whereas infection of a drug-susceptible strain would not. This sort of assay may be able to provide information on drug susceptibilities in 24–48 h after obtaining a primary culture.

In addition to being a surrogate marker for the activity of known antituberculosis agents, luciferase–reporter systems can also be used as a simple and rapid screening procedure to identify compounds with antimicrobial activity against *M. tuberculosis*. Such screening systems have been developed using (i) a mycobacteriophage expressing firefly luciferase (106), (ii) a plasmid expressing firefly luciferase (71), and (iii) plasmids expressing the *Vibrio* luciferase genes (72,55); these systems show great promise for rapidly identifying candidates for new antituberculosis drugs.

### *4.5. Virulence Determinants*

A commonly used genetic approach to elucidating virulence mechanisms of bacterial pathogens is to isolate and characterize mutants that are less virulent. For the mycobacteria, attempts to isolate less virulent strains began shortly after the identification of *M. tuberculosis* as the causative agent of tuberculosis. The basic approach was based on Pasteur's attenuation of viruses and involved passing a virulent strain serially on laboratory medium and periodically assaying virulence (reviewed in Ref. 107). After 13 years of such serial culturing, Calmette and Guerin isolated an avirulent strain of *M. bovis* which was designated BCG (Bacillus–Calmette–Guerin). Similarly, in 1934, Steenken et al. (108,109) described the isolation of an attenuated strain (designated H37Ra) of the H37R isolate of *M. tuberculosis.* Numerous biochemical studies of these mutants did not reveal the molecular basis of their lack of virulence, and the limited genetic systems of mycobacteria hampered attempts to identify the mutations responsible for the lack of virulence.

One way to identify the mutation responsible for the loss of virulence starts with isolating recombinant DNA clones that restore the virulence of the mutated strain. For example, Pascopella et al. (62) constructed a recombinant DNA library of *M. tuberculosis* strain H37Rv (the virulent strain) DNA in a cosmid vector which allowed integration of the foreign DNA into the chromosome at the mycobacteriophage L5 attachment site. They then electroporated the library into *M. tuberculosis* strain H37Ra (the avirulent strain), infected mice with a pool of the transformed H37Ra bacilli, and recovered bacilli that had multiplied in the mice. (The last step should enrich for transformants that carry a piece of the H37Rv genome that complemented the defect in H37Ra because H37Ra grows very poorly in mice.) In this way, they identified an approximate 25,000-base-pair region of the H37Rv genome that could partially restore the virulence of H37Ra. Further analysis of this region, designated *ivg,* should identify the gene(s) responsible for the loss of virulence.

This sort of complementation approach should be feasible with any pair of virulent and avirulent strains. Unfortunately, the attenuation of mycobacterial pathogens by serial passage on laboratory medium is of limited utility because many years of serial passing of the slowly growing mycobacteria might be required to produce an attenuated strain, and because such attenuated strains often contain several mutations, some of which are unrelated to virulence. The latter might explain why the *ivg* locus only partially restored the virulence of H37Ra. Alternate molecular genetic strategies to generate attenuated mutants are discussed below.

### *4.6. Intracellular Survival*

One of the key aspects of the virulence of *Mycobacterium tuberculosis* is that it survives and replicates within the host's phagocytic cells, primarily within macrophages. Several strategies have been hypothesized for how mycobacteria avoid being killed by macrophages, including (i) prevention of the oxidative burst in phagocytosing cells, (ii) inhibition of phagosome–lysosome fusion, (iii) resistance to lysosomal enzymes, (iv) secretion of inhibitors or inactivators of bactericidal agents, (v) exudation of lipids or capsules, and (vi) escape from the phagosome into the cytoplasm (reviewed in Refs. 110 and 111). A key step in sorting out these possibilities will be the identification and characterization of the mycobacterial genes and gene products required for intracellular survival and replication. Three commonly used molecular genetic approaches to this are (i) characterization of mutants defective in intracellular survival, (ii) analysis of mycobacterial genes that confer the ability to resist the bactericidal activities of macrophages onto normally susceptible cells, and (iii) identification of genes that are specifically expressed during intracellular replication (reviewed in Ref. 46). Each approach might identify overlapping but distinct groups of genes and gene products. For example, one class of mutants identified by the first approach will probably be those defective in scavenging required nutrients (e.g., purines) from the host cell. Such genes are unlikely to be found by the second approach.

Mutants defective in intracellular survival can be isolated following serial passing on laboratory medium as described above. Alternatively, a more systematic and comprehensive approach to investigating the genes required for intracellular survival is to first isolate a large number of randomly mutated strains and then to test individual strains for intracellular survival. The key to this approach is to use a mutagenesis system that causes mutations at random locations throughout the genome and that allows the selection of strains that have undergone a mutagenic event. For example, Fields et al. (112) used a transposon-mediated mutagenesis system to isolate mutants of *Salmonella typhimurium* (mutated cells were selected by virtue of the genetic marker carried by the transposon) and then screened 9516 mutants individually for the ability to survive in macrophages (screening approximately 9000 independent mutants should saturate the genome). In this way, they identified 83 transposon-generated mutants of *Salmonella typhimurium* that were less resistant to killing by macrophages than the wild-type strain. These mutations affect many different genes and processes, including purine biosynthesis, amino acid auxotrophy, serum sensitivity, lipopolysaccharide biosynthesis, or colony morphology. One group of mutations affects *phoP*, an inducible regulatory gene, and displays an increased sensitivity to defensins (113). Many of the transposon-generated mutants affect survival or killing processes that have not yet been characterized.

A similar approach has been proposed to investigate the intracellular survival of *M. tuberculosis* (46). Insertional mutants could be generated using illegitimate recombination (103) or a transposon (77). For the latter, a transposon carrying a selectable genetic marker (e.g., kanamycin resistance) might be delivered to the recipient mycobacterium on a plasmid that has a temperature-sensitive origin of replication (77). Transformants would be isolated using the selectable marker of the transposon at the temperature permissive for plasmid replication and then transferred to the nonpermissive temperature. At the nonpermissive temperature, the only way for the mycobacterium to continue to express the drug-resistance marker carried by the transposon would be for the transposon to have inserted into the bacterial chromosome and, hence, create a chromosomal mutation. Although these mutagenesis systems are feasible, they are relatively inefficient, and further developments will be needed to allow the generation of a large number of independent mutants.

A second strategy for identifying the genes involved in intracellular survival is to use a recombinant DNA approach to identify the genes of a pathogen that will confer a particular virulence-associated feature onto a nonpathogenic strain. For example, Isberg and Falkow (114) constructed a cosmid library of genomic DNA from *Yersinia pseudotuberculosis* in *E. coli,* and by passing the library through the epithelial cell line Hep 2, they were able to identify a cosmid that conferred the ability to invade Hep 2 cells onto the recipient *E. coli.* Subsequently, this invasion phenotype was shown to be due to a single gene, *inv,* which encodes a protein of ~100 kDa. By using similar systems, invasion genes have been isolated from *Y. enterocolitica, S. typhi, S. typhimurium,* and *Shigella* (reviewed in Ref. 115).

Because mycobacteria can also invade cultured cells, including HeLa, Hep 2, and fibroblast cells (116), Arruda et al. (117) passed a recombinant DNA library of *M. tuberculosis* H37Ra genomic DNA in *E. coli* through HeLa cells and isolated an *E. coli* transformant that could invade HeLa cells. This invasive phenotype was associated with the expression of a 20,000-Da protein from an approximate 1500-base-pair fragment of the *M. tuberculosis* H37Ra genome. Studies are in progress to define the mechanism of invasion and properties of the expressed protein.

Of course, in the infected human, macrophages are the primary host cell for *M. tuberculosis* and *M. leprae.* So, to identify mycobacterial genes and gene products that are required for entry and survival within the normal host cell, recombinant DNA libraries of *M. leprae* genomic DNA in *E. coli* have been passed through macrophages to enrich clones carrying genes that endow the normally susceptible *E. coli* bacteria with an enhanced ability to survive within macrophages (118). Following 3 cycles of enrichment, 15 independent clones were isolated. Three recombinants were characterized in detail, and each conferred significantly enhanced survival on *E. coli* cells carrying them. Two of the cloned DNAs also conferred enhanced survival onto *M. smegmatis* cells. Further characterization of these genes and gene products should provide insights into the survival of *M. leprae* within macrophages.

The third molecular genetic approach is based on the idea that genes are expressed only when their functions are needed. For example, *E. coli* cells produce

$\beta$-galactosidase only when lactose is to be metabolized. Thus, the gene products required for the intracellular survival and replication of *M. tuberculosis* are likely to be expressed only when the bacilli are growing within macrophages. If so, the identification of genes that are expressed only during intracellular growth should identify important virulence factors of *M. tuberculosis,* and the in-depth analysis of differentially expressed promoters, their associated genes and gene products, and biologic functions should generate insights into the molecular mechanisms of the intracellular survival and virulence of *M. tuberculosis.* Both genetic and physical biochemical techniques have been used to identify such differentially expressed genes.

Mahan et al. (119) recently described an elegant genetic approach to identifying in vivo expressed genes, which basically involves cloning promoters in front of a gene whose product is required for growth in vivo, transforming the recombinants into a pathogenic bacterium that does not express the required gene product, and passing the recombinant through an animal host. The recombinants recovered from the animal host should contain promoters that directed expression of the required gene during growth in vivo. Because the approach targets functional promoters, it is usually necessary to use the identified promoters as hybridization probes to isolate the corresponding full-length gene which is then characterized by standard molecular biologic techniques. To use this approach with *M. tuberculosis,* one would need to (i) identify and clone a gene whose product was required for the growth of *M. tuberculosis* in vivo, (ii) construct a promoterless version of this gene, (iii) isolate a strain of *M. tuberculosis* that does not express this gene, and (iv) have an animal model of tuberculosis available. These tools are still being developed, although a recent report describes a potentially suitable mutant of *M. tuberculosis* that requires leucine for growth in vitro and does not grow in vivo (46).

An alternate molecular genetic approach involves an easily measured enzymatic activity rather than a selectable marker. The use of a reporter enzyme circumvents the need to identify and construct a promoterless essential gene and corresponding mutant of *M. tuberculosis.* A particularly useful reporter enzyme for this approach is firefly luciferase, which has been used successfully as a reporter system to determine the activities of drugs against *M. tuberculosis* (71,106). An additional advantage of this reporter system is that its substrate (luciferin) can enter intact cells, which greatly simplifies the enzymatic assay by obviating the need to lyse the mycobacteria. The basic strategy is to (i) clone pieces of the genome of a virulent *M. tuberculosis* strain in front of a promoterless luciferase gene, (ii) electroporate the recombinants into *M. tuberculosis* H37Rv, and (iii) assay individual transformants for reporter activity when they are growing in liquid media or inside macrophages. As a first step in this approach, Marston et al. (120) constructed a luciferase-reporter plasmid vector (designated pCL5) which has (i) an *E. coli* origin of replication from pUC18, (ii) an aminoglycoside 3′-phosphotransferase gene from Tn5, (iii) a modified integrase gene and attachment site from mycobacteriophage L1, (iv) the *E. coli* rRNA transcription terminators T1T2, (v) a promoterless firefly luciferase gene, and (vi) a unique *BamH*I cloning site immediately upstream of the luciferase gene. Next, small fragments of *M. tuberculosis* H37Rv genomic DNA were generated by partial digestion with the restriction enzyme *Sau*3a and inserted into the unique *BamH*I site of pCL5. These recombinant plasmids were purified and electroporated into *M. tuberculosis* H37Rv to produce a library of transformants carrying integrated pCL5 recombinants. An initial analysis revealed that ~14% of the *M. tuberculosis* transformants produced significant luciferase activity while growing in liquid media. The next step in these studies will be to assay luciferase production in individual *M. tuberculosis* H37Rv transformants growing inside macrophages. Recombinants containing promoters that are expressed only during intracellular growth should express luciferase activity while growing inside macrophages but not while growing in liquid medium.

Biophysical methods such as subtractive hybridization can also be used to characterize differences in gene expression. For example, mRNA subtractive hybridization can clearly detect differences in gene expression by two populations of cells that are due to differences in the transcription of a single gene (121). Briefly, this technique starts with the isolation of RNA from two strains or from the same strain grown under different conditions (121,122). Next, reverse transcriptase is used to generate complementary single-stranded DNA from the target RNA, which is then hybridized to a large excess of RNA from a second population. Double-stranded nucleic acid is removed by passing the sample over a hydroxylapatite column, and RNA is removed by treatment with RNase to generate a "subtracted" population of single-stranded cDNA. This cDNA can be recycled through the subtraction process to improve the enrichment or can be converted to double-stranded DNA, amplified using a PCR step, and used as Southern blot probes or cloned into suitable vectors. Minor variations in this approach have been used to identify genes that are expressed (i) by H37Rv that are not expressed by H37Ra (122,123), (ii) by tubercle bacilli in a sputum specimen that are not expressed by the bacilli growing in liquid medium (124), and (iii) by *M. avium* bacilli growing in macrophages but not by the bacilli growing in liquid medium (125).

## 5. Summary and Future Directions

Although the immunology and physiology of the mycobacteria have been intensively investigated for more than 100 years, studies of the molecular genetics of mycobacteria really began only about 10 years ago with the cloning of *M. tuberculosis* and *M. leprae* genes into *E. coli*, and molecular genetic manipulation of mycobacterial genes became feasible only about 7 years ago with the development of a reliable transfection system for mycobacteria. Many sophisticated molecular genetic tools and techniques for the mycobacteria have been developed, although additional research is needed to develop a gene replacement method for the pathogenic mycobacteria, additional selectable markers and vectors for co-transformation studies, and an efficient transposon-mediated mutagenesis system for making random mutations in the *M. tuberculosis* chromosome. Nonetheless, the currently available molecular genetic tools for mycobacteria have set the stage for rapid advances in (i) characterizing genes, gene products, and their functions, (ii) understanding the regulation of gene expression and protein processing, (iii) identifying the molecular basis of drug resistance and developing rapid assays to determine drug susceptibilities of isolates, (iv) elucidating the mechanisms responsible for the ability of mycobacteria to survive and multiply within professional phagocytic cells, (v) defining critical virulence determinants and the basis of the pathogenicity of the mycobacteria, and (vi) expressing foreign genes in mycobacteria and developing *M. bovis* BCG as a multivalent vaccine vehicle. Finally, a better understanding of the molecular biology and genetics of mycobacteria should greatly improve the diagnosis, treatment, and prevention of infections by these deadly pathogens that have plagued humans since prehistoric times.

**References**

1. Minikin DE. (1991) Chemical principles in the organization of lipid components in the mycobacterial cell envelope. Res. Microbiol. 142:423–427.
2. Bloom BR, ed. (1994) Tuberculosis: Pathogenesis, Protection, and Control. Washington DC: American Society for Microbiology Press.
3. Kubica GP, Wayne LG, eds. (1984) The Mycobacteria: A Sourcebook. New York: Marcel Dekker.
4. Hansen GA (1880) *Bacillus leprae.* Virchows Archiv 79:32–42.
5. Koch R (1882) Die Aetiologie der Tuberkulose. Berliner Klin Wochenschr 19:221–230.
6. Koch R (1891) Weitere Mittheilung uber das Tuberkulin. Dtsch Med Wochenschr 17:1189–1192.
7. Seibert FB, Munday B (1932) The chemical composition of the active principle of tuberculin. XV. A precipitated purified tuberculin protein suitable for the preparation of a standard tuberculin. Am Rev Tuberc 25:724–737.
8. Daniel TM, Janicki BW (1978) Mycobacterial antigens: A review of their isolation, chemistry, and immunologic properties. Microbiol Rev 42:84–113.
9. Anderson AB, Brennan P (1994) Proteins and antigens of *Mycobacterium tuberculosis.* In: Bloom BR, ed. Tuberculosis: Pathogenesis, Protection, and Control, pp. 307–332. Washington DC: American Society for Microbiology Press.
10. Anderson RJ (1939) The chemistry of the lipoids of the tubercle bacillus and certain other organisms. Fortschr Chem Organisch Naturstoffe 3:145–202.
11. Besra GS, Chatterjee D (1994) Lipids and carbohydrates of *Mycobacterium*

*tuberculosis.* In: Bloom BR, ed. Tuberculosis: Pathogenesis, Protection, and Control, pp. 285–306. Washington DC: American Society for Microbiology Press.

12. Brennan PJ, Draper P (1994) Ultrastructure of *Mycobacterium tuberculosis.* In: Bloom BR, ed. Tuberculosis: Pathogenesis, Protection, and Control, pp. 271–284. Washington DC: American Society for Microbiology Press.
13. Clark-Curtiss JE (1990) Genome structure of mycobacteria. In: McFadden JJ, ed. Molecular Biology of the Mycobacteria, pp. 77–96. London: Harcourt Brace Jovanovich Publishers.
14. Clark-Curtiss J, Jacobs W, Docherty M, Ritchie L, Curtiss R (1985) Molecular analysis of DNA and construction of genomic libraries of *Mycobacterium leprae.* J Bacteriol 161:1093–1102.
15. Jacobs WR, Docherty M, Curtiss R, Clark-Curtiss JE (1986) Expression of *Mycobacterium leprae* genes from a *Streptococcus mutans* promoter in *Escherichia coli* K-12. Proc Natl Acad Sci USA 83:1926–1939.
16. Young RA, Mehra V, Sweetser D, Buchanan T, Clark-Curtiss J, Davis RW, Bloom BR (1985) Genes for the major protein antigens of the leprosy parasite *Mycobacterium leprae.* Nature 316:450–452.
17. Young RA, Bloom BR, Grosskinsky CM, Ivanyi J, Thomas D, Davis RW (1985) Dissection of *Mycobacterium tuberculosis* antigens using recombinant DNA. Proc Natl Acad Sci USA 82:2583–2587.
18. Young RA, Davis RW (1983) Efficient isolation of genes by using antibody probes. Proc Natl Acad Sci USA 80:1194–1198.
19. Young DB, Mehlert A (1989) Serology of mycobacteria: Characterization of antigens recognized by monoclonal antibodies. Rev Infect Dis 11:S431–S435.
20. Young RA (1990) Stress proteins and immunology. Annu Rev Immunol 8:401–420.
21. Shinnick TM (1991) Heat shock proteins as antigens of bacterial and parasitic pathogens. Curr Top Microbiol Immunol 167:145–160.
22. Shinnick T (1995) Mycobacterial heat shock proteins. In: Rom W, Garay S, eds. Tuberculosis. Boston: Little, Brown and Co.
23. Winfield JB, Jarjour WN (1991) Stress proteins, autoimmunity, and autoimmune disease. Curr Top Microbiol Immunol 167:161–189.
24. van Eden W (1991) Heat-shock proteins as immunogenic bacterial antigens with the potential to induce and regulate autoimmune arthritis. Immunol Rev 121:5–28.
25. Rinke de Wit TF, Bekelie S, Osland A, Miko TL, Hermans PWM, van Soolingen D, Drijfhout JW, Schoningh R, Janson AAM, Thole JER (1992) Mycobacteria contain two *groEL* genes: The second *Mycobacterium leprae groEL* gene is arranged in an operon with *groES.* Mol Microbiol 6:1995–2007.
26. Kong TH, Coates ARM, Butcher PD, Hickman CJ, Shinnick TM (1993) *Mycobacterium tuberculosis* expresses two chaperonin-60 homologs. Proc Natl Acad Sci USA 90:2608–2612.
27. Mazodier P, Guglielmi G, Davies J, Thompson CJ (1991) Characterization of the *groEL*-like genes in *Streptomyces albus.* J Bacteriol 173:7382–7386.

28. Hughes AL (1993) Contrasting evolutionary rates in the duplicate chaperonin genes of *Mycobacterium tuberculosis* and *M. leprae.* Mol Biol Evol 10:1343–1359.

29. Sela S, Clark-Curtiss JE, Bercovier H (1989) Characterization and taxonomic implications of the rRNA genes of *Mycobacterium leprae.* J Bacteriol 171:70–73.

30. Miller LM, Crawford JT, Shinnick TM (1994) The *Mycobacterium tuberculosis rpoB* gene. Antimicrob Agents Chemother 38:805–811.

31. Colston MJ, Davis EO (1994) Homologous recombination, DNA repair, and mycobacterial *recA* genes. In: Bloom BR, ed. Tuberculosis: Pathogenesis, Protection, and Control, pp. 217–226. Washington DC: American Society for Microbiology Press.

32. Colston MJ, Davis EO (1994) The ins and outs of protein splicing elements. Molec Microbiol 12:359–363.

33. Greenberg J, Woodley CL (1984) Genetics of mycobacteria. In: Kubica GP, Wayne LG, eds. The Mycobacteria: A Sourcebook, pp. 629–639. New York: Marcel Dekker.

34. Mizuguchi Y (1984) Mycobacteriophages. In: Kubica GP, Wayne LG, eds. The Mycobacteria: A Sourcebook, pp. 641–662 New York: Marcel Dekker.

35. Katanuma N, Nakasato H (1954) A study of the mechanism of the development of streptomycin resistant organisms by addition of deoxyribonucleic acid prepared from resistant bacilli. Kekkaku 29:19–22.

36. Bloch H, Walter A, Yamamura Y (1959) Failure of desoxyribonucleic acid from mycobacteria to induce bacterial transformation. Am Rev Respir Dis 80:911.

37. Tokunaga T, Sellers MI (1964) Infection of *Mycobacterium smegmatis* and D29 phage DNA. J Exp Med 119:139–149.

38. Mizaguchi Y, Tokunaga T (1968) Spheroplast of mycobacteria. II. Infection of phage and its DNA on glycine-treated mycobacteria and spheroplast. Med Biol 77:57–60.

39. Jacobs WR, Tuckman M, Bloom BR (1987) Introduction of foreign DNA into mycobacteria using a shuttle phasmid. Nature 327:532–535.

40. Snapper SB, Lugosi L, Jekkel A, Melton RE, Kieser T, Bloom BR, Jacobs WR (1988) Lysogeny and transformation in mycobacteria: Stable expression of foreign genes. Proc Natl Acad Sci USA 85:6987–6991.

41. Jacobs WR, Kalpana GV, Cirilo JD, Pascopella L, Snapper SB, Udani RA, Jones W, Barletta RG, Bloom BR (1991) Genetic systems for mycobacteria. Methods Enzymol 204:537–555.

42. Snapper SB, Melton RE, Mustapha S, Kieser T, Jacobs WR (1990) Isolation and characterization of efficient plasmid transformation mutants of *Mycobacterium smegmatis.* Mol Microbiol 11:1911–1919.

43. Hatfull GF, Jacobs WR (1994) Mycobacteriophages: Cornerstones of mycobacterial research. In: Bloom BR, ed. Tuberculosis: Pathogenesis, Protection, and Control, pp. 165–183. Washington DC: American Society for Microbiology Press.

44. Hatfull GF (1993) Genetic transformation of mycobacteria. Trends Microbiol 1:310–314.

45. Berlein JE, Stover CK, Offutt S, Hanson MS (1994) Expression of foreign genes in mycobacteria. In: Bloom BR, ed. Tuberculosis: Pathogenesis, Protection, and Control, pp. 239–252. Washington DC: American Society for Microbiology Press.

46. Jacobs WR, Bloom BR (1994) Molecular genetic strategies for identifying virulence determinants of *Mycobacterium tuberculosis.* In: Bloom BR, ed. Tuberculosis: Pathogenesis, Protection, and Control, pp. 253–268. Washington DC: American Society for Microbiology Press.

47. Falkinham JO, Crawford JT (1994) Plasmids In: Bloom BR, ed. Tuberculosis: Pathogenesis, Protection, and Control, pp. 185–198. Washington DC: American Society for Microbiology Press.

48. Hatfull GF (1994) Mycobacteriophage L5: A toolbox for tuberculosis. ASM News 60:255–260.

49. Anes E, Portugal I, Moniz-Pereira J (1992) Insertion into the *Mycobacterium smegmatis* genome of the *aph* gene through lysogenization with the temperate mycobacteriophage Ms6. FEMS Microbiol Lett 95:21–25.

50. Crawford JT (1996) Unpublished results.

51. Ranes MG, Rauzier J, Lagranderie M, Gheroghiu M, Gicquel B (1990) Functional analysis of pAL5000, a plasmid from *Mycobacterium fortuitum:* Construction of a 'Mini' mycobacterium–*Escherichia coli* shuttle vector. J Bacteriol 172:2793–2797.

52. Garbe TR, Barathi J, Barnini S, Zhang Y, Abou-Zeid C, Tang D, Mukherjee R, Young DB (1994) Transformation of mycobacterial species using hygromycin resistance as selectable marker. Microbiology 140:133–138.

53. Donnelly-Wu MK, Jacobs WR, Hatfull GF (1993) Superinfection immunity of mycobacteriophage L5: Applications for genetic transformation of mycobacteria. Molec Microbiol 7:407–417.

54. Qin MH, Taniguchi H, Mizuguchi Y (1994) Analysis of the replication region of a mycobacterial plasmid, pMSC262. J Bacteriol 176:419–425.

55. David M, Lubinsky-Mink S, Ben-Zvi A, Ulitzur S, Kuhn J, Suissa M (1992) A stable *Escherichia coli–Mycobacterium smegmatis* plasmid shuttle vector containing the mycobacteriophage D29 origin. Plasmid 28:267–271.

56. Singer MEV, Finnerty WR (1988) Construction of an *Escherichia coli–Rhodococcus* shuttle vector and plasmid transformation in *Rhodococcus* spp. J Bacteriol 170:638–645.

57. Plikaytis BB, Shinnick TM. (1996) Unpublished observations.

58. Radford AJ, Hodgson ALM (1991) Construction and characterization of a *Mycobacterium–Escherichia coli* shuttle vector. Plasmid 25:149–153.

59. Zainuddin ZF, Kunze ZM, Dale JW (1989) Transformation of *Mycobacterium smegmatis* with *Escherichia coli* plasmids carrying a selectable resistance marker. Molec Microbiol 3:29–34.

60. Gormley EP, Davies J (1991) Transfer of plasmid RSF1010 by conjugation from *Escherichia coli* to *Streptomyces lividins* and *Mycobacterium smegmatis.* J Bacteriol 173:6705–6708.

61. Hinselwood S, Stoker NG (1992) An *Escherichia coli–Mycobacterium* shuttle cosmid vector. Gene 110:115–118.
62. Pascopella L, Collins FM, Martin JM, Lee MH, Hatfull GF, Stover CK, Bloom BR, Jacobs WR (1994) Use of in vivo complementation in *Mycobacterium tuberculosis* to identify a genomic fragment associated with virulence. Infect Immun 62:1313–1319.
63. Hermans J, Martin C, Huijberts GNM, Goosen T, de Bont JAM (1991) Transformation of *Mycobacterium aurum* and *Mycobacterium smegmatis* with the broad host-range gram-negative cosmid vector pJRD215. Molec Microbiol 5:1561–1566.
64. Lee MH, Pascopella L, Jacobs WR, Hatfull GF (1991) Site specific integration of mycobacteriophage L5: Integration-proficient vectors for *Mycobacterium smegmatis,* BCG, and *M. tuberculosis.* Proc Natl Acad Sci USA 88:3111–3115.
65. Martin C, Mazodier P, Mediola MV, Gicquel B, Smokvina T, Thompson CJ, Davies J (1991) Site-specific integration of the *Streptomyces* plasmid pSAM2 in *Mycobacterium smegmatis.* Molec Microbiol 5:2499–2502.
66. Haeseleer F, Pollet JF, Haumont M, Bollen A, Jacobs P (1993) Stable integration and expression of the *Plasmodium falciparum* circumsporozoite protein coding sequence in mycobacteria. Molec Biol Parasit 57:117–126.
67. Stover CK, de la Cruz VF, Fuerst TR, Burlein JE, Benson LA, Bennett LT, Bansal GP, Young JF, Lee MH, Hatfull GF, Snapper SB, Barletta RG, Jacobs WR, Bloom BR (1991) New use of BCG for recombinant vaccines. Nature 351:456–460.
68. Matsuo K, Yamaguchi R, Yamazaki A, Tasaka H, Terasaka K, Totsuka M, Kobayashi K, Yukitake H, Yamada T (1990) Establishment of a foreign antigen secretion system in mycobacteria. Infect Immun 58:4049–4054.
69. Barletta RG, Kim DD, Snapper SB, Bloom Br, Jacobs WR (1992) Identification of expression signals of the mycobacteriophages Bxb1, L1, and TM4 using the *Escherichia–Mycobacterium* shuttle plasmids pYUB75 and pYUB76 designed to create translational fusions to the *lacZ* gene. J Gen Microbiol 138:23–30.
70. Timm J, Perilli MG, Duez C, Trias J, Orefici G, Fattorini L, Amicosante G, Oratore A, Joris B, Frere JM, Pugsley AP, Gicquel B (1994) Transcription and expression analysis, using *lacZ* and *phoA* gene fusions, of *Mycobacterium fortuitum* $\beta$-lactamase genes cloned from a natural isolate and a high-level $\beta$-lactamase producer. Molec Microbiol 12:491–504.
71. Cooksey RC, Crawford JT, Jacobs WR, Shinnick TM (1993) A rapid method for screening antimicrobial agents for activities against a strain of *Mycobacterium tuberculosis* expressing firefly luciferase. Antimicrob Agents Chemother 37:1348–1352.
72. Andrew PW, Roberts IS (1993) Construction of a bioluminescent mycobacterium and its use for assay of antimycobacterial agents. J Clin Microbiol 31:2251–2254.
73. Das Gupta SK, Bashyam MD, Tyagi AK (1993) Cloning and assessment of mycobacterial promoters by using a plasmid shuttle vector. J Bacteriol 175:5186–5192.

74. Guilhot C, Gicquel B, Davies J, Martin C (1992) Isolation and analysis of IS6120, a new insertion sequence from *Mycobacterium smegmatis.* Molec Microbiol 6:107–113.

75. Cirillo JD, Barletta RG, Bloom BR, Jacobs WR (1991) A novel transposon trap for mycobacteria: Isolation and characterization of IS1096. J Bacteriol 173:7772–7780.

76. Guilhot C, Gicquel B, Martin C (1992) Temperature sensitive mutants of the *Mycobacterium* plasmid pAL5000. FEMS Microbiol Lett 98:181–186.

77. Guilhot C, Otal I, Vanrompaey I, Martin C, Gicquel B (1994) Efficient transposition in mycobacteria—Construction of *Mycobacterium smegmatis* insertional mutant libraries. J Bacteriol 176:535–539.

78. Husson RN, James BE, Young RA (1990) Gene replacement and expression of foreign DNA in mycobacteria. J Bacteriol 172:519–524.

79. Connelly N, Bloom BR, Jacobs WR. Unpublished observations cited in Ref. 46.

80. Curcic R, Deretic V (1994) Promoter probe vectors for mycobacteria based on the *xylE*gene. Abstract U51. 94th General Meeting of the American Society for Microbiology, p. 181.

81. Eisenach KD, Crawford JT, Bates JH (1986) Genetic relatedness among strains of the *Mycobacterium tuberculosis* complex—Analysis of restriction fragment heterogenicity using cloned DNA probes. Am Rev Respir Dis 133:1065–1068.

82. Clark-Curtiss JE, Docherty MA (1989) A species-specific repetitive sequence in *Mycobacterium leprae* DNA. J Infect Dis 159:7–15.

83. McFadden JJ, Butcher PD, Chiodini R, Hermon-Taylor J (1987) Crohn's disease-isolated mycobacteria are identical to *Mycobacterium paratuberculosis* as determined by DNA probes that distinguish between mycobacterial species. J Clin Microbiol 25:796–801.

84. Hermans PWM, van Sooligan D, van Embden JDA (1992) Characterization of a major polymorphic tandem repeat in *Mycobacterium tuberculosis* and its potential use in the epidemiology of *Mycobacterium kansasii* and *Mycobacterium gordonae.* J Bacteriol 174:4157–4165.

85. Crawford JT (1996) Molecular approaches to the detection of mycobacteria. In: Gangadharam PRJ, Jenkins PA, eds. Mycobacteria. Volume 1—Basic Aspects. New York: Chapman & Hall (this volume).

86. Clark-Curtiss JE, Walsh GP (1989) Conservation of genomic sequences among isolates of *Mycobacterium leprae.* J Bacteriol 171:4844–4851.

87. Kapur V, Whittam TS, Musser JM (1994) Is *Mycobacterium tuberculosis* 15,000 years old? J Infect Dis 170:1348–1349.

88. Fox GE, Stackebrandt E (1987) The application of 16S rRNA cataloguing and 5S rRNA sequencing in bacterial systematics. Methods Microbiol 19:405–458.

89. Rogall T, Wolters J, Flohr T, Bottger EC (1990) Towards a phylogeny and definition of species at the molecular level within the genus *Mycobacterium.* Int J Syst Bacteriol 40:323–330.

90. Kirschner P, Meier A, Bottger EC (1993) Genotypic identification and detection of

mycobacteria: Facing novel and uncultured pathogens. In: Persing DH, Tenover F, White TJ, Smith TF, eds. Diagnostic Molecular Microbiology: Principles and Applications, pp. 173–190. Washington DC: American Society for Microbiology Press.

91. Woese CR (1992) Prokaryote systematics: The evolution of a science. In: Balows A, Truper HG, Dworkin M, Harder W, Schleifer KH, eds. The Prokaryotes: A Handbook on the Biology of Bacteria: Ecophysiology, Isolation, Identification, Applications, vol. 1, p. 3–18. New York: Springer-Verlag.

92. Pitulle C, Dorsch M, Kazda J, Wolters J, Stackebrandt E (1992) Phylogeny of rapidly growing members of the genus *Mycobacterium.* Int J Syst Bacteriol 42:337–343.

93. Stahl DA, Urbance JW (1990) The division between fast- and slow-growing species corresponds to natural relationships among the mycobacteria. J Bacteriol 172:116–124.

94. Musial CE, Tice LS, Stockman L, Roberts GD (1988) Identification of mycobacteria from culture by using Gen-Probe Rapid Diagnostic System for *Mycobacterium avium* complex and *Mycobacterium tuberculosis* complex. J Clin Microbiol 26:2120–2123.

95. Rogall T, Flohr T, Bottger EC (1990) Differentiation of *Mycobacterium* species by direct sequencing of amplified DNA. J Gen Microbiol 136:1915–1920.

96. Cole ST, Smith DR (1994) Toward mapping and sequencing the genome of *Mycobacterium tuberculosis.* In: Bloom BR, ed. Tuberculosis: Pathogenesis, Protection, and Control, pp. 227–238. Washington DC: American Society for Microbiology Press.

97. Bergh S, Cole ST (1994) MycDB: An integrated mycobacterial database. Molec Microbiol 12:517–534.

98. Shinnick TM (1987) The 65-kilodalton antigen of *Mycobacterium tuberculosis.* J Bacteriol 169:1080–1088.

99. Hatfull GF, Sarkis GJ (1993) DNA sequence, structure, and gene expression of mycobacteriophage L5: A phage system for mycobacterial genetics. Molec Microbiol 7:395–405.

100. Falkow S (1988) Molecular Koch's postulates applied to microbial pathogenicity. Rev Infect Dis 10:S274–S276.

101. Telenti A, Imboden P, Marchesi F, Lowrie D, Cole S, Colston MJ, Matter L, Schopfer K, Bodmer T (1993) Detection of rifampin-resistance mutations in *Mycobacterium tuberculosis.* Lancet 341:647–650.

102. Jun Jin D, Gross CA (1988) Mapping and sequencing of mutations in the *Escherichia coli rpoB* gene that lead to rifampin resistance. J Molec Biol 202:45–58.

103. Kalpana GV, Bloom BR, Jacobs WR (1991) Insertional mutagenesis and illegitimate recombination in mycobacteria. Proc Natl Acad Sci USA 88:5433–5437.

104. Aldovini A, Husson RN, Young RA (1993) The *uraA* locus and homologous recombination in *Mycobacterium bovis* BCG. J Bacteriol 175:7282–7289.

105. Gangadharam PRJ. Drug resistance in tubercle bacillic In: Gangadharam PRJ,

Jenkins PA, eds. Mycobacteria. Volume 1—Basic Aspects. New York: Chapman & Hall. (This volume).

106. Jacobs WR, Barletta RG, Udani R, Chan J, Kalkut G, Sosne G, Kieser T, Sarkis GJ, Hatfull GF, Bloom BR (1993) Rapid assessment of drug susceptibilities of *Mycobacterium tuberculosis* by means of luciferase reporter phages. Science 260:819–822.

107. Guerin C (1980) The history of BCG. In: Rosenthal SR, ed. BCG vaccine: Tuberculosis–Cancer, pp. 35–43. Littleton, MA: PSG Publishing Co.

108. Steenken W, Oatway WH, Petroff SA (1934) Biological studies of the tubercle bacillus. III. Dissociation and pathogenicity of the R and S variants of the human tubercle bacillus (H37). J Exp Med 60:515–540.

109. Steenken W, Gardner LU (1946) History of H37 strain of tubercle bacillus. Am Rev Tuberc 54:62–66.

110. Horwitz MA (1988) Intracellular parasitism. Curr Opin Immunol 1:41–46.

111. Rastogi N, ed. (1990) Killing intracellular mycobacteria: dogmas and realities. Fifth Forum in Microbiology. Res Microbiol 141:191–270.

112. Fields P, Swanson R, Haidaris C, Heffron F (1986) Mutants of *Salmonella typhimurium* that cannot survive within the macrophage are avirulent. Proc Natl Acad Sci USA 83:5189–5193.

113. Fields PI, Grossman EA, Heffron F (1989) A *Salmonella* locus that confers resistance to microbicidal proteins from phagocytic cells. Science 243:1059–1062.

114. Isberg RR, Falkow S (1985) A single genetic locus encoded by *Yersinia pseudotuberculosis* permits invasion of cultured animal cells by *Escherichia coli* K-12. Nature 317:262–264.

115. Finlay BB, Falkow S (1989) Common themes in microbial pathogenicity. Microbiol Rev 53:210–230.

116. Shepard CC (1958) A comparison of the growth of selected mycobacteria in HeLa, monkey kidney, and human amnion cells in tissue culture. J Exp Med 107:237–246.

117. Arruda S, Bomfim G, Knights R, Hiuma-Byron T, Riley LW (1993) Cloning of an *M. tuberculosis* DNA fragment associated with entry and survival inside cells. Science 261:1454–1457.

118. Mundayoor S, Shinnick TM (1994) Identification of genes involved in the resistance of mycobacteria to killing by macrophages. Ann NY Acad Sci 730:26–36.

119. Mahan MJ, Slauch JM, Mekalanos JJ (1993) Selection of bacterial virulence genes that are specifically induced in host tissues. Science 259:686–688.

120. Marston BJ, King CH, Quinn FD, Shinnick TM (1994) Differentially expressed genes of *Mycobacterium tuberculosis.* Abstract in Proceedings of the 29th U.S.–Japan Conference on Tuberculosis and Leprosy, pp. 143–147.

121. Utt EA, Brousal JP, Kikuta-Oshima LC, Quinn FD (1995) mRNA subtractive hybridization identifies variations in bacterial gene expression. Nucleic Acids Res.

122. Kikuta-Oshima LC, King CH, Shinnick TM, Quinn FD (1994) Methods for the

identification of virulence genes expressed in *Mycobacterium tuberculosis* strain H37Rv. Ann NY Acad Sci 730:263–265.

123. Kinger AK, Tyagi JS (1993) Identification and cloning of genes differentially expressed in the virulent strain of *Mycobacterium tuberculosis*. Gene 131:113–117.

124. Quinn FD, King CH, Kikuta-Oshima LC, Utt EA, Shinnick TM (1993) Subtractive hybridization methods for the identification of *M. tuberculosis* genes expressed specifically in virulent strains and clinical specimens. Abstract in Proceedings of the 28th U.S.–Japan Conference on Tuberculosis and Leprosy, pp. 184–188.

125. Plum G, Clark-Curtiss JE. Induction of *Mycobacterium avium* gene expression following phagocytosis by human macrophages. Infect Immun 62:476–483.

# 4

# Molecular Approaches to the Detection of Mycobacteria

*Jack T. Crawford*

## 1. Introduction

The application of molecular biology to the detection of mycobacteria began with nucleic acid hybridization techniques for the identification of the various species of mycobacteria. The first commercially available probe tests were offered in the mid-1980s by Gen-Probe (San Diego, California), and these have since been replaced by the current Accuprobe system. The use of DNA probe assays for identification of isolates has become standard practice and is described elsewhere in the volume (Chapter 13 Heifets and Jenkins). Two problems with the probe methods are the limited coverage (*Mycobacterium tuberculosis* complex, *M. avium* complex, *M. kansasii,* and *M. gordonae*) and the limited sensitivity that prevents their use directly with clinical specimens. The first problem is primarily one of practicality; it is unlikely that there would be sufficient demand for an expanded panel of probes to justify production. The limited sensitivity of the probes, however, is a more fundamental problem. Sensitivity for the DNA probe test for *M. tuberculosis* is usually stated as 100%, meaning that all isolates of the *M. tuberculosis* complex react with the probe. In the current context, sensitivity means the limit of detection of probes in numbers of bacilli. Although the Accuprobe test is a quite sensitive probe assay and was once considered for direct detection, the number of bacilli in many clinical specimens is below the level of detection with probes. The limited sensitivity of the probes is overcome by application only to cultures in which the number of organisms present in the clinical sample has been increased by growth.

Nucleic acid amplification methods provide the means for direct detection of mycobacteria in clinical specimens. Instead of amplifying the number of organisms, these procedures increase the amount of nucleic acid target by in vitro enzymatic means. Polymerase chain reaction (PCR) was the first amplification method described and is the method that has been used for most research applications. Other amplification methods have been developed subsequently. These include transcription-mediated amplification (TMA), strand displacement amplification (SDA), ligase chain reaction (LCR), and others. All such methods are designed to increase the amount of a specific target sequence, allowing, in theory, detection of a single organism in a clinical specimen.

## 2. Specimen Processing

All of the amplification assays begin with a specimen-processing step. Because sputum is the most common clinical specimen submitted for detection of *M. tuberculosis,* most published reports have dealt primarily with sputum or other respiratory specimens. Amplification assays do not require highly purified nucleic acid as a starting material, but concentration of the nucleic acid and elimination of inhibitors is desirable. Standard digestion and decontamination procedures used for processing sputum for mycobacterial culture, such as the sodium hydroxide/*N*-acetyl-L-cysteine method, provide a concentrated pellet of bacilli and eliminate much of the material that would inhibit amplification. A portion of the sample can then be used for the preparation of smears and cultures, and another portion can be used for amplification. Some additional processing, such as washing the sediment with buffer, might be required for the amplification assay. Lysis of the bacilli to release the nucleic acid has been accomplished using sonication, shaking with beads, and treatment with enzymes, detergents, alkali, and heat. Further nucleic acid purification can improve the sensitivity of the assay, but the methods are generally considered incompatible with routine use in the clinical laboratory. Other clinical specimens, including blood, urine, stool, tissue, pleural fluid, and spinal fluid, can be assayed with appropriate modifications in the specimen-processing procedure.

## 3. PCR Assays

The polymerase chain reaction uses a DNA polymerase and oligonucleotide primers to reproduce a specific target segment of DNA. Cycles of denaturation, primer binding, and elongation of primers by polymerase produce an exponential increase in the target sequence yielding $10^5$ or more copies in a few hours. Many reports describing the application of PCR methods for detection of mycobacteria, especially *M. tuberculosis,* have been published in the past few years (1–19). Only a few assays have been evaluated with a sufficient number of clinical samples and adequate clinical data to allow a critical evaluation of sensitivity and specificity.

A number of characterized genes and various other sequences have been used as targets for PCR assays. One approach is to target a sequence which is found in many or all mycobacteria but which has species-specific regions. The first published report of a PCR assay for *M. tuberculosis* described an assay that detected the *hsp60* (65-kDa) heat shock protein gene (1,2,20). Portions of the gene are conserved and primers were selected that produce a 383-base-pair (bp) amplified product from DNA of all mycobacteria. An oligonucleotide probe for a sequence within this 383-bp region specific for various species was used to detect product from *M. tuberculosis.* Other species of mycobacteria, such as *M. avium,* could be detected in a similar manner using other probes.

The alternative approach is to use an *M. tuberculosis*-specific target for amplification. The *hsp60* target has been used for species-specific amplification by using primers for the unique sequences (3). Assays for *M. tuberculosis* and *M. leprae* have been described. The most widely studied target has been the insertion sequence IS6110 which has been demonstrated only in isolates of the *M. tuberculosis* complex, including *M. tuberculosis, M. bovis, M. africanum,* and *M. microti* (21). Several assays targeting this sequence have been described; all produce amplification product only with DNA from the *M. tuberculosis* complex (4–12). In addition to being specific for the *M. tuberculosis* complex, this target is repeated 8–20 times within the genome in most *M. tuberculosis* isolates. This repetitive target increases the sensitivity of the assay allowing detection of smaller numbers of bacilli. However, a significant number of *M. tuberculosis* isolates have only a few copies, and isolates with no copies have been reported, especially in Southeast Asia.

Most of the assays developed in research laboratories, sometimes referred to as "home-brew" PCR assays, use simple detection systems, often gel electrophoresis, for detecting PCR product. Most do not have internal controls and few have included mechanisms for preventing carryover of PCR product. These assays can be recommended for routine diagnostic use only when they are performed under carefully controlled conditions. The commercial PCR assay currently being evaluated (Amplicor, Roche Molecular Systems, Somerville, New Jersey) is based on amplification of a 584-bp region of the gene encoding 16S rRNA. The assay uses primers specific for conserved regions in the target gene to produce amplified product from all mycobacteria. Biotinylated primers are used to label the amplicon. Following amplification, the amplicon is captured using an *M. tuberculosis*-specific oligonucleotide attached to wells in a microtiter plate. A colorimetric system is used to detect the biotin label in the captured amplicon. This approach could permit simultaneous testing for several species of mycobacteria in a single assay using a series of species-specific oligonucleotide probes in separate wells. Carryover of amplicon which results in false positive results is a major problem for amplification assays (22). In the Amplicor system, dUTP is substituted for TTP in the PCR, rendering the product susceptible to digestion with uracil-*N*-glycosylase. This enzyme is incorporated into the master mix for the PCR, resulting in destruction of any contaminating amplicon without affecting native target DNA in the sample.

## 4. Evaluations of PCR Assays

A number of studies evaluating PCR assays for detection of *M. tuberculosis* have been published. In most of the studies, the assays used were developed by independent researchers. These studies suggest the types of results that are likely to be obtained using commercial products. In one early study, Eisenach et al. reported results of 178 sputum samples tested with an *M. tuberculosis*-specific assay based on detection of IS6110 and including an internal control for inhibition (5). Specimens were processed for smears and culture, and a portion of the concentrated sample was tested using the PCR assay. All smear-positive samples from patients with tuberculosis, as defined by clinical and other laboratory results, were positive by PCR. Specimens from patients with nontuberculous mycobacterial infections, including strongly smear-positive samples, were negative, demonstrating the specificity of the assay. Specimens from patients whose final diagnosis was not mycobacterial infection were also negative. There were insufficient smear-negative/culture-positive samples to allow any calculation of the overall sensitivity of the assay. This study was performed in a research laboratory; however, other larger studies using similar assays but with simpler specimen processing procedures have demonstrated that amplification assays can be performed on a routine basis in clinical laboratories (10–12). These studies demonstrated the excellent specificity of PCR assays based on IS6110. Smear-positive samples yielded a sensitivity of >90% relative to culture. Sensitivity with smear-negative/culture-positive specimens was lower. The PCR methods used in these studies can detect DNA equivalent to that found in a single bacillus. The modest sensitivity with samples containing smaller numbers of bacilli reflects the inherent difficulty in preparing clinical samples for amplification assays. Lysis of tubercle bacilli to release the target DNA is difficult, and only a small sample of the specimen can be used for amplification. Also, material in the clinical specimen might inhibit the amplification reaction. Purification and concentration of the nucleic acid adds laborious steps to the procedure.

Recent publications have described evaluations of the Roche Amplicor *M. tuberculosis* assay using respiratory specimens (23–25). The assay can be completed in about 8 h after receipt of the specimen. As was the case with the research assays, the sensitivity with smear-positive specimens was excellent, around 95%, but with smear-negative/culture-positive specimens, the sensitivity was lower.

However, some culture-negative samples are positive by PCR. Analysis of these samples suggests that most are not false positive. It is known that digestion/decontamination procedures kill many mycobacteria in the specimen; these dead bacilli can still be detected by amplification. The specificity of the assay was greater than 97%.

## 5. TMA

The transcription-based amplification assay produced by Gen-Probe was the first assay to be marketed. The target for amplification is the 16S rRNA and, thus, it takes advantage of the high copy number of the rRNA. As with the PCR assay targeting the rRNA gene, product is produced from all mycobacteria, and probe assays are used to determine the species. Reverse transcriptase is used to produce a DNA copy from the rRNA and incorporate a specific promoter region. RNA polymerase then produces RNA transcripts to provide the amplification. The amplification step is isothermal; that is, it does not require thermal cycling. The RNA product is detected using the well-proven chemiluminescent hybridization protection assay system currently used for the Accuprobe culture confirmation assay.

Recent publications have described evaluations of the Gen-Probe Mycobacterium Tuberculosis Direct Test (MTD) (26–29). About 5 h are required for completion of the assay. In one larger study, 758 sputum samples from 235 patients were tested (26). Following resolution of discrepant results, the sensitivity and specificity of the assay were 82% and 99% versus 88% and 100%, respectively, for culture. Again, the assay showed higher sensitivity with smear-positive specimens. The slightly lower sensitivity of the amplification assay is offset by the speed of the assay relative to culture.

## 6. Other Amplification Methods

A strand displacement amplification (SDA) assay specific for *M. tuberculosis* has been developed (Becton Dickinson Microbiology Systems, Cockeysville, Maryland) (30,31). This is an isothermal amplification procedure. Yields of SDA have been significantly increased by the use of thermal-stable enzymes and higher incubation temperatures. Several other methods, including ligase chain reaction, cycling probe reaction, and *QB* replicase amplification, have been used for mycobacterial assays. A novel approach is a method that amplifies the detection signal rather than the target. All of these amplification approaches have strengths and weaknesses in terms of sensitivity, specificity, ease of use, cost of reagents and equipment, and technical expertise required for the assay. All of these factors and others, such as automation of the assay, will determine which of the competing methods will prove successful in the clinical laboratory. However, it is clear that specimen processing can be the most difficult aspect of assay development for mycobacteria and is probably the limiting factor for sensitivity in all of the assays.

## 7. Role of Amplification Assays

Assays based on nucleic acid amplification methods have the potential to greatly facilitate the diagnosis of tuberculosis by providing same-day detection of *M. tuberculosis* in clinical samples. How these assays will be integrated into the routine of the mycobacteriology laboratory and what role the results of such assays will play in the diagnosis of mycobacterial disease is still unclear. Even more uncertain is the question of whether these highly sophisticated and costly tests will be useful in those parts of the world where tuberculosis is a major problem but resources are limited. Aside from the cost and technical requirements of the assays, there are still questions about the performance of the existing assays and the advantages that they provide over standard methods.

Amplification assays are most useful for detection of organisms that are difficult or slow to culture. A second consideration is the correlation between the presence of the organism and disease. On both of these points, tuberculosis seems to be an ideal target for an amplification assay. Detection of *M. tuberculosis* in clinical samples from a patient with symptoms consistent with tuberculosis is generally considered diagnostic for tuberculosis. Amplification assays have proven to be highly specific, and detection of *M. tuberculosis* nucleic acid using an amplification assay should be diagnostic as well. One major concern has been the high level of false-positive results that can occur with amplification assays. This problem can be controlled with properly designed methods and careful work, but, even then, the rate of false-positive assays can equal true positives in very low-incidence situations. Unfortunately, false-positive cultures resulting from cross-contamination or laboratory error are now recognized as a major problem also. In both cases, the laboratory results must be evaluated to determine if they are compatible with the clinical picture. A second concern has been the lower sensitivity of the assays which with smear-negative samples compared with culture. Whereas increased sensitivity would be desirable, the ability to detect *M. tuberculosis* in most samples in a few hours rather than in 2–4 weeks appears to be a considerable advance. Samples that are negative with an amplification test can still prove to be positive by culture. Of course, a significant percentage of patients diagnosed with tuberculosis are never culture confirmed. A negative result with either culture or amplification should not be interpreted as ruling out tuberculosis. A highly specific, rapid assay could reliably detect those patients who are most likely to be infectious. In this regard, amplification assays could be considered an alternative to smears but with much higher specificity. It is probable that all samples that are positive with an amplification assay will still be cultured for confirmation, drug susceptibility testing, and epidemiologic purposes.

## 8. Detection of Drug Resistance

Another promising application of molecular methods is the detection of drug resistance. The increasing occurrence of drug resistance makes susceptibility testing of all initial isolates desirable. Standard methods are based on the growth of the isolate, either on solid medium or in a liquid culture system. Recent advances in our knowledge of the mechanisms of drug resistance in *M. tuberculosis* and the development of new methods for detecting mutations have made it possible to detect resistance at the genetic level. A second and quite different approach is the use of luciferase-reporter bacteriophage to assay drug action. Both of these approaches have shown potential for expediting detection of resistance.

## 9. Molecular Analysis of Drug-Resistance Mutations

The development of an assay to detect drug-resistance mutations requires identification of the gene or genes involved and determination of the specific mutations that impart resistance. The most straightforward example of this is rifampin resistance. Analysis of *E. coli* and other bacteria demonstrated that rifampin binds to the $\beta$ subunit of RNA polymerase to inhibit transcription. Rifampin-resistance mutants are located in the *rpo*B gene that encodes the $\beta$ subunit. Because this is an essential enzyme, only mutations that result in amino acid substitutions or insertion or deletion of amino acids are possible. Using the sequence of *rpo*B from *E. coli* it was possible to identify and sequence the corresponding gene in *M. leprae* and *M. tuberculosis* (32,33). Although *rpoB* is a rather large gene with and open reading frame of 3534 bp, most mutations imparting resistance to rifampin are located in a small region (34–36). PCR amplification of that small region combined with direct DNA sequencing has been used to analyze a large number of rifampin-resistant isolates of *M. tuberculosis.* The majority of mutations occur in two codons, but numerous other less frequent mutations have been demonstrated, including small in-frame deletions and insertions.

A similar approach was used to demonstrate mutations that impart streptomycin resistance in *M. tuberculosis* (37,38). A single base change in the gene encoding the S12 ribosomal protein (*rpsL*) accounts for most high-level streptomycin resistance. A second, less common mutation in the adjacent codon in *rpsL* has also been reported. Lower-level streptomycin resistance is imparted by mutations in two sites in 16S rRNA (*rrs*). About 25% of streptomycin resistant isolates do not have mutations in either or these targets and the mechanism is unknown. Fluoroquinolones are known to bind to DNA gyrase, and resistance mutations in *M. tuberculosis* have been shown to be clustered in one region of *gyrA* (39).

Demonstration of the mechanisms of resistance to isonicotinic acid hydrazide (INH) has been more complicated because the drug is highly active only on *M. tuberculosis.* The first resistance mechanism for INH to be identified was loss of catalase–peroxidase activity as a result of deletions in the gene designated *katG* (40). It was subsequently demonstrated that point mutations in various sites in *katG,* in particular, residues Ser315 and Arg463, impart reduced activity and various levels of resistance to INH, and these mutations are more common than deletions (41). Apparently, catalase–peroxidase is not the primary target of action but is involved in activating INH.

A gene designated *inhA* has been shown to be involved in INH resistance and also low-level ethionamide resistance (42). This gene codes for an enzyme involved in mycolic acid synthesis that is one target of INH. The gene was initially identified using a mutant that carried a point mutation that imparted INH resistance. Analysis of clinical isolates indicates that this mutation is not common and most low-level INH-resistant mutations have been found in regulatory regions upstream from the *inhA* structural gene, apparently resulting in overproduction of the enzyme.

## 10. Rapid Assays for Detecting Drug Resistance

The mutations involved in drug resistance cannot be detected easily by simple probe assays but must be identified by determination of the actual DNA sequence or by the use of assays designed specifically to detect mutations (43). The first step is amplification of the target sequence. This has been accomplished using PCR, but other amplification methods probably could be adapted for some of the approaches. Most published studies have been performed with isolates, but, ideally, the starting material would be a clinical specimen. The primer sites flank the region in which the mutations occur and both wild-type and mutant sequences are amplified. For most of the published assays, the homologous sequences from nontuberculous mycobacteria are also amplified. When the mutations are clustered, such as *rpoB,* a single amplification is sufficient. For larger regions, such as *katG,* multiple or multiplex amplification is needed.

Various methods have been used to detect mutations in amplified DNA. The most widely used methods are DNA sequencing, hybridization with oligonucleotide probes, single-strand conformation polymorphism analysis, heteroduplex analysis, and chemical cleavage of mismatches. The most reliable method is direct sequencing. Using cycle sequencing and an automated sequencer, it is possible to sequence multiple PCR products at one time and obtain very high accuracy. In some instances, sequencing in only one direction is sufficient. Using sequencing, all mutations can be detected, including those that have not been recognized previously. The presence of a new mutation does not automatically indicate drug resistance, although it appears that sequences in *M. tuberculosis* are highly conserved. Sequencing is most useful when there are multiple possible mutations clustered in a small region, such as with *rpoB,* where a single sequencing reaction is sufficient. The complexity of the procedure, cost, and throughput for sequencing are the primary limitations. Although sequencing might become routine in some clinical laboratories, it is unlikely to be used routinely for drug-resistance testing on a wide scale.

The use of oligonucleotide probes is the simplest approach for detection of mutations, but this is practical only when there are few mutations and they are clustered in a small region. The most popular approach for detection of mutations has been single-strand conformational polymorphism (SSCP) analysis (44). For this analysis, the PCR product is heat denatured and the single-strand molecules are electrophoresed on a nondenaturing acrylamide gel. The molecules are separated on the basis of secondary structure which is altered by single base changes. Each amplicon produces two bands on the gel. Migration of the test DNA is compared to product from known wild-type and mutant strains. The size of the DNA that can be analyzed is limited to a few hundred bases. SSCP analysis is frequently performed using PCR products labeled with $^{32}P$ or $^{35}S$, and electrophoresis is performed on large sequencing gels. After electrophoresis, the gels are dried and the bands detected by radioautography. In one study, 17 rifampin-resistant isolates with different *rpoB* mutations were examined (45). A specific SSCP pattern was associated with each mutation, although some of these patterns are very difficult to distinguish from the wild-type pattern. Reliable analysis would require a library of recognized mutants and their corresponding SSCP patterns and very careful technique. These authors also evaluated an automated analysis procedure in which the amplicon was labeled with fluorescein and analyzed using a Pharmacia ALF automated sequencer. Specific patterns were demonstrated to various mutations.

Nonradioactive (cold) SSCP methods in which the DNA is detected by ethidium bromide staining offer many advantages over radioactive labeling (46,47). In addition to eliminating the hazards of isotopes, this procedure yields much sharper bands that are easier to analyze.

Molecular analysis has also been applied to INH resistance. PCR/SSCP analysis was used to characterize the *katG* gene in 36 INH-resistant isolates (48). Mutations in *katG* occur in many sites and this required amplification and analysis of 12 separate segments. Direct sequencing was used to characterize the *inhA* region in these isolates. Of 36 isolates, 21 were mutant in *katG,* 10 were mutant in *inhA,* 4 had mutations in both genes, and no mutations were detected in 9 isolates. This last number is disturbing because it means that despite this extensive analysis, resistance in these nine isolates would have been missed.

Although molecular analysis of drug-resistance mutations is promising, there are several pitfalls to be dealt with. It will be necessary to perform at least one assay for each drug. At least two genes are involved in INH resistance, and analyzing these two genes does not detect all INH resistance. A similar situation exists for streptomycin resistance. The assays are technically complex and are unfamiliar to many clinical laboratory personnel. Another concern is the certainty of correlating mutation with resistance. As additional sequence data are accumulating it becomes clear that *M. tuberculosis* sequences are highly conserved, decreasing the probability of detecting silent mutations. The possibility of mixed cultures must also be considered. Because the amplification reactions are generally species-specific, nontuberculous mycobacteria in the sample will amplify and could be mistaken for mutant *M. tuberculosis.*

Perhaps the most serious problem will be recognizing emerging resistance. Conventional susceptibility testing using the method of proportions defines resistance as 1% resistant mutants in population. Most of the assays described to date cannot detect low levels of mutations in a background of wild-type sequences.

The best prospect for molecular analysis is rifampin resistance. The analysis is the most straightforward, and it has been suggested that this assay alone could be used to rapidly predict resistance. Standard susceptibility testing could be performed on any isolate that is mutant in *rpoB* or on all isolates. Ideally, the starting material would be a clinical sample for direct detection and demonstration of drug resistance.

## 11. Luciferase-Reporter Phage

The use of a reporter gene system to assay drug action provides a novel alternative to genetic detection of resistance mutations (49). There are two clear advantages of this approach over the mutation-detection approach—the reporter gene assay measures actual drug activity at the phenotypic level and the same assay can be applied simultaneously to numerous drugs, including new drugs for which the mechanism of action is not known. The firefly luciferase gene has been developed as a reporter system for mycobacteria. This enzyme catalyzes the production of light from luciferin and ATP. This reaction can be detected using a luminometer and current instruments allow detection in various formats including microtiter plates. Expression of the luciferase gene in mycobacteria has been accomplished by insertion of a mycobacterial promoter sequence from *hsp60.* To use the luciferase gene as a reporter in an assay with clinical isolates of *M. tuberculosis,* it is necessary to introduce the gene in to the bacilli. Bacteriophages which adsorb to bacteria and deliver their DNA with high efficiency provide the vector. Jacobs and co-workers have cloned the luciferase gene into the genome of a mycobacteriophage and obtained expression of the gene in *M. tuberculosis* following infection with the phage. The luciferase-reporter mycobacteriophage is used for susceptibility testing in the following manner. The *M. tuberculosis* isolate is inoculated into medium containing the test drug. The culture is incubated for one to several days to allow drug action, and then cultures are infected with the reporter mycobacteriophage. After several hours, the cultures are assayed for luciferase activity (i.e., for light production after addition of luciferin). If the isolate is susceptible to the drug, phage replication and luciferase production are inhibited and no light is produced. Low levels of intracellular ATP will also block light production. If the isolate is resistant to the drug, light is produced. Although this assay is not ready for actual clinical use, the basic features have been defined. Ideally, a mycobacteriophage specific for *M. tuberculosis* could be used and this would allow direct application to clinical samples as a combination detection and drug-susceptibility assay.

## References


1. Hance AJ, Grandchamp B, Levy-Frebault V, Lecossier D, Rauzier J, Bocart D, Gicquel B (1989) Detection of and identification of mycobacteria by amplification of mycobacterial DNA. Molec Microbiol 3:843–849.
2. Brisson-Noel A, Gicquel B, Lecossier D, Levy-Frebault V, Nassif X, Hance AJ (1989) Rapid diagnosis of tuberculosis by amplification of mycobacterial DNA in clinical samples. Lancet ii:1069–1071.
3. Plikaytis BB, Gelber RH, Shinnick TM (1990) Rapid and sensitive detection of *Mycobacterium leprae* using a nested-primer amplification assay. J Clin Microbiol 28:1913–1917.
4. Eisenach KD, Cave MD, Bates JH, Crawford JT (1990) Polymerase chain reaction amplification of a repetitive DNA sequence specific for *Mycobacterium tuberculosis.* J Infect Dis 161:977–981.
5. Eisenach KD, Sifford MD, Cave MD, Bates JH, Crawford JT (1991) Detection of *Mycobacterium tuberculosis* in sputum samples using a polymerase chain reaction. Am Rev Respir Dis 144:1160–1163.
6. Hermans P, van Soolingen D, Dale J, Schuitema A, McAdams R, Catty D, van Embden JDA (1990) Insertion element IS986 from *Mycobacterium tuberculosis:* A useful tool for diagnosis and epidemiology of tuberculosis. J Clin Microbiol 28:2051–2058.
7. Thierry D, Brisson-Noel A, Vincent-Levy-Frbault V, Nguyen S, Guesdon JL, Gicquel B (1990) Characterization of a *Mycobacterium tuberculosis* insertion sequence, IS6110, and its application in diagnosis. J Clin Microbiol 28:2668–2673.
8. Kolk AHJ, Schuitema ARJ, Kuijper S, Van Leeuwen J, Hermans PWM, van Embden JDA, Hartskeerl RA (1992) Detection of *Mycobacterium tuberculosis* in clinical samples by using a polymerase chain reaction and a nonradioactive detection system. J Clin Microbiol 30:2567–2575.

9. Shawar RM, El-Zaatari FAK, Nataraj A, Clarridge JE (1993) Detection of *Mycobacterium tuberculosis* in clinical samples by two-step polymerase chain reaction and nonisotopic hybridization methods. J Clin Microbiol 31:61–65.
10. Forbes BA, Hicks KES (1993) Direct detection of *Mycobacterium tuberculosis* in respiratory specimens in a clinical laboratory by polymerase chain reaction. J Clin Microbiol 31:1688–1694.
11. Nolte FS, Metchock B, McGowan JE Jr, Edwards A, Okwumabua O, Thurmond C, Mitchell PS, Plikaytis B, Shinnick T (1993) Direct detection of *Mycobacterium tuberculosis* in sputum by polymerase chain reaction and DNA hybridization. J Clin Microbiol 31:1777–1782.
12. Clarridge JE III, Shawa RM, Shinnick TM, Plikaytis BB (1993) Large-scale use of polymerase chain reaction for detection of *Mycobacterium tuberculosis* in a routine mycobacteriology laboratory. J Clin Microbiol 31:2049–2056.
13. Patel RJ, Fries JWU, Piessens WF, Wirth DF (1990) Sequence analysis and amplification by polymerase chain reaction of a cloned DNA fragment for identification of *Mycobacterium tuberculosis.* J Clin Microbiol 28:513–518.
14. Hermans PWM, Schuitema ARJ, van Soolingen D, Verstynen CPHJ, Bik EM, Thole JER, Kolk AH, van Embden JDA (1990) Specific detection of *Mycobacterium tuberculosis* complex strains by polymerase chain reaction. J Clin Microbiol 28:1204–1213.
15. Pao CC, Benedict Yen TS, You J-B, Maa J-S, Fiss EH, Chang C-H (1990) Detection and identification of *Mycobacterium tuberculosis* by DNA amplification. J Clin Microbiol 28:1877–1880.
16. De Wit D, Steyn L, Shoemaker S, Sogin M (1990) Direct detection of *Mycobacterium tuberculosis* in clinical specimens by DNA amplification. J Clin Microbiol 28:2437–2441.
17. Cousins DV, Wilton SD, Francis BR, Gow BL (1992) Use of polymerase chain reaction for rapid diagnosis of tuberculosis. J Clin Microbiol 30:255–258.
18. Soini H, Skurnik M, Liippo K, Tala E, Viljanen MK (1992) Detection and identification of mycobacteria by amplification of a segment of the gene coding for the 32-kilodalton protein. J Clin Microbiol 30:2025–2028.
19. Altamirano M, Kelly MT, Wong A, Bessuille ET, Black WA, Smith JA (1992) Characterization of a DNA probe for detection of *Mycobacterium tuberculosis* complex in clinical samples by polymerase chain reaction. J Clin Microbiol 30:2173–2176.
20. Shinnick TM (1987) The 65-kilodalton antigen of *Mycobacterium tuberculosis.* J Bacteriol 169:1080–1088.
21. Thierry D, Cave MD, Eisenach KD, Crawford JT, Bates JH, Gicquel B, Guesdon JL (1990) IS6110, an IS-like element of *Mycobacterium tuberculosis* complex. Nucleic Acids Res. 18:188.
22. Noordhoek GT, Kolk AHJ, Bjune G, Catty D, Dale JW, Fine PEM, Godfrey-Faussett P, Cho S-N, Shinnick T, Svenson SB, Wilson S, van Embden IDA (1994) Sensitivity

and specificity of PCR for detection of *Mycobacterium tuberculosis:* A blind comparison study among seven laboratories. J Clin Microbiol 32:277–284.

23. Vuorinen P, Miettinen A, Vuento R, Hallstom O (1995) Direct detection of *Mycobacterium tuberculosis* complex in respiratory specimens by Gen-Probe Amplified Mycobacterium tuberculosis Direct Test and Roche Amplicor Mycobacterium Tuberculosis Test. J Clin Microbiol 33:1856–1859.
24. Beavis KG, Lichty MB, Jungkind DL, Giger O (1995) Evaluation of Amplicor PCR for direct detection of *Mycobacterium tuberculosis* from sputum specimens. J Clin Microbiol 33:2582–2586.
25. Moore DF, Curry JI (1995) Detection and identification of *Mycobacterium tuberculosis* directly from sputum sediments by Amplicor PCR. J Clin Microbiol 33:2686–2691.
26. Jonas V, Alden MJ, Curry JI, Kamisango K, Knott CA, Lankford R, Wolfe JM, Moore DF (1993) Detection and identification of *Mycobacterium tuberculosis* directly from sputum sediments by amplification of rRNA. J Clin Microbiol 31:2410–2416.
27. Abe C, Hirano K, Wada M, Kazumi Y, Takahashi M, Fukasawa Y, Yoshimura T, Miyagi C, Goto S (1993) Detection of *Mycobacterium tuberculosis* in clinical specimens by polymerase chain reaction and Gen-Probe Amplified Mycobacterium Tuberculosis Direct Test. J Clin Microbiol 31:3270–3274.
28. Miller N, Hernandez SG, Cleary TJ (1994) Evaluation of Gen-Probe Amplified Mycobacterium Tuberculosis Direct Test and PCR for direct detection of *Mycobacterium tuberculosis* in clinical specimens. J Clin Microbiol 32:393–397.
29. Pfyffer GE, Kissling P, Wirth R, Weber R (1994) Direct detection of *Mycobacterium tuberculosis* complex in respiratory specimens by a target-amplified test system. J Clin Microbiol 32:918–923.
30. Walker GT, Frasier MS, Schram JL, Little MC, Nadeay JG, Malinowski DP (1992) Strand displacement amplification—An isothermal, in vitro DNA amplification technique. Nucleic Acids Res 20:1691–1696.
31. Spargo CA, Haaland PD, Jurgensen SR, Shank DD, Walker GT (1993) Chemiluminescent detection of strand displacement amplified DNA from species comprising the *Mycobacterium tuberculosis* complex. Molec Cell Probes 7:395–404.
32. Honore NS, Bergh S, Chanteau S, Doucet-Populaire F, Eiglmeier K, Garnier T, Georges C, Launois P, Limpaiboon T, Newton S, Niang K, Del Portill P, Ramesh GR, Reddi P, Ridel PR, Sittisombut N, Wu-Hunter S, Cole ST (1993) Nucleotide sequence of the first cosmid from the *Mycobacterium leprae* genome project: Structure and function of the rif-str regions. Molec Microbiol 7:207–214.
33. Miller LP, Crawford JT, Shinnick TM (1994) The *rpoB* gene of *Mycobacterium tuberculosis.* Antimicrob Agents Chemother 38:805–811.
34. Telenti A, Imboden P, Marchesi F, Lowrie D, Cole S, Colston MJ, Matter L, Schopfer K, Bodmer T (1993) Detection of rifampicin-resistance mutations in *Mycobacterium tuberculosis.* Lancet 341:647–650.
35. Telenti A, Imboden P, Marchesi F, Lowrie D, Cole S, Colston MJ, Matter L, Schopfer

K, Bodmer T (1993) Detection of rifampicin-resistance mutations in *Mycobacterium tuberculosis.* Lancet 341:647–650.

36. Williams DL, Waguespack C, Eisenach K, Crawford JT, Portaels F, Salfinger M, Nolan CM, Abe C, Sticht-Groh V, Gillis TP (1994) Characterization of rifampin-resistance in pathogenic mycobacteria. Antimicrob Agents Chemother 38:2380–2386.
37. Douglass J, Steyn LM (1993) A ribosomal gene mutation in streptomycin-resistant *Mycobacterium tuberculosis* isolates. J Infect Dis 167:1505–1507.
38. Finken M, Kirschner P, Meier A, Bottger EC (1993) Molecular basis of streptomycin resistance in *Mycobacterium tuberculosis:* Alterations of the ribosomal protein S12 gene and point mutations within a functional 16S ribosomal RNA pseudoknot. Molec Microbiol 9:1239–1246.
39. Takiff HE, Salazar L, Guerrero C, Philipp W, Huang WM, Kreiswirth B, Cole ST, Jacobs WR Jr, Telenti A (1994) Cloning and nucleotide sequence of *Mycobacterium tuberculosis gyrA* and *gyrB* genes and detection of quinolone resistance mutations. Antimicrob Agents Chemother 38:773–780.
40. Zhang Y, Heym B, Allen B, Young D, Cole S (1992) The catalase–peroxidase gene and isoniazid resistance of *Mycobacterium tuberculosis.* Nature 358:591–593.
41. Heym B, Zhang Y, Poulet S, Young D, Cole ST (1993) Characterization of the *katG* gene encoding a catalase–peroxidase required for the isoniazid susceptibility of *Mycobacterium tuberculosis.* J Bacteriol 175:4255–4259.
42. Banerjee A, Dubnau E, Quemard A, Balausbramanian KS, Um, Wilson T, Collins D, De Lisle G, Jacobs WR Jr (1994) *inhA,* a gene encoding a target for isoniazid and ethionamide in *Mycobacterium tuberculosis.* Science 263:227–230.
43. Grompe M (1993) The rapid detection of unknown mutations in nucleic acids. Nature Genetics 5:111–117.
44. Orita M, Owajama H, Kanazawa H, Hayashi K, Sekiya T (1989) Detection of polymorphisms of human DNA by gel electrophoresis as single-strand conformation polymorphisms. Proc Natl Acad Sci USA 86:2766–2770.
45. Telenti A, Imboden P, Marchesi F, Schmidheini T, Bodmer T (1993) Direct, automated detection of rifampin-resistant *Mycobacterium tuberculosis* by polymerase chain reaction and single-strand conformation polymorphism. Antimicrob Agents Chemother 37:2054–2058.
46. Yap EP, McGee JO (1992) Nonisotopic SSCP detection in PCR products by ethidium bromide staining. Trends Genet 8:49.
47. Hongyo T, Buzard GS, Calvert RJ, Weghorst CM (1993) "Cold SSCP": A simple, rapid and non-radioactive method for optimized single-strand conformation polymorphism analyses. Nucleic Acids Res 21:3637–3642.
48. Heym B, Alzari PM, Honore N, Cole ST (1995) Missense mutations in the catalase–peroxidase gene, *katG,* are associated with isoniazid resistance in *Mycobacterium tuberculosis.* Molec Microbiol 15:235–245.
49. Jacobs WR Jr, Barletta RG, Udani R, Chan J, Kalkut G, Sosne G, Kieser T, Sarkis G, Hatfull GF, Bloom BR (1993) Rapid assessment of drug susceptibility of *Mycobacterium tuberculosis* by means of luciferase reporter phages. Science 260:819–822.

# 5

# Pathogenesis of Mycobacterial Disease

*John M. Grange*

The mycobacteria causing disease in animals and human beings can be divided into two major groups: the obligate pathogens and the environmental saprophytes that occasionally cause opportunist disease. The principal species in both groups are listed in Table 5.1. Some diseases caused by opportunist species have characteristic features and, thus, have individual names (e.g., Buruli ulcer, swimming pool granuloma), whereas others are responsible for much more variable presentations of disease, often resembling the many manifestations of tuberculosis.

Even within those mycobacterial diseases bearing specific names, there is an enormous variation in their clinical features, course, and eventual outcome which is almost entirely attributable to differences in host immunity. Several mycobacterial diseases clearly show a variability related to time after initial infection. This is particularly evident in tuberculosis, in which the immunopathological features and clinical presentation of disease following primary infection differ in many ways from disease due to endogenous reactivation or exogenous reinfection later in life. Thus, despite a great variability in clinical presentations, Wallgren was able to discern a common pattern or "timetable" of events in childhood tuberculosis (Table 5.2) (1). An extension of this "timetable" is needed in regions where most tuberculosis is of the reactivation type in older persons (2).

Mycobacterial disease also shows variability at any given time according to the nature of the host's immune responsiveness. The classical example is leprosy, in which there is a well-defined "spectrum" of clinical, immunological, and pathological features with inappropriately strong immune reactivity at one pole of the spectrum and a specific anergy to the leprosy bacillus at the other pole (3).

Three key features of host immunity must be considered in any discussion of the pathogenesis of mycobacterial disease. First, host immunity to mycobacteria is usually very good. In the absence of an immunocompromising disorder, only a minority of infections by the leprosy or tubercle bacilli lead to clinically evident

Table 5.1. The principal species of mycobacteria pathogenic for human beings.

| | | | |
|---|---|---|---|
| **1. Specific named diseases** | | | |
| Tuberculosis | *M. tuberculosis* | *M. bovis* | *M. africanum* |
| Leprosy | *M. leprae* | | |
| Swimming pool granuloma | *M. marinum* | | |
| Buruli ulcer | *M. ulcerans* | | |
| **2. Non-specific opportunist diseases: slowly growing species** | | | |
| *M. avium** | *M. intracellulare** | *M. kansaii* | *M. xenopi* |
| *M. scrofulaceum* | *M. malmoense* | *M. szulgai* | *M. gordonae* |
| *M. simiae* | *M. asiaticum* | *M. genavense* | *M. celatum* |
| **3. Non-specific opportunist diseases: rapidly growing species** | | | |
| *M. chelonae* | *M. fortuitum* | | |

*Often grouped together as *M. avium-intracellulare* or the *M. avium* complex.

Table 5.2. The 'time-table' of tuberculosis in children*

| Time after infection | Main clinical features |
|---|---|
| 3–8 weeks | Conversion to tuberculin positivity |
| | Erythema nodosum |
| | Phlyctenular conjunctivitis |
| 3 months | Tuberculous meningitis |
| | Miliary tuberculosis |
| 3–6 months | Pleural effusion |
| 3–9 months | Local complications from enlarged lymph nodes |
| | Endobronchial tuberculosis |
| | Segmental lesions and/or collapse |
| 1–3 years | Bone and joint lesions |
| | First appearance of post-primary lesions (mostly in post-pubertal children) |
| >5 years | Renal tuberculosis |

*Modified from Wallgren.[1]

disease. Likewise, human beings are exposed daily to environmental mycobacteria, yet overt disease in otherwise healthy people is extremely rare. Second, unlike the great majority of other bacteria, obligate (and, perhaps, other) mycobacterial pathogens can persist in tissues in a poorly understood state of dormancy and cause reactivation disease years or decades later. Third, immune responsiveness varies qualitatively as well as quantitatively. There is, after decades of debate, strong evidence that immune reactivity leading to resolution of disease is qualitatively different from that leading to tissue necrosis and progression of disease (4,5).

## 1. Mycobacterial Pathogenicity and Virulence

The pathogenesis and immunopathology of any infectious disease are a result of characteristics of both the infecting organism and the infected host—"the seed and the soil." The mycobacterial species and, in some cases, strains within a species clearly vary in their pathogenicity and virulence, but the actual causes of such variation are shrouded in mystery. Very little difference has, for example, been found between virulent *Mycobacterium tuberculosis* and the attenuated Bacille–Calmette–Guérin (BCG) strain. One of the problems in ascertaining the cause of virulence is that most studies on the properties of mycobacteria have been conducted *in vitro.* It is possible that determinants of virulence are induced *in vivo.*

A number of putative determinants of virulence in *M. tuberculosis* have been described, including certain toxic lipids in the cell wall and susceptibility to killing by hydrogen peroxide (6). The problem is that virulent strains lacking a given determinant, and attenuated strains with the determinant, are not infrequently found, indicating that virulence is not due to any single factor so far described. The ability to shuttle small amounts of DNA from one mycobacterial cell to another provides a means of analyzing virulence at a molecular level, but, to date, no clear determinant has thereby been identified. It may well be that rather than a single characteristic, the way in which an antigen-presenting cell interprets a complex array of antigens and adjuvants determines further immune events and, hence, pathogenicity (7).

## 2. Immune Responses Contributing to Protection and Pathogenesis in Mycobacterial Disease

The immunology of mycobacterial disease is discussed in detail in Chapters 8 and 9; therefore, only those immune responses relevant to pathogenesis are briefly summarized here.

The "classical" theory of mycobacterial immunity proposed by Mackaness (8) is based on antigen recognition by helper/inducer T cells followed by their clonal proliferation and secretion of gamma-interferon which, by activating macrophages, leads to the formation of the characteristic lesion of mycobacterial disease—the granuloma. This consists of a compact aggregate, many layers thick, of activated macrophages around the pathogen and a peripheral zone containing lymphocytes responsible for macrophage activation. The closely interdigitated macrophages bear a morphological resemblance to epithelial cells and are, therefore, termed "epithelioid cells." Some of the macrophages fuse to form multinucleate giant cells (Langhans' cells) which, although not unique to tuberculosis, strongly support the histological diagnosis.

One of the problems faced by experimental immunologists is the inability to create granulomas in *in vitro* systems. Much work on the role of the macrophage in immunity to mycobacterial disease has been based on monolayers of cultivated macrophages, often derived from blood rather than from tissues. In such systems, activated human macrophages inhibit replication of tubercle bacilli, but the demonstration of actual killing has been elusive. It is likely, however, that the aggregation of macrophages in the granuloma is a much more effective killing system than the isolated macrophage. Being metabolically very active, the epithelioid cells consume oxygen and nutrients diffusing into the granuloma. Consequently, the interior region becomes anoxic and necrotic, thereby providing an environment which inhibits the growth of mycobacteria and probably kills the great majority.

The theory of mycobacterial immunity outlined above now requires extensive revision, particularly as it does not account for immune responses that cause gross tissue necrosis and progression of disease. Such a revision requires a summary of the principal cells involved in protective and non-protective immune responses in mycobacterial disease; namely, the antigen-presenting cells (such as the dendritic cells in lymphoid tissue), T lymphocytes, natural killer (NK) cells, and macrophages. In some circumstances, B lymphocytes can, by producing antibody, also contribute to immunopathology.

The T cells are divided into subsets according to their possession of certain cell surface markers and by the nature of the cytokines they release. On the basis of cell surface markers, most T cells are divided into those that are CD4 or CD8 positive. Initially, it was thought that there was a clear functional difference between these two types of cell, with CD4+ cells having inducer/helper functions and CD8+ cells having suppressor/cytotoxic functions. This distinction is now not so clear-cut, as, for example, cytotoxic CD4+ cells have been described (9).

Mycobacterial antigens are taken up and processed by antigen-presenting cells and presented to antigen-specific T lymphocytes in close association with products of the major histocompatibility complex (MHC; Fig. 5.1). Antigen derived from mycobacteria within phagosomes of macrophages is presented by HLA-D molecules (MHC Class II) and principally activates CD4+ helper/inducer cells that are responsible for enhancement of bactericidal activity. In other cells, such as effete phagocytes, the bacilli escape from the phagosomes and replicate freely in the cytoplasm. Antigen from this site is presented by HLA-A and -B molecules (MHC Class I) and activates, and serve as targets for, cytotoxic CD8+ T cells. This enables cells that cannot cope with the bacilli within them to be lysed, thereby permitting the bacilli to be engulfed by immunologically competent phagocytes. Experiments with "knockout" mice lacking functional CD8+ T cells

Figure 5.1. Antigen presentation by MHC molecules on the surface of the antigen presenting cell. Mycobacterial antigen derived from the phagosome is presented by MHC Class II molecules to CD4+ T cells (mostly helper/inducer cells) whereas antigen from bacilli within the cell cytoplasm is presented by MHC Class I molecules to CD8+ T cells (mostly cytotoxic/suppressor cells).

show that this cytolytic activity is crucial for a protective response in tuberculosis (10), and there is evidence of a similar involvement of these cells in human beings (11). CD8+ T cells can also have a suppressing effect in immune responses and, as outlined below, these cells predominate over CD4+ cells in the more anergic forms of leprosy and tuberculosis.

Whereas most T cells have antigen receptors consisting of two protein chains, alpha and beta, a minority have receptors composed of gamma and delta chains and are thus termed gamma/delta ($\gamma\delta$) cells. They appear early in ontogeny and have structural and functional similarities with the natural killer cells described below. A major subset of $\gamma\delta$ cells recognize certain epitopes of a class of proteins known as heat shock proteins (HSPs, see below), which are expressed on the surface of stressed cells (12). There is evidence that these cells, or at least a subpopulation thereof, recognize HSPs presented in ways other than by the MHC molecules (13). They are abundant in early tuberculous lesions but, later on, the alpha/beta ($\alpha\beta$) cells dominate.

The exact role of $\gamma\delta$ T cells in mycobacterial disease is unknown, but patients with active tuberculosis have a much lower percentage of circulating $\gamma\delta$ T cells responding to *M. tuberculosis* than healthy infected (tuberculin-positive) persons and immunologically reactive leprosy patients have more of these cells than anergic (lepromatous) patients, suggesting a protective role. It has been postulated that $\gamma\delta$ cells recognizing HSPs on effete, stressed, mycobacteria-laden, macrophages have a similar function in early tuberculous lesions to the antigen-specific CD8+ cytotoxic cells mentioned above, affording protection before the specific immune response develops (14).

Natural killer (NK) cells also appear early in ontogeny and have a nonspecific cytolytic activity. Their precise function, if any, in protection against, or pathogenesis of, mycobacterial disease is unknown. Some authors report increased nonspecific cytolytic activity in tuberculosis (15), whereas others report a decreased activity (16), although the exact identity of the cells involved was not determined in the latter study. More recently, it has been shown that patients with chronic multidrug-resistant tuberculosis have greatly reduced NK cell activity compared with patients with less chronic, drug-sensitive, disease (17). It was postulated that this defect favored a high bacterial load and poor response to therapy, thereby leading to the emergence of drug resistance.

## 3. Delayed Hypersensitivity and the Koch Phenomenon

A cardinal feature of infection by, and disease due to, mycobacteria is the development of dermal reactivity to injected preparations of mycobacterial antigens. Although such reactivity forms the basis of tuberculin and lepromin testing, the nature of this reaction and its relation to pathogenesis and protection has been the source of confusion for many decades. The tuberculin test is cited as a typical example of delayed-type hypersensitivity (DTH), a term which is often used synonymously with cell-mediated immunity (CMI). It is now evident that there is more than one type of DTH: some types are associated with tissue necrosis and others are not (18).

The necrotic type of DTH was first described by Koch in 1891 in the guinea pig and became known as the Koch phenomenon (19). Thus, necrotizing, or incipiently necrotizing, dermal reactions to mycobacterial antigens in man have been termed Koch-type reactions. Koch injected virulent tubercle bacilli intradermally into the flanks of guinea pigs and observed the ensuing reaction. A small nodule developed at the injection site after 10–14 days; this nodule ulcerated and remained open, and contained viable tubercle bacilli, until the death of the animal. The regional lymph nodes were grossly involved and the disease became disseminated, leading to death 3 or 4 months later. Animals given a further injection of tubercle bacilli at another site 4–6 weeks after the first injection developed a quite different reaction at the second site. Within 24–48 h, an area of skin 0.5–1 cm at the injection site became darkened and necrotic and eventually sloughed off, leaving a shallow ulcer that healed. No viable tubercle bacilli were found in this ulcer and the regional lymph nodes were not involved, suggesting that the second challenge had been overcome. An identical dermal reaction was elicited by injection of either killed tubercle bacilli or a bacteria-free filtrate of the medium in which the bacilli had been cultivated. This filtrate was termed tuberculin.

On the belief that tuberculin induced a protective immune response, Koch administered this agent systemically to patients with tuberculosis. Unfortunately, many patients developed severe allergic reactions ("tuberculin shock") and there were a few deaths. Although some patients with chronic skin tuberculosis (*lupus vulgaris*) made dramatic improvements, the therapeutic results were, in general, unimpressive and this mode of treatment was largely abandoned. It was the subsequent use of tuberculin by von Pirquet as a skin test reagent to indicate infection by the tubercle bacillus that kept Koch's discovery alive to the present day (20).

For decades, there was disagreement as to whether the characteristic tissue-necrotizing hypersensitivity reaction occurring in tuberculosis was an exaggerated form of the classical protective, macrophage activating, immune reaction or whether it was a distinct and antagonistic phenomenon (21). Lefford, with prophetic insight, postulated that the different reactions were attributable either to separate T-cell populations or to multifunctional cells at different stages of maturation (22). He did not, however, consider this as being of practical importance, as he thought it unlikely that infection or immunization could cause a single T-cell function to predominate.

One of the problems is that the tuberculin test appears to indicate more than one immune phenomenon. There are several components to the tuberculin reaction: a nonspecific infiltration of blood-derived white cells around the capillaries (as seen in any acute inflammation), a more specific migration of T cells into the interstitial dermal tissue and the clinically observable dermal swelling (23). The latter is cytokine mediated and is independent of the intensity of the cellular infiltrate: clinically negative reactions may, on biopsy, show numerous cells.

A great light has been shed on this argument by the discovery that CD4+ helper T cells are divisible into functionally different subsets according to the cytokines they liberate. Thus, the major cytokines produced by T-helper-1 ($T_H1$) cells are interleukin 2 (Il-2) and gamma interferon (INF$\gamma$), whereas those produced by T-helper-2 ($T_H2$) cells are interleukins 4 and 10 (24). It is now apparent that cytotoxic CD8+ T cells are likewise divisible into these two types so the terms Type 1 and Type 2, rather than $T_H1$ and $T_H2$, are increasingly used.

These two functional types are derived by different maturation pathways from uncommitted ($T_HO$) precursor cells. The cytokines produced by the committed cells are mutually antagonistic, so that the presence of $T_H1$ cytokines down-regulates maturation to $T_H2$ cells and *vice versa.* Thus, infection or immunization could cause a single T-cell function to predominate. Furthermore, contact with the mycobacterial antigen, even if insufficient to cause a measurable immune response, may, depending on the dose and route of infection, "imprint" the immune system with a tendency to respond with a $T_H1$- or $T_H2$-mediated reaction (25,26). Thus, the multiplicity of epitopes in a complex antigen such as a mycobacterium induces a unified pattern of immune responses, a phenomenon which Bretscher has termed coherence (25). This concept provides an explanation of the profound effects of prior exposure to environmental mycobacteria on the efficacy of subsequent BCG vaccination. It also underpins the development of rational immunotherapy as described in Chapter 9. The T-cell maturation pathway is controlled by steroid hormones and the balance between them in the lymphoid tissue where the maturation occurs (27). Glucocorticoids such as cortisol promote $T_H2$ maturation, whereas dehydroepiandrosterone promotes $T_H1$ maturation and opposes the former.

## 4. The Role of Tumor Necrosis Factor in Immunopathogenesis of Mycobacterial Disease

Tumor necrosis factor (TNFα) has been shown to play a key role in immune responses in mycobacterial disease, notably tuberculosis. Treatment of mice infected with BCG or *M. tuberculosis* with an antibody to TNFα led to a rapid progression of disease (28). It has been shown that macrophages and other cells infected with virulent strains of *M. tuberculosis* are rendered exquisitely sensitive to killing by TNFα (29). It is likely that this provides a means, in addition to cell-mediated cytotoxicity, to release tubercle bacilli from immunologically effete or incompetent cells. Mycobacterial infection also elicits a more generalized effect of TNFα on tissues, analogous to the Shwartzman phenomenon. In this phenomenon, injection of a gram-negative endotoxin into the skin primes the site for a necrotic reaction when the same endotoxin is given intravenously 24 h later. A similar phenomenon occurs in mycobacterial lesions: as Koch originally demonstrated, systemic injection of tuberculin led to necrosis around tuberculous foci, with their elimination if they were on the skin. It appears that TNFα is involved, as such necrosis is accompanied by a massive systemic release of this cytokine (4). (In this context, it is of note that the extreme cachexia seen in advanced, untreated tuberculosis is due to systemic release of TNFα; in the earlier literature, cachectin was a synonym for TNFα.)

It has been shown that this mycobacterial analogue of the Shwartzman phenomenon differs from that induced by gram-negative endotoxin in that it is T-cell dependent (30). Furthermore, it is critically dependent on the T-cell maturation pathway, as mixed $T_H1/T_H2$ cytokines prime the tissues for necrosis, whereas pure $T_H1$ cytokines do not (Fig. 5.2) (31). This important finding explains how the T-cell type critically affects the pattern of immune responsiveness, protective or tissue destroying, induced by mycobacteria and indicates that the ideal immunoprophylactic and immunotherapeutic agent would be a $T_H1$ adjuvant.

An interesting and useful correlate of the necrotizing reaction described above is the elevated level of agalactosyl IgG [Gal(0)] (i.e., immunoglobulin G lacking

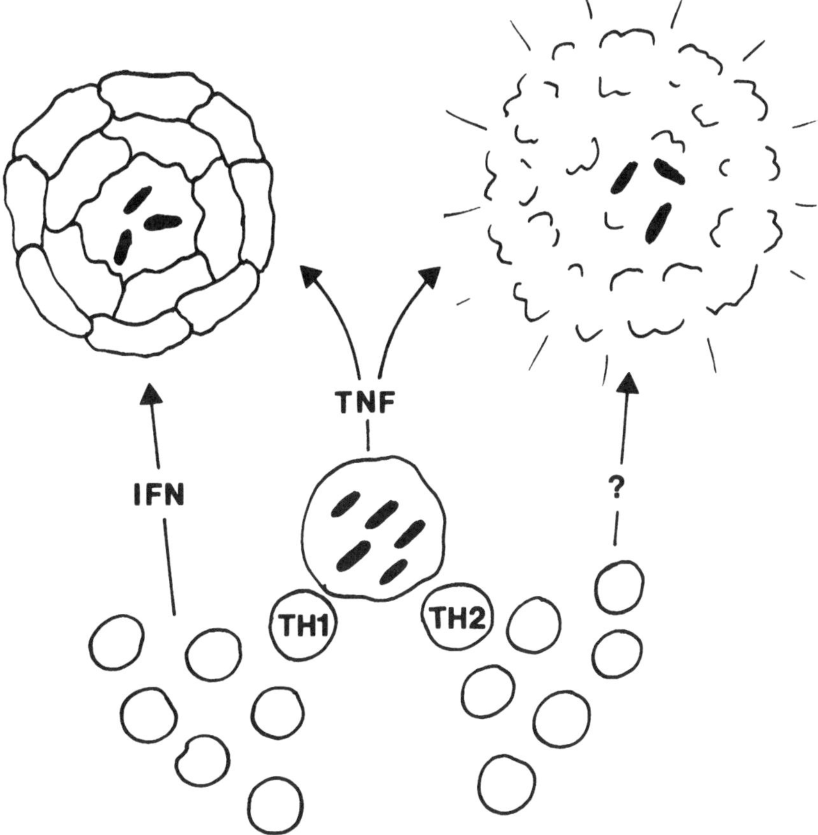


Figure 5.2. Role of tumour necrosis factor (TNF) in protective immunity and immunopathology in tuberculosis and the role of TH1 and TH2 helper T cells in determining the nature of the immune response. In the TH1 response, TNF acts in synergy with gamma interferon (INF) to induce granuloma formation but in the TH2 response, tissues are rendered extremely sensitive to necrosis by TNF.

a terminal galactose on the sugar chain in the conserved $CH_2$ domain of the heavy chain). In health, Gal(0) accounts for 20–25% of the total circulating IgG, rising to 40% or more in the elderly. An increase in circulating Gal(0) occurs in a number of diseases, including active tuberculosis and some autoimmune diseases, in which tissue necrotizing reactions occur (32). The function of Gal(0) remains unknown, but the lack of the terminal galactose exposes the bacteriomimetic sugar *N*-acetyl glucosamine, which is expressed on the surface of cells undergoing programmed cell death (apoptosis) and thereby facilitates their endocytosis by adjacent healthy cells. It can, therefore, contribute to the pathogenesis of necrotic lesions by triggering excessive cell death. Whatever its function, the reduction of Gal(0) is a marker for the efficacy of immunotherapy for tuberculosis (see Chapter 9).

## 5. Recognition of Mycobacterial Antigens in Health and Disease

Much work has been carried out on the recognition of purified mycobacterial antigens by T-cell clones, but patterns of antigen recognition, easily determined by multiple skin testing, are of greater clinical relevance (33,34). Such testing has revealed that healthy but infected (tuberculin or lepromin positive) persons show considerable cross-reactivity, including reaction to mycobacterial species not encountered in their environment. This indicates the recognition of common mycobacterial antigens. By contrast, patients with active disease show much less cross-reactivity, indicating reduced responsiveness to common antigens. A similar diminished reactivity to common mycobacterial antigens has been observed in patients seropositive for the HIV infection (35) and those with South American trypanosomiasis (Chagas' disease) (36), both conditions being due to unrelated intracellular pathogens.

Many of the common mycobacterial antigens belong to a class of highly conserved proteins termed heat shock proteins (HSPs) (37). These are present in all living cells, usually in small amounts confined to the cytoplasm. Is states of stress, however, they appear in large amounts on the cell surface and could lead to the detection of stressed, pathogen-laden cells and their lysis by cytotoxic CD4+, CD8+, and $\gamma\delta$ T cells. Thus, diminution in recognition of HSPs could contribute to the progression of disease.

## 6. Pathogenesis and Healing in Tuberculosis

The granulomatous lesions of tuberculosis show a natural tendency to heal by fibrous scarring, often with subsequent contraction and calcification. Such healing does not require microbiological quiescence: indeed, it might be an important contributory factor to the onset of quiescence. Some primary pulmonary lesions show a laminated appearance with successive layers of calcified scar tissue indicative of repeated phases of healing and progression of disease. In the prechemotherapy era, a minority of cavities underwent spontaneous closure and resolution, presumably as the result of contracting scar tissue. It is likely that encapsulation of lesions by fibrous scarring plays a role in the prevention of lymphatic and hematogenous dissemination of tubercle bacilli from postprimary lesions and from primary lesions in BCG-vaccinated subjects. Thus, there might be a relation between such fibrosis and the immune response, including the necrotic Koch-type reactivity.

There are similarities between wound repair and granuloma formation in chronic infections. Macrophages are the predominant cells in wounds 48 h after injury and, by releasing cytokines including interleukin-1 and the tumor necrosis factor, play a key role in subsequent healing and fibrosis (38). Healing involves obliteration of capillaries and contraction of fibrous tissue: one agent involved in this process is endothelin-1 (ET-1), an extremely potent vasoconstrictor (39). Expression of mRNA of ET-1 is induced by TNF$\alpha$ which, as outlined above, is produced in tuberculous granulomas. It is possible that ET-1 contributes not only to healing and to encapsulation of the disease process but also to the obliteration of capillaries and lymphatics in postprimary lesions and thereby to the extensive tissue necrosis. It would, therefore, be of interest if ET-1 production is affected by the predominant T-cell type—$T_H1$ or $T_H2$. In this context, studies of blood flow in tuberculin reactions show that there is considerable slowing of the blood flow in the center of indurated test sites, especially those showing frank necrosis, but increased blood flow in the softer, nonindurated test sites (40).

## 7. Initial Infection by Mycobacteria

This brief summary of immune phenomena in mycobacterial disease permits an analysis of the events occurring in initial infection by a mycobacterial pathogen. The word *infection* is often used rather loosely and it has subtly different meanings to the physician, epidemiologist, microbiologist, and immunologist. In the case of mycobacteria, infection can mean entry of the bacilli into the body by inhalation, ingestion, or inoculation; interaction with the immune system leading to altered immune reactivity, usually detected by skin testing; establishment of bacilli in the tissues with the induction of some form of lesion, or persistence in the tissues without causing overt disease. In the case of tuberculosis, it is often stated that entry of tubercle bacilli into the lungs of a previously uninfected person leads to the establishment of a primary complex (see below). Whether, in fact, this always occurs is unknown. It is possible that some inhaled bacilli are engulfed by alveolar macrophages and are either destroyed or removed from the lung, possibly on the ciliary escalator, before they can become established.

## 8. The Primary Lesions Caused by Mycobacteria

There is enormous variation in the nature and evolution of the primary lesion of any mycobacterial disease, but there are three basic components: a focus of disease at the site of entry of bacilli, involvement of the regional lymph nodes, and spread by the lymphatics and bloodstream leading to distant lesions. The extent to which these three components occur, or are evident, varies according to the causative organism, the site of its entry, and the host's immune response.

The evolution of this pattern of disease is most clearly seen in pulmonary tuberculosis, which is usually caused by the inhalation of small clumps of tubercle bacilli, about 5 $\mu$m in diameter, in the cough spray. These reach the alveoli or small bronchioles where they can establish a local focus of disease termed the Ghon focus. There is limited information on the initial events following infection in the human lung, owing to the rarity of opportunities to study them. The classical studies of Lurie and Dannenberg on the rabbit provide a probable account (41,42). Any tubercle bacilli not initially destroyed by the alveolar macrophage replicate and, after killing the cell, are released and elicit a local inflammatory reaction. This consists of an accumulation of blood-derived white cells, principally polymorphonuclear leukocytes and macrophages. The resulting granuloma is of the "foreign body" type characterized by a low turnover of cells. At this stage, the specific immune response has not developed, but there is evidence that gamma–delta T cells (see above) respond nonspecifically to certain components of the bacillus and cause macrophage aggregation and activation (14). They may also, as described above, lyse effete macrophages, enabling the bacilli within them to be engulfed by more aggressive phagocytes.

If these early, nonspecific, protective reactions do not overcome the infection, bacilli are transported in the lymphatics to the lymph nodes at the hilum of the lung, resulting in the formation of further lesions. The lesion consisting of the Ghon focus and the hilar lymphadenopathy is termed the primary complex. During the first few weeks, specific cell-mediated immune responses are induced and after about 2 months the characteristic high-turnover granulomas of immune origin are evident.

Similar complexes occur if tubercle bacilli are ingested, the usual cause being milk containing *M. bovis*. In this case, the initial focus, which is usually too small to be detected clinically, occurs in the tonsil or the intestinal wall and there is involvement of the tonsillar nodes at the angle of the jaw and the mesenteric nodes, respectively. Another form of complex results from traumatic inoculation of the skin, which manifests as a warty lesion at the site of injury and involvement of lymph nodes at the elbow or in the axilla. This was once an occupational hazard of anatomists, pathologists, and butchers; hence, the local lesion was termed prosector's (or butcher's) wart (43). Rene Laenecc, the French physician who invented the stethoscope, acquired such a lesion after cutting his left index finger while sawing through the vertebrae of a patient who had died of spinal tuberculosis.

Early post-inoculation tuberculous lesions show a relatively large bacillary load and a mixed granulomatous and pyogenic appearance known as suppurative granulomatous dermatitis (44). Subsequently, the pyogenic element gives way to a pure granulomatous reaction and a reduction in bacterial load. A similar sequence of events is seen in *M. marinum* infection (swimming pool granuloma, see below).

Primary complexes, particularly those in the lung, usually go unnoticed by the patient and resolve, but, occasionally, local complications can arise from the Ghon focus or the enlarged lymph nodes. Thus, the Ghon focus may rupture into the pleural cavity, initiating tuberculous pleurisy or, if the immune response is inadequate, there may be alternative phases of progression and encapsulation resulting in a large lesion, sometimes showing concentric laminations. Such lesions may enter the postprimary phase without the usual latent interval (Fig. 5.3) and eventually erode into a bronchus with formation of a cavity (Fig. 5.4). Grossly enlarged lymph nodes can compress one of the main bronchi and thereby cause collapse of a lobe of the lung. Alternatively, a lymph node can erode through the bronchus, leading to endobronchial tuberculosis with eventual bronchial stricture or to a spreading disease in the lobe aerated by that bronchus. Lymph nodes can also erode into the pericardial cavity, thereby causing tuberculous pericarditis.

Lymph nodes in the neck can ulcerate through the skin with the formation of draining sinuses and secondary skin lesions which may become very chronic

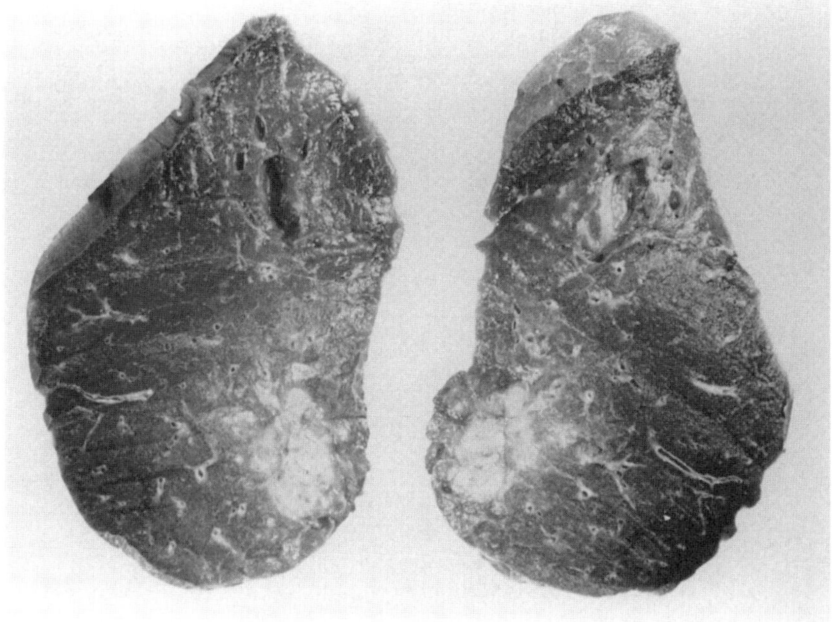

Figure 5.3. Progressive primary tuberculosis with tuberculoma formation in the lung of a 12 year old boy.

(lupus vulgaris). This combination of tuberculous lymphadenitis (scrofula) and overlying skin lesions is termed scrofuloderma. Mesenteric lymph nodes can rupture into the abdominal cavity and cause a spreading tuberculous peritonitis.

In addition to forming the localized lesions described above, tubercle bacilli disseminate further by the lymphatic stream and the bloodstream and lodge in many organs of the body. Primary tuberculosis is, therefore, a systemic infection, but preferential sites for overt disease are the brain and meninges, bones, notably the spine, and the kidney. Bacilli also reach other parts of the lung and can manifest as radiologically visible lesions (Simon's foci) in the apical regions. These may be the precursors of postprimary lesions.

The site and nature of the primary lesion of leprosy is a mystery. Originally thought to be transmitted by skin contact, it is now believed that the leprosy bacilli are inhaled or ingested (45). The first observable lesion in leprosy is a small skin lesion with rather nonspecific features of inflammation and containing a few acid-fast bacilli. This is termed indeterminate (Idt) leprosy. It is often assumed that these bacilli reach the skin directly via the hematogenous route, but it is possible

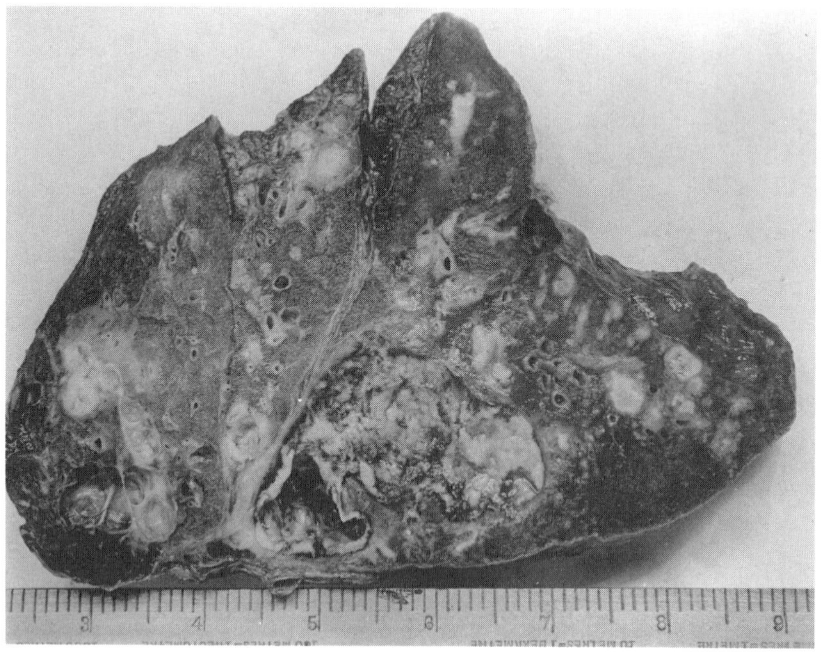


Figure 5.4. Tuberculous cavities and spreading endobronchial disease in the lung of a 17 year old girl.

that this early skin lesion is secondary to one in the nerve trunk serving that segment of skin (46).

The distinction between primary and postprimary lesions is usually not evident in disease due to environmental mycobacteria (EM), with the exception of *M. ulcerans* infection (Buruli ulcer, see below). Cervical lymphadenopathy due to EM, usually *M. avium-intracellulare,* in young children (see below) is a probable example of a primary complex.

## 9. Mycobacterial Persistence

The most important characteristic of mycobacteria contributing to the pathogenesis of disease is the ability of bacterial cells to enter a state of dormancy, or near dormancy, and to persist in tissues in such a quiescent state for years or decades. The nature of this "persistor" state is shrouded in mystery. Various explanations have been advanced and some *in vivo* and *in vitro* experimental models have been developed (47). Thus, for example, mycobacteria remain viable but dormant for long periods under anaerobic conditions. Other explanations include the existence of elaborate life cycles or microspores (Much's granules) (48). Dormancy must to some extent be determined by the host's immune defense mechanisms, as factors leading to a weakening of such mechanisms predispose to reactivation of mycobacterial disease.

## 10. Postprimary Disease: The Immunopathological Spectrum in Leprosy

Ridley and Jopling showed that the enormous variation in the clinical and histological manifestations of leprosy was the result of a "spectrum" of immunological reactivity (3). At one end of the spectrum, the disease is characterized by a small number of discrete lesions, containing massive tuberculosis-like granulomas but very few bacilli. It has been suggested that the reaction is, at least in part, to bacterial debris that is not easily cleared from the site of disease (49). At the other end of the spectrum, there is an absence of specific immune reactivity to *M. leprae,* the lesions are widespread and diffuse, and they contain numerous bacilli. Between these polar forms, termed tuberculoid (TT) and lepromatous (LL) leprosy, respectively, there is a range of intermediate forms which, for convenience, are classified as borderline tuberculoid (BT), mid-borderline (BB), and borderline lepromatous (BL). The immunological features of the leprosy spectrum are summarized in Table 5.3. The nature of the spectrum of leprosy has been reviewed in detail by Ridley (49).

It is often stated that because TT leprosy is limited and usually self-healing, it is a manifestation of protective immunity, but the true protective immunity is that which prevents early, indeterminate leprosy from progressing to one of the above

Table 5.3. The spectrum of leprosy

| Feature | Point on TT | spectrum BT | BB | BL | LL |
|---|---|---|---|---|---|
| Bacilli in lesions | − | + | + | + + | + + + |
| Bacilli in nasal discharge | − | − | − | + | + + |
| Granuloma formation | + + + | + + | + | − | − |
| Macrophage maturity | + + + | + + | + | + | − |
| *In vitro* correlates of CMI* | + + + | + + | + | + | − |
| Reactivity to lepromin | + + | + | − | − | − |
| Specific antibody IgG | + | + | + | + + | + |
| Specific antibody IgM | + | + | + | + | + + + |

*Cell mediated immunity

forms. The characteristic feature of TT and BT leprosy is the occurrence of massive non-necrotizing granulomas, far out of proportion to the apparent bacterial load, which, as space-occupying lesions, can cause serious nerve damage. Thus, even at this reactive pole of the spectrum, immunopathology as well as protection occurs.

The most bizarre feature of leprosy is the total lack of T-cell responsiveness to *M. leprae* at the LL pole. The mechanism by which the bacillus can suppress immune reactivity to its numerous antigens, including many that are shared by other mycobacteria and, indeed, by many other bacterial genera, remains a mystery. Certainly, patients with LL leprosy can respond, sometimes very strongly, in skin tests to tuberculin (50). The defect does not appear to be due to an absence of CD4+ T-helper cells responding to antigens of *M. leprae* but rather to a suppression of the activity of these cells (51). When challenged by individual purified antigens of *M. leprae,* T-helper cells from LL patients respond well in lymphocyte transformation assays, but they fail to respond to whole bacilli. Thus, there must be some component of the leprosy bacillus that switches off T-cell recognition, but it is not known why this only happens in certain patients. Lesions in LL leprosy contain relatively few T cells: those within the lesion are principally CD8+ and there are a few CD4+ cells in the periphery of the lesions (52). By contrast, TT lesions contain many more T cells with a preponderance of CD4+ cells throughout the lesions.

There is evidence that the immunological spectrum in leprosy is, at least to some extent, related to the balance between $T_H1$ and $T_H2$ T cells. Thus, the circulating *M. leprae*-specific T cells in patients with TT leprosy are mostly of the $T_H1$ type, and $T_H1$-associated cytokines are detectable in the lesions, but there appears to be a mixed $T_H1$ and $T_H2$ cytokine response in LL leprosy (53). As outlined earlier, the non-necrotic protective response in tuberculosis is $T_H1$ mediated and necrotizing immunopathology is $T_H2$ or mixed $T_H1/T_H2$ mediated. Yet, paradoxically, typical necrotizing reactions do not occur in LL leprosy and this is probably the result of specific suppression of this response by the CD8+ T cells (50).

In addition to the underlying disease, leprosy patients can undergo reactions which may cause extensive damage to nerves and other vital structures. These are divisible into two types (45). Type 1 reactions are sometimes termed upgrading (or reversal) reactions as they appear to be associated with a shift in immunity toward the reactive (TT) end of the spectrum. The influx of immunologically competent cells and the onset of an inflammatory response in a major nerve trunk can lead to gross swelling, which, if untreated, can lead rapidly to permanent loss of motor or sensory function. The cytokines in Type 1 reactions are mainly of the $T_H1$ pattern, suggesting that such reactions are triggered by a $T_H2$ to $T_H1$ shift (50).

The other type of leprosy reaction, Type 2 [also termed erythema nodosum leprosum (ENL)], usually occurs in patients near the LL pole of the spectrum who have a high bacillary load and high levels of antibody. The cytokine patterns in the lesions are of a mixed $T_H1$ and $T_H2$ pattern which may merely reflect the maturation pattern of leprosy-specific T cells of the patient. ENL is usually regarded as being due to antigen–antibody complex formation, but there is evidence that TNF-mediated tissue necrosis is involved, probably due to a relaxation of suppression by CD8+ cells (54). The immune complex formation may merely focus the cell-mediated reaction (50) or it can be a secondary phenomenon due to deposition of antibody on damaged tissue. Glomerulonephritis, due to deposition of immune complexes in the glomeruli, occurs more frequently in patients with multibacillary disease (LL and BL) than in those with paucibacillary disease, but opinions differ as to whether it is more common in patients with ENL (55).

The suggested association between cytokine patterns and the position of the patient on the immune spectrum has implications for immunotherapy. It has been shown that the injection of small amounts of gamma interferon (a $T_H1$ cytokine) into the skin of LL patients led to an influx of T cells, principally of the CD4+ type, granuloma formation and a destruction of leprosy bacilli, strongly suggesting a local shift toward the TT end of the spectrum (56). On the other hand, systemic immunotherapy based on cytokine administration might well induce extensive granuloma formation, leading to serious nerve damage.

## 11. Postprimary Tuberculosis—Its Threefold Nature

With the exception of the uncommon cases of progressive primary tuberculosis, postprimary tuberculosis is due to either endogenous reactivation of persisting bacilli or to exogenous reinfection. For many years, it was generally and rather dogmatically held that exogenous reinfection was prevented by the immune response induced by the primary disease. It is now clear from several epidemiological studies and from DNA "fingerprinting" studies (57), that exogenous reinfection does occur. As a general rule, most postprimary tuberculosis in regions where the disease is uncommon and in decline is due to endogenous reactivation, whereas in regions where there are many infectious patients, exogenous reinfection occurs relatively more frequently (58).

The most usual site of reactivation tuberculosis is the upper part of the lung. As foci of primary tuberculosis do not have a predilection for this site, it is assumed that bacilli reach the upper part of the lung by the lymphatic or hematogenous route, either at the time of the primary infection or sometime during the latent phase. In the early part of the 20th century, there was much disagreement as to whether postprimary tuberculosis following nonpulmonary primary infection, such as tonsillar or intestinal infection by *M. bovis,* occurred in the lung. This question is still not completely resolved. In the years 1977–1990, about 40% of postprimary tuberculosis due to *M. bovis* in southeast England occurred in the lung, but the possibility that the primary focus of infection in these patients was in the lung cannot be excluded (59).

Postprimary tuberculosis is characterized by extensive tissue necrosis which, in the lung, can lead to very large lesions, termed tuberculomas. These consist of a mass of caseous necrosis surrounded by granulomatous tissue containing activated macrophages (epithelioid cells), giant cells, and lymphocytes and surrounded by a varying degree of fibrosis. The lumen of blood vessels and lymphatics entering these lesions are often obliterated. As in primary lesions, the center of the tuberculoma is acidic and anoxic, providing a poor environment for growth and survival of tubercle bacilli. The caseous material is eventually softened or liquified by proteases released from activated macrophages, and the expanding necrotic process often erodes into a bronchus, permitting the liquified contents to escape, thereby forming a cavity (Fig. 5.4). Oxygen and carbon dioxide enter the cavity and the acidity is neutralized, providing a very good environment for the replication of the tubercle bacillus.

Under the conditions within the cavity, the number of actively replicating bacilli increases enormously (Fig. 5.5), but they are almost all extracellular and behave more like saprophytes on the necrotic wall of the cavity than as primary pathogens. They almost certainly, however, secrete antigens and other substances that fuel the pathological process. Some of the bacilli become detached from the cavity wall and enter the sputum so that the patient becomes "open" or infectious. As a useful general rule, infectious patients have enough tubercle bacilli in their sputum for them to be detected by standard microscopical techniques (i.e., at least 5000 bacilli/ml of sputum) (60).

Some of the bacilli escaping from the cavity reach other parts of the same and opposite lung and establish secondary foci of disease. Bacilli can lodge in the

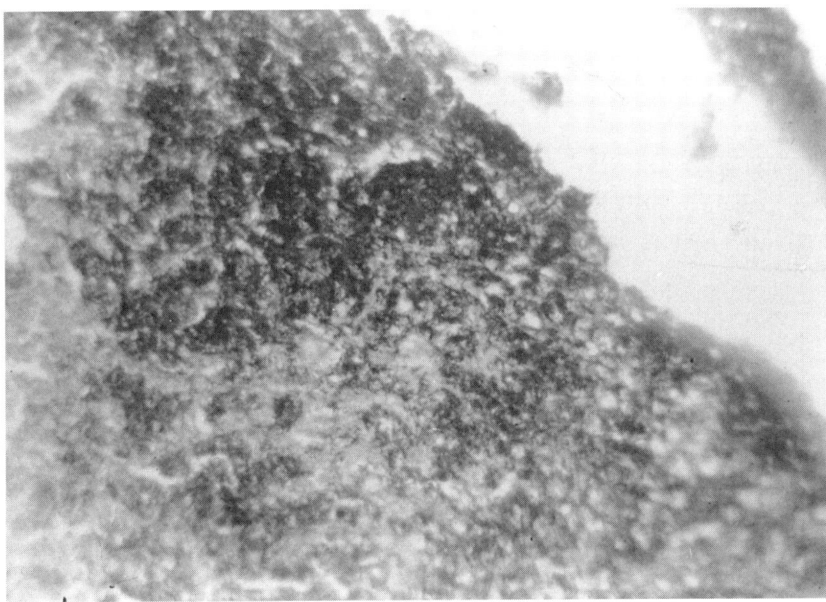

Figure 5.5. The wall of a tuberculous cavity with numerous acid-fast rods (dark stained) growing in the aerated wall composed of friable, caseous necrotic tissue.

larynx, causing tuberculous laryngitis, or they can be swallowed and infect the intestinal wall, causing indurating ulcers and anal abscesses and fistulae.

Cavity formation therefore often heralds a progression of disease, both in terms of bacillary load and spread within the lung, although even at this stage, spontaneous resolution occurs in about one-fifth of untreated patients. In the prechemotherapeutic era, the only effective treatment for cavitating pulmonary tuberculosis was surgical removal or closure of the cavity. The latter was achieved by reducing the volume of the lung, either by introducing air into the pleural space or by thoracoplasty, in which large sections of the chest wall were removed.

In contrast to primary tuberculosis, the characteristic spread to the local lymph nodes and hematogenous dissemination is rarely seen in postprimary disease unless the patient is immunocompromised. In this sense, the immunopathological process in this form of tuberculosis serves the useful purpose of walling off the infection. As the major protective effect of BCG vaccination is the prevention of the serious nonrespiratory forms of primary tuberculosis, it may act by inducing such a walling-off process.

Cavitating pulmonary tuberculosis can be seen as three superimposed conditions: persisting dormant or near-dormant bacilli in a probable endosymbiotic relation to the host, pathogenic bacilli causing tissue destruction and inflammation but, in so doing, generating an unfavorable environment for themselves, and saprophytic bacilli actively replicating in the favorable environment of the cavity wall. This three-in-one process has important implications for modern short-course chemotherapy which is designed to rapidly kill the actively replicating bacilli in the cavity wall, then to kill the less actively replicating bacilli in the acidic inflammatory lesions, and finally to eliminate the near-dormant persisting bacilli (see Volume II, Chapter 2).

## 12. The Immune Spectrum in Postprimary Tuberculosis

Although there is a range of immune reactivity in human tuberculosis, this bears only a superficial resemblance to the immunopathological spectrum of leprosy.

Ridley and Ridley have divided tuberculosis into three groups, each with two subgroups, according to the histological features of 54 patients (Table 5.4) (61). As in leprosy, the number of bacilli and the maturity of the macrophages in the lesions were inversely related.

All patients in Group 1 had cutaneous tuberculosis and showed some features in common with tuberculoid leprosy. Thus, the lesions showed well-organized granulomas composed of mature epithelioid cells and some giant cells, little or

Table 5.4. The spectrum of tuberculosis

| Group | Main cell type | Type of necrosis | Giant cells | Bacilli seen |
|---|---|---|---|---|
| 1a | Organized mature epithelioid cells | None | + | None |
| 1b | Unorganized mature and immature epithelioid cells | Patchy fibrinoid | + | Rare |
| 2a | Immature epitheloid cells | Caseation, no nuclear debris | + | Scanty |
| 2b | Immature epithelioid cells or undifferentiated histiocytes | Necrosis and nuclear debris with polymorphs | + | + |
| 3a | Scanty macrophages | Extensive, basophilic, coarse nuclear debris | − | + + |
| 3b | Very few macrophages | Extensive, eosinophilic, scanty nuclear debris | − | + + + |

*Data from Ridley and Ridley[64]

no tissue necrosis, and no detectable bacilli. When, however, such lesions are stained by the immunoperoxidase method, the mycobacterial antigen is detectable within macrophages and is more diffusely distributed within the lesions. This suggests that, as in tuberculoid leprosy, the immune response can be directed at residual and nonviable mycobacterial components that are not readily cleared from the site of disease.

The majority of the patients were in Group 2 and had a wide range of respiratory and nonrespiratory forms of tuberculosis. All patients in Group 3 had disseminated disease. Even within disseminated forms of tuberculosis, there is a range of immunological reactivity. In miliary tuberculosis, each focus of infection is surrounded by a granuloma, giving it a millet seed-like appearance (Latin *milium*—a millet seed). Although often progressive, some cases of miliary tuberculosis are relatively chronic, indicating a successful immune response. In other cases, disseminated bacilli do not elicit such a granulomatous response and the tissues contain numerous minute foci teeming with bacilli (cryptic disseminated tuberculosis—see below).

Although lepromatous leprosy appears to be the result of a very specific anergy to the causative organism, often with normal or even exaggerated immune responses to other mycobacteria, anergic tuberculosis can be the result of a much more generalized immunosuppression due to other causes. Since the early 1980s, HIV has become a prevalent and increasing cause of such anergic tuberculosis as described below. Although CD8 + cells are found in lesions of severe, progressive tuberculosis (62), it is uncertain whether they are a cause or a result of diminished immune responsiveness.

As mentioned above, chronic skin tuberculosis (lupus vulgaris) has some features in common with tuberculoid leprosy. The skin can also be involved in disseminated disease, giving rise to a much more acute condition termed *tuberculosis cutis miliaris disseminata.*

The occurrence of immune reactions in tuberculosis analogous to the two types of leprosy reactions is controversial. Two types of skin reactions occur in tuberculosis: erythema nodosum and the tuberculides. The former is a nodular vasculitis occurring in the subdermal connective tissue and is most often seen during streptococcal infections. In tuberculosis, it usually occurs in children with primary infection at the time of tuberculin conversion and it may be associated with phlyctenular conjunctivitis. In a rare chronic form, erythema induratum or Bazin's disease, the lesions sometimes ulcerate. The pathogenesis of nodular vasculitis in primary tuberculosis is unknown.

Mystery likewise shrouds the group of skin reactions termed tuberculides (63). Many forms have been described and bear rather quaint descriptive latin epithets. The two main types are lichen scrofulosorum and papulonecrotic tuberculide. The former commences as a perivascular and periappendicular infiltration of white cells in the dermis and develops into non-necrotic granulomas with epithelioid and giant cells. The latter is similar, but there is necrosis due to an obliterative vasculitis. On occasions, this necrosis is very extensive, leading to gangrene of the extremities. It is possible that the reactions represent exaggerated non-necrotizing and necrotizing hypersensitivity reactions to blood-borne whole tubercle bacilli or debris thereof.

## 13. Mycobacterial Disease and Immunosuppression

It has long been known that various causes of immunosuppression favour the development of primary tuberculosis following infection as well as reactivation of latent disease. In recent years, HIV has emerged as a very important predisposing factor in tuberculosis (Chapter 8, this volume). In addition, HIV predisposes to disease due to environmental mycobacteria, notably *M. avium-intracellulare* (see below).

Tuberculosis in immunosuppressed persons differs from that in the more immunocompetent (64). Not only is there a failure to develop the characteristic high-turnover granuloma of immunogenic origin, there is also a suppression of tissue-necrotizing reactions and scar formation that would otherwise limit the spread of infection. Thus, discrete pulmonary lesions and cavity formation are both less common in immunosuppressed patients. Instead, there may be a radiologically rather nonspecific spreading pulmonary lesion. Nonpulmonary lesions due to unrestricted bacillary dissemination are frequent in such patients. Dissemination can present as one or more solitary lesions, as widespread lymphadenopathy or as multiorgan involvement. The latter differs from miliary tuberculosis because the discrete granulomas do not develop. Instead, organs contain minute necrotic foci teeming with acid-fast bacilli. These might not be visible radiologically and are only detected by biopsy or at postmortem. Thus, this form of the disease has been termed *cryptic disseminated tuberculosis* (65).

There is no doubt that HIV infection has a profound effect on tuberculosis and there is evidence for the reverse. It has been observed that, even if successfully treated, tuberculosis in an HIV-positive person has a very deleterious effect on future health. Indeed, tuberculosis appears to drive the patient into the full picture of AIDS with considerable shortening of life. Although the details are not clear, there is evidence that TNFα and other immunological mediators released in tuberculosis lead to transactivation of the HIV provirus and its subsequent replication (66). In addition, tuberculosis causes a CD4+ T-cell lymphopenia which may synergize with that induced by the HIV (67). Whatever the cause, the occurrence of active tuberculosis in the HIV-positive patient has very serious consequences, demanding strenuous efforts to prevent such disease by programmes of chemoprophylaxis or, preferably, immunoprophylaxis (68).

In contrast to tuberculosis, evidence for an interaction between HIV and leprosy is limited and conflicting. In general, clinical leprosy is no more frequent in HIV-positive than in HIV-negative persons and, among leprosy patients, there is no relationship between HIV status and the bacterial load. It has been observed that certain infections, notably those due to parasites and worms, are not adversely affected by HIV infection (69), and it has been postulated that leprosy might be among these "missing infections" (70). If this is the case, there must be very basic differences in the immune responses between leprosy and tuberculosis.

## 14. Genetic Factors Affecting Pathogenesis of Mycobacterial Disease

The existence of genes that determine resistance to tuberculosis have long been suspected. In the mouse, a *bcg* gene confers resistance to early stages of infection by BCG and other intracellular pathogens, apparently by affecting the innate ability of the macrophages to inhibit or kill the pathogens. There is suggestive evidence for a similar gene determining disease susceptibility in man (71).

Attempts to find linkages between susceptibility to tuberculosis and the Class I HLA genes (HLA-A and -B) have been unsuccessful. Studies of Class II (HLA-D) genes have been more promising and have revealed that the HLA-DR2 gene appears to predispose to the development of tuberculosis, particularly radiologically advanced, smear-positive disease (72,73). The HLA-DR2 specificity can affect antigen recognition, as persons of this genotype have higher levels of antibody to epitopes on a 38-kDa protein unique to *M. tuberculosis* than those lacking this genotype (74).

Class II genes also affect reactivity to skin testing with tuberculin. Thus, skin testing with a range of mycobacterial sensitins revealed that persons lacking the HLA-DR3 gene tend to respond poorly to all sensitins, whereas those of HLA-DR4 phenotype respond relatively strongly to species-specific antigens of *M. tuberculosis* (75).

## 15. Mycobacterial Disease and Autoimmunity

Considerable interest has been expressed in the relation between infection by mycobacteria and autoimmune disease. It has been shown that Freund's complete adjuvant (a suspension of killed mycobacteria in oil) induces autoimmune arthritis in rats and that this arthritis is adoptively transferrable by T cells reacting to mycobacterial 65-kDa heat shock protein (76). Low levels of a range of autoantibodies are detectable in sera from leprosy and tuberculosis patients and a small minority of patients with tuberculosis develop an arthritis termed Poncet's disease. Also, patients with rheumatoid arthritis, in common with those with tuberculosis, have elevated levels of agalactosyl IgG (77). There is also an association between rheumatoid arthritis and the HLA-DR4 phenotype which, as mentioned earlier, is associated with strong dermal reactivity to species-specific antigens of *M. tuberculosis* (75).

Despite these findings, the evidence for overt autoimmune phenomena contributing to the pathogenesis of mycobacterial disease is limited and conflicting (28). Likewise, the ability of mycobacteria to act as triggers for autoimmune diseases such as rheumatoid arthritis remains unproven. Although, in theory, mycobacterial heat shock proteins could, by their cross-reactivity with their human counterparts, induce autoimmune phenomena (78), they are probably usually prevented from doing so by an elaborate regulatory system (79).

## 16. Pathogenesis of Disease Due to Environmental Mycobacteria (EM)

Such disease is divisible into four main groups: post-inoculation; localised lymphadenitis; pulmonary and disseminated.

### *16.1. Post-inoculation Disease*

Two main types are seen: superficial warty lesions resembling primary inoculation tuberculosis and deeper lesions following injections and more extensive accidental or surgical wounds (43). In addition, *M. ulcerans* causes a very specific post-inoculation disease and is described separately below.

The most frequent superficial mycobacterial disease is termed swimming pool granuloma, fish tank granuloma, or fish fancier's finger (80). It is caused by the waterborne species *M. marinum* and enters superficial cuts and abrasions, usually on knees and elbows of swimmers and the hands of tropical fish enthusiasts. The local lesion initially shows a mixed granulomatous and suppurative response (suppurative granulomatous dermatitis), but older lesions have a more distinct granulomatous appearance. Lymphatic spread often occurs and, if the primary lesion is on the hand, one or more proximal lesions can develop along the superficial lymphatics of the arm. This is termed sporotrichoid spread as a similar phenomenon occurs in the fungal infection sporotrichosis. Disease due to *M. marinum,* although chronic, is usually benign and self-limiting, but involvement of joints in patients with rheumatoid arthritis and disseminated disease in immunocompromised patients have been described. On rare occasions, other mycobacterial species, including *M. chelonae* and *M. kansasii,* cause similar lesions.

Post-injection abscesses and deep-wound infections are almost exclusively caused by the rapidly growing species *M. chelonae* and *M. fortuitum.* (*M. terrae* is a rare cause of infection of wounds contaminated by soil.) The abscesses show rather nonspecific inflammation and pyogenic tissue destruction. They appear between 1 and 12 months after the injection or injury; they may be as large as 3 in. in diameter and discharging sinuses often develop. Local and systemic spread is uncommon, although a spreading cellulitis sometimes occurs in insulin-dependent diabetics. More serious infections have followed surgical procedures, notably the insertion of prosthetic heart valves (81).

### *16.2. Lymphadenitis*

Typically, this occurs in otherwise healthy children under the age of 5 years with involvement of a single node in the neck. Other age groups and sites are occasionally involved. Histologically, the nodes show caseating granulomas similar to those seen in tuberculosis. In the absence of immune suppression, excision is usually curative. Various species have been isolated, but the most usual causes are *M. avium-intracellulare* and *M. scrofulaceum.* It seems likely that the lymphadenopathy seen in young children are analogous to the primary tuberculosis complex following tonsillar infection by *M. bovis,* as EM have been isolated from excised tonsils of children. Such disease increased in frequency in Sweden after the cessation of neonatal BCG vaccination (82). Lymph nodes may also be involved in disseminated mycobacterial disease (see below).

### *16.3. Pulmonary Disease*

This is caused by many of the species listed in Table 5.1, but the most frequent causes worldwide are *M. avium-intracellulare* and *M. kansasii.* Despite the fact that pulmonary disease is one of the commonest forms of disease due to EM, there have been very few studies on its pathogenesis and evolution. In many cases, there is an underlying localized or systemic predisposing factor, including old tuberculous cavities, industrial dust disease, rheumatoid lung, cancer, cystic fibrosis, and various congenital, acquired and iatrogenic causes of immunosuppression, including renal transplantation and HIV infection. In a significant minority of patients, however, there is no detectable predisposing cause. Care must be made in diagnosis on the basis of culture, as these mycobacteria can occur in sputum as a result of transitory contamination of the mouth and pharynx or of secondary saprophytic colonization of diseased lung tissue.

### *16.4. Disseminated Disease*

Before the early 1980s, disseminated disease due to environmental mycobacteria was very rare. Most cases occurred in children or young adults with various congenital defects in cell-mediated immunity and in recipients of renal transplants. In addition, there were rare reports of solitary non-respiratory lesions, usually in bones and the genitourinary tract, presumably due to haematogenous spread from a primary focus. As in the case of pulmonary disease, many different species were involved.

Since that time, the situation has changed dramatically owing to the advent of the HIV/AIDS pandemic. Environmental mycobacteria are important opportunist AIDS-related pathogens, although, for unknown reasons, such disease is mostly caused by members of the *M. avium-intracellulare* complex, particularly serotypes that belong to *M. avium* (83). In the United States and Holland, up to 50% of AIDS patients eventually develop disease due to *M. avium,* whereas in Sweden, only about 10% of patients develop such disease and it is relatively uncommon in Africa. In contrast to tuberculosis, such disease usually occurs late in the course of HIV infection, almost always after other AIDS-defining conditions have presented. In the United States, it has been estimated that AIDS patients with disease due to *M. avium* have, on average, 7 months to live (84). Treatment, although rarely curative, is indicated, as symptoms are alleviated and the quality of life is enhanced.

AIDS-related *M. avium* disease is often disseminated at the time of diagnosis and bacilli are present in large numbers in the blood, bone marrow, and various internal organs and are mostly within macrophages (85). Involvement of the intestinal wall is one of the causes of the chronic and debilitating diarrhea that often afflicts these patients and bacilli are often present in the faeces, even in the absence of intestinal symptoms.

The origin of the *M. avium* is uncertain. Some authors state that the bacilli are recent acquisitions from the environment; others postulate that the disease results from reactivation of dormant foci of infection in the gut-associated lymphoid tissue, possibly acquired in childhood. It has been postulated that neonatal BCG vaccination which, as discussed above, reduced the incidence of *M. avium* lymphadenopathy in children explains the low incidence of AIDS-related *M. avium* disease in Sweden, relative to the United States where such patients would not have been vaccinated neonatally (86). Arguments in favor of both sources of infection have been advanced and it is possible that, as in tuberculosis, both endogenous reactivation and exogenous infection occur (87).

## 17. Disease Due to *Mycobacterium ulcerans* (Buruli Ulcer)

This disease, which occurs in limited regions within the tropics, is quite unique among the mycobacterioses with respect to its clinical features and underlying immunopathology (88). The primary lesion is a firm nodule in the skin resulting from traumatic inoculation of the bacilli, usually by spiky vegetation. In some cases, this primary lesion resolves, but in others there is progression. *Mycobacterium ulcerans* is unique among pathogenic mycobacteria in that a major determinant of virulence is a toxin (89). This, in the progressive cases, causes widespread colliquitive necrosis of the subcutaneous fat. Secondary necrosis of the overlying skin results in deeply undermined ulcers (Fig. 5.6). In the progressive

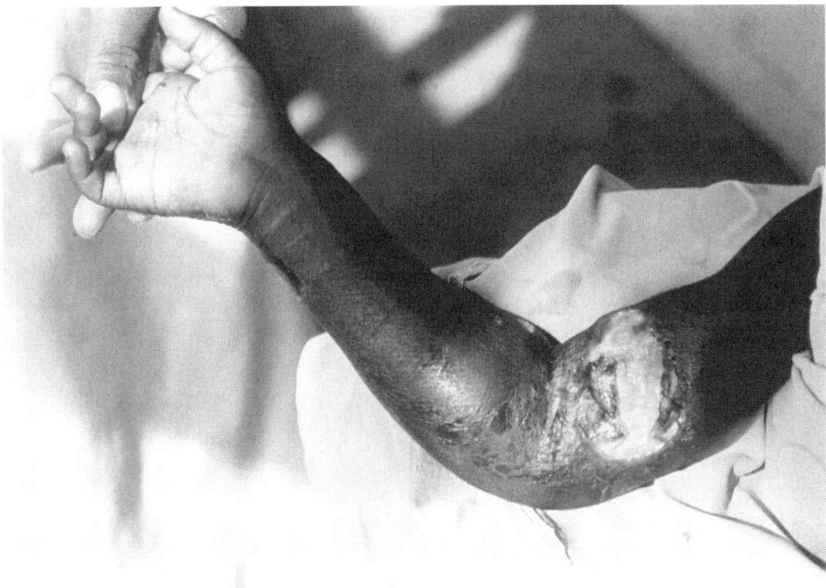

Figure 5.6. Buruli ulcer (*M. ulcerans* infection). Ulcerative lesion on the upper arm of a child. (Courtesy of Dr. Alan Knell, Wellcome Tropical Institute.)

stage of the disease, there is an immunological anergy to the bacillus, which is less specific than in leprosy, as patients fail to react to tuberculin as well as to *burulin,* a skin test reagent prepared from *M. ulcerans (89).* The lesions contain numerous bacilli and there is little or no evidence of granuloma formation. The reason for this anergy is unknown, although there is some evidence that T cells responding to mycobacterial antigen are trapped in lymph nodes.

The ulcer can reach an enormous size, sometimes involving a whole limb. Although *M. ulcerans* is susceptible to rifampicin and some other antimicrobial agents *in vitro,* chemotherapy has proved disappointing, and the only effective treatment in the anergic phase is radical surgery. If untreated, a stage is, however, reached when immune reactivity returns spontaneously, the patient becomes burulin and tuberculin positive, and there is a cellular infiltrate in the lesion resulting in granuloma formation and a disappearance of the bacilli. Healing then takes place, although, because of the size of the ulcers, the gross fibrosis can lead to extensive scarring and deformities.

The mechanism of the characteristic shift from anergy to immune reactivity in Buruli ulcer remains a mystery, but, if it was understood, it might pave the way to immunotherapeutic intervention at an early stage, before gross ulceration occurs, and it might have application to other mycobacterial diseases. A lesion can sometimes show anergy at one part and healing in another, suggesting that local factors, possibly cytokine release, affect the immune reactivity. It would be of interest to determine whether this shift was accompanied by change in cytokine patterns related to $T_H2$ and $T_H1$ T cells.

## 18. Conclusions

Although there are considerable differences between the clinical features and the underlying pathogenesis of the various mycobacterial diseases, some common trends are discernable. In most cases, immune responses to mycobacteria are very effective: in the absence of HIV infection or other predisposing factor, disease appears to be the result of inappropriate immune reactivity. This, in turn, is reflected in altered patterns of cytokine production and release, as well as factors affecting the responsiveness of tissues to these cytokines. An understanding of the variable responses to infection by mycobacteria has considerable implication for the understanding of pathogenesis as well as to improved tools for the prevention, diagnosis, and therapy of a group of diseases that probably cause more adult morbidity and mortality than all other infectious diseases together.

## References


1. Wallgren A (1948) The "time-table" of tuberculosis. Tubercle 29:245–251.
2. Rubilar M, Sime PJ, Moudgil H, Chilvers ER, Leitch AG (1994) Time to extend "the timetable of tuberculosis"? Respir Med 88:481–482.
3. Ridley DS, Jopling WW (1966) Classification of leprosy according to immunity: A five group system. Int J Leprosy 34:255–273.
4. Rook GAW, Al Attiyah R (1991) Cytokines and the Koch phenomenon. Tubercle 72:13–20.
5. Stanford JL (1991) Koch's phenomenon: Can it be corrected? Tubercle 72:241–249.
6. Goren MB, Grange JM, Aber VR, Allen BW, Mitchison DA (1982) Role of lipid content and hydrogen peroxide susceptibility in determining the guinea-pig virulence of *Mycobacterium tuberculosis.* Br J Exp Pathol 63:693–700.
7. Rook GAW (1991) Mobilising the appropriate T cell subset: The immune response as taxonomist? Tubercle 72:253–254.
8. Mackaness GB (1967) The immunology of antituberculous immunity. Am Rev Respir Dis 97:337–344.
9. Ottenhoff THM, Kale AB, van Embden JDA, Thole JER, Liessling R (1988) The recombinant 65 KD heat shock protein of *Mycobacterium bovis* BCG/*M. tuberculosis*

is a target molecule for CD4+ cytotoxic T lymphocytes that lyse human monocytes. J Exp Med 168:1947–1952.

10. Flynn JL, Goldstein MM, Triebold KJ, Koller B, Bloom BR (1992) Major histocompatibility complex class 1-restricted T cells are required for resistance to *M. tuberculosis* infection. Proc Natl Acad Sci USA 98:12013–12017.
11. Bothamley GH, Festenstein F, Newland A (1992) Protective role for CD8 cells in tuberculosis. Lancet 339:315–316.
12. Born W, Hall L, Dallas A, Boymel J, Shinnick T, Young D, Brennan P, O'Brien R (1990). Recognition of a peptide antigen by heat shock reactive gd T lymphocytes. Science 249:67–69.
13. Srivastava PK (1991) Stress-induced protein in immune response to cancer. Curr Top Microbiol Immunol 167:109–123.
14. Barnes PF, Grisso CL, Abrams JS, Band H, Rea TH, Modlin RL (1992) Gamma delta T lymphocytes in human tuberculosis. J Infect Dis 165:506–512.
15. Yoneda T, Mikami R, Sakaguchi Y, Shirai F (1987) The relationship between natural killer cell activity and delayed-type hypersensitivity reaction to 2,4-dinitrochlorobenzene in the spectrum of chronic intractable pulmonary tuberculosis. Tubercle 68:59–64.
16. Restrepo LM, Barrera LF, Garcia LF (1990) Natural killer cell activity in patients with pulmonary tuberculosis and in healthy controls. Tubercle 71:95–102.
17. Ratcliffe LT, MacKenzie CR, Lukey PT, Ress SR (1992) Reduced natural killer cell activity in multi-drug resistant pulmonary tuberculosis. Scand J Immunol 11(Suppl):167–170.
18. Rook GAW, Stanford JL (1979) The relevance to protection of three forms of delayed skin test response evoked by *M. leprae* and other mycobacteria in mice: Correlation with the classical work in the guinea-pig. Parasite Immunol 1:111–123.
19. Koch R (1891) Weitere Mitteilungem über ein Heilmittel gegen Tuberculose. Dtsch Med Wschr 17:101–102.
20. von Pirquet C (1907) Demonstration zur Tuberculindiagnose durch Hautimpfung. Berl Klin Wschr 48:699.
21. Bothamley GH, Grange JM (1991) The Koch phenomenon and delayed hypersensitivity: 1891–1991. Tubercle 72:7–11.
22. Lefford MJ (1975) Delayed hypersensitivity and immunity in tuberculosis. Am Rev Respir Dis 111:243–246.
23. Beck JS, Morley SM, Gibbs JH, Potts RC, Ilias MI, Kardjito T, Grange JM, Stanford JL, Brown RA (1986) The cellular responses of tuberculosis and leprosy patients and of healthy controls in skin tests to new tuberculin and healthy subjects. Clin Exp Immunol 64:484–494.
24. Mosmann TR (1991) Regulation of immune responses by T cells with different cytokine secretor phenotypes. Role of a new cytokine: cytokine synthesis inhibitory factor (Il-10). Int Arch Allergy Appl Immunol 94:110–115.

25. Bretscher PA (1992) A strategy to improve the efficacy of vaccination against tuberculosis and leprosy. Immunol Today 13:342–345.

26. Grange JM, Stanford JL, Rook GAW, Onyebujoh P, Bretscher PA (1994) Tuberculosis and HIV: light after darkness. Thorax 49:537–539.

27. Rook GAW, Onyebujoh P, Stanford JL (1993) THI → TH2 switch and loss of CD4 cells in chronic infections: An immuno-endocrinological hypothesis not exclusive to HIV. Immunol Today 16:568–569.

28. Rook GAW, Bloom BR (1994) Mechanisms of pathogenesis in tuberculosis. In Bloom BR, ed Tuberculosis: Pathogenesis, Protection, and Control, pp. 485–501. Washington DC: American Society for Microbiology.

29. Filley EA, Bull HA, Dowd PM, Rook GAW (1992) The effect of *Mycobacterium tuberculosis* on the susceptibility of human cells to the stimulatory and toxic effects of tumour necrosis factor. Immunology 77:505–509.

30. Al Attiyah R, Moreno C, Rook GAW (1992) TNF-alpha-mediated tissue damage in mouse footpads primed with mycobacterial preparations. Res Immunol 143:6011–610.

31. Hernandez-Pando R, Rook GAW (1994) The role of TNFα in T cell-mediated inflammation depends on the Th1/Th2 cytokine balance. Immunology 82:591–595.

32. Rook GAW, Onyebujoh P, Wilkins E, Ly HM, Al Attiyah R, Bahr G, Corrah T, Hernandez H, Stanford JL (1994) A longitudinal study of per cent agalactosyl IgG in tuberculosis patients receiving chemotherapy, with and without immunotherapy. Immunology 81:149–154.

33. Stanford JL, Nye PM, Rook GAW, Samuel NM, Fairbank A (1981) A preliminary investigation of the responsiveness or otherwise of patients and staff of a leprosy hospital to groups of shared or species specific antigens of mycobacteria. Leprosy Rev 52:321–327.

34. Kardjito T, Beck JS, Grange JM, Stanford JL (1986) A comparison of the responsiveness to four new tuberculins among Indonesian patients with pulmonary tuberculosis and healthy subjects. Eur J Respir Dis 69:142–145.

35. Khoo SH, Wilkins EGL, Fraser I, Hamour AA, Stanford JL (1996) Lack of skin test reactivity to common mycobacterial antigens in human immunodeficiency virus infected individuals with high CDA counts. Thorax 51:932–935.

36. Bottasso OA, Ingledew N, Keni M, Morini J, Pividori JF, Rook GAW, Stanford JL (1994) The cellular immune response to common mycobacterial antigens is suppressed in persons seropositive for *Trypanosoma cruzi*. Lancet 344:1540–1541.

37. Cohen IR, Young DB (1991) Autoimmunity, microbial immunity and the immunological homunculus. Immunol Today 12:105–110.

38. Appleton I (1994) Wound repair: The role of cytokines and vasoactive mediators. J R Soc Med 87:500–502.

39. Thiemermann C, Corder R (1992) Is endothelin-1 the regulator of myofibroblast contraction during wound healing? Lab Invest 67:677–679.

40. Potts RC, Beck JS, Gibbs JH, Grange JM, Kardjito T, Stanford JL (1992) Measurements of blood flow and histometry of the cellular infiltrate in tuberculin skin

test responses of the Koch-type and Listeria-type in pulmonary tuberculosis patients and apparently healthy controls. Int J Exp Pathol 73:565–572.

41. Lurie MB (1964) Resistance to Tuberculosis: Experimental Studies in Native and Acquired Defensive Mechanisms. Cambridge, MA: Harvard University Press.
42. Dannenberg AM (1993) Immunopathogenesis of pulmonary tuberculosis. Hosp Pract 28:51–58.
43. Grange JM, Noble WC, Yates MD, Collins CH (1988) Inoculation mycobacterioses. Clin Exp Dermatol 13:211–220.
44. Savin JA, Wilkinson DS (1986) Mycobacterial infections including tuberculosis. In Rook A, Wilkinson DS, Ebling FJG, Champion RH, Burton JL, eds. Textbook of Dermatology, pp. 791–822. Oxford: Blackwell.
45. Jopling WH (1984) Handbook of Leprosy, 3rd ed. London: William Heinemann.
46. Mishra B, Mukherjee A, Girdhar A, Husain S, Malaviya GN, Girdhar BK (1993) Evolution of early lesions in leprosy. Leprosy Rev 64:259–266.
47. Grange JM (1992) The mystery of the mycobacterial persistor. Tubercle Lung Dis 73:249–251.
48. Stanford JL (1987) Much's granules revisited. Tubercle 68:241–242.
49. Ridley DS (1988) Pathogenesis of Leprosy and Related Diseases. London: Wright.
50. Stanford JL (1994) The history and future of vaccination and immunotherapy for leprosy. Trop Geogr Med 46:93–107.
51. Bloom BR, Salgame P, Diamond B (1992) Revisiting and revising suppressor T cells. Immunol Today 13:131–136.
52. Modlin RL, Hofman FM, Taylor CR, Rea TH (1983) T-lymphocyte subsets in the skin lesions of patients with leprosy. J Am Acad Dermatol 8:182–189.
53. Yamamura M, Uyemura K, Deans RJ, Weinberg K, Rea TH, Bloom BR, Modlin RL (1991) Defining protective responses to pathogens: Cytokine profiles in leprosy patients. Science 254:277–279.
54. Filley E, Andreoli A, Steele J, et al. (1989) A transient rise in agalactosyl IgG correlating with free interleukin-2 receptors during episodes of erythema nodosum leprosum. Clin Exp Immunol 76:343–347.
55. Eastwood JB, Dilly SA, Grange JM (1995) Tuberculosis, leprosy and other mycobacterial diseases. In: Cattell WR, ed. Infections of the Kidney and Urinary Tract. Oxford: Oxford University Press.
56. Samuel NM, Grange JM, Samuel S, Lucas S, Owilli OM, Adalla S, Leigh IM, Navarrette C (1987) A study of the effects of intradermal administration of recombinant gamma interferon in lepromatous leprosy patients. Leprosy Rev 58:389–400.
57. Dwyer B, Jackson K, Raios K, Sievers A, Wilshire E, Ross B (1993) DNA restriction fragment analysis to define an extended cluster of tuberculosis in homeless men and their associates. J Infect Dis 167:490–494.
58. Styblo K (1980) Recent advances in epidemiological research in tuberculosis. Adv Tuberc Res 20:1–63.

59. Grange JM, Yates MD (1994) Zoonotic aspects of *Mycobacterium bovis* infection. Vetin Microbiol 40:137–151.

60. Rouillon A, Perdrizet S, Parrot R (1976) Transmission of tubercle bacilli: The effects of chemotherapy. Tubercle 57:275–299.

61. Ridley DS, Ridley MJ (1987) Rationale for the histological spectrum of tuberculosis. A basis for classification. Pathology 19:186–192.

62. Ainslie GM, Solomon JA, Bateman ED (1992) Lymphocyte and lymphocyte subset numbers in blood and in bronchoalveolar lavage and pleural fluid in various forms of human pulmonary tuberculosis at presentation and during recovery. Thorax 47:513–518.

63. Grange JM (1989) Tuberculosis and environmental (atypical) mycobacterioses: Bacterial, pathological and immunological aspects. In Harahap M, ed. Mycobacterial Skin Diseases, pp. 1–32. Dordrecht: Kluwer Academic Publishers.

64. Festenstein F, Grange JM (1991) Tuberculosis and the acquired immune deficiency syndrome. J Appl Bacteriol 71:19–30.

65. Proudfoot AT (1971) Cryptic disseminated tuberculosis. Br J Hosp Med 5:773–780.

66. Osborn L, Kunkel S, Nabel GJ (1989) Tumour necrosis factor and interleukin 1 stimulate the human immunodeficiency virus enhancer by activation of the nuclear factor kappa B. Proc Natl Acad Sci USA 86:2236–2240.

67. Beck JS, Potts RC, Kardjito T, Grange JM (1985) T4 lymphopenia in patients with active pulmonary tuberculosis. Clin Exp Immunol 60:49–54.

68. Stanford JL, Stanford CA, Rook GAW, Carswell JW, Grange JM, Pozniak A (1993) Chemoprophylaxis for tuberculosis. Respir Med 87:398–399.

69. Lucas SB (1990) Missing infections in AIDS. Trans R Soc Trop Med Hyg 84(Suppl 1):34–38.

70. Lucas SB (1993) Human immunodeficiency virus and leprosy. Leprosy Rev 64:97–103.

71. Schurr E, Malo D, Radzioch D, Buschman E, Morgan K, Gros P, Skamene E (1991) Genetic control of innate resistance to mycobacterial infections. Immunol Today 12:A42–A45.

72. Brahmajothi V, Pitchappan RM, Kakkanaiah VN, et al. (1991) Association of pulmonary tuberculosis and HLA in South India. Tubercle 72:123–132.

73. Khomenko AG, Litvinov VI, Chukanova VP, Pospelov LE (1990) Tuberculosis in patients with various HLA phenotypes. Tubercle 71:187–192.

74. Bothamley GH, Beck JS, Geziena M, et al. (1989) Association of tuberculosis and *M. tuberculosis*-specific antibody levels with HLA. J Infect Dis 159:549–555.

75. Ottenhoff THM, Torres P, de las Aguas JT, et al. Evidence for an HLA-DR4-associated immune response gene for *Mycobacterium tuberculosis*. Lancet 2:310–312.

76. van Eden W, Thole JER, van der Zee R, Noordzij A, van Embden JDA, Yamamura EJ, Hensen EJ, Cohen IR (1988) Cloning of the mycobacterial epitope recognised by T lymphocytes in adjuvant arthritis. Nature 331:171–173.

77. Rook GAW (1988) Rheumatoid arthritis, mycobacterial antigens and agalactosyl IgG. Scand J Immunol 28:487–493.
78. Das PK, Grange JM (1993) Mycobacteria in relation to tissue immune response and pathogenesis. Rev Med Microbiol 4:15–23.
79. Cohen IR, Young DB (1991) Autoimmunity, microbial immunity and the immunological homunculus. Immunol Today 12:105–110.
80. Collins CH, Grange JM, Noble WC, Yates MD (1985) *Mycobacterium marinum* infections in man. J Hyg 94:135–149.
81. Grange JM (1992) Mycobacterial infections following heart valve replacement. J Heart Valve Dis 1:102–109.
82. Romanus V (1988) Swedish experiences 12 years after the cessation of general BCG vaccination of new borns in 1975. Bull Int Tuberc Lung Dis 63(4):34–38.
83. Pitchenik AE, Fertel D (1992) Medical management of AIDS patients: Tuberculosis and nontuberculous mycobacterial disease. Med Clin North Am 76:121–171.
84. Horsburg CR, Selik RM (1989) The epidemiology of disseminated mycobacterial infection in the acquired immunodeficiency syndrome (AIDS). Am Rev Respir Dis 139:4–7.
85. Kiehn TE, Edwards FF, Brennon P, et al. Infections caused by *Mycobacterium avium* complex in immunocompromised patients: Diagnosis by blood culture and fecal examination, antimicrobial susceptibility tests, and morphological and seroagglutination characteristics. J Clin Microbiol 21:168–173.
86. Kallenius G, Hoffner SE, Svenson SB (1989) Does vaccination with Bacille–Calmette–Guérin protect against AIDS? Rev Infect Dis 11:349–351.
87. Grange JM (1994) Is the incidence of AIDS-associated *Mycobacterium avium-intracellulare* disease affected by previous exposure to BCG, *Mycobacterium tuberculosis* or environmental mycobacteria? Tubercle Lung Dis 75:234–236.
88. van der Werf TS, van der Graaf, Groothuis DG, Knell AJ (1989) *Mycobacterium ulcerans* infection in Ashanti region, Ghana. Trans R Soc Trop Med Hyg 83:410–413.
89. Hockmeyer WT, Krieg RE, Reich M, Johnson RD (1978) Further characterization of *Mycobacterium ulcerans* toxin. Infect Immun 21:124–128.

# 6

# Transmission of Mycobacteria

*Joseph O. Falkinham III*

## 1. Introduction

Members of the bacterial genus *Mycobacterium,* including *Mycobacterium tuberculosis, Mycobacterium bovis, Mycobacterium kansasii, Mycobacterium marinum, Mycobacterium ulcerans, Mycobacterium xenopi, Mycobacterium paratuberculosis,* and *Mycobacterium avium* demonstrate a wide variety of transmission modes. Rapidly growing mycobacteria, including *Mycobacterium fortuitum, Mycobacterium chelonae,* and *Mycobacterium abscessus,* have also been shown to infect patients by a number of different routes. These include transmission via (1) aerosols, (2) water, (3) soil, (4) dust, (5) food products, and (6) equipment. For some *Mycobacterium* species, transmission via a variety of modes (e.g., *M. avium* via aerosols and water) has been demonstrated.

Understanding these different modes of mycobacterial transmission is important in this era of emerging drug-resistant mycobacterial strains. Both multidrug-resistant *M. tuberculosis* and the characteristically drug-resistant nontuberculous mycobacteria (i.e., *M. avium*) have challenged our ability to treat mycobacterial infections. In the absence of effective antibiotic therapy, reduction of the morbidity and mortality of infections due to mycobacteria can be achieved by prevention of transmission. This is especially important in situations where there is a high proportion of immunocompromized individuals who are particularly susceptible to mycobacterial infections (1–3).

Prevention of transmission of *M. tuberculosis* is a particularly important problem, in light of the virulence of this pathogen. Transmission of tuberculosis has been reported in situations where persons are exposed within confined spaces, like submarines (4), schools (5), and aircraft (6). Nosocomial tuberculosis is due, in part, to the presence of undiagnosed patients (7,8). This is a particular concern because of the reported delay in diagnosis of pulmonary tuberculosis in hospitals due to infrequent use of tuberculin skin tests, misinterpretation of unusual chest roentgenograms, and waiting for culture results in patients with negative acid-fast smears (9). Failures by patients to comply with treatment regimens (10) and the fact that treatment does not immediately halt transmission (11) also contribute to the spread of tuberculosis. Consequently, measures to reduce or eliminate *M. tuberculosis* transmission will have a direct effect on reducing the chance of infection in other individuals.

An understanding of mycobacterial transmission modes will contribute to our developing understanding of the epidemiology of the rise in tuberculosis and other mycobacterial diseases in the United States. For example, two reports have established that between 33% and 40% of persons with active tuberculosis were recently infected and that tuberculosis was not a consequence of reactivation of a previous infection (12,13).

The initial section of this chapter is a review of the sources of members of the genus *Mycobacterium.* Its purpose is to illustrate the great variety of mycobacterial habitats and, hence, potential sources of transmission and infection. Following that section, mycobacterial transmission will be presented from two different perspectives. First, different modes of transmission employed by the mycobacteria will be described. Second, modes of transmission demonstrated by individual species will follow. The first is organized by mode of transmission, the second by mycobacterial species. One objective is to identify shared and unique features of mycobacterial transmission. Further, by approaching the subject from two different perspectives, it is hoped that the material will be accessible to readers of different interests and backgrounds. Following those two sections, experimental studies of mycobacterial transmission are reviewed to provide an understanding of methods for detecting routes of transmission, of the mechanism of mycobacterial transmission, and of the microbial, environmental, and host factors influencing mycobacterial transmission. Based on that background, a discussion of methods for the surveillance of mycobacterial transmission and preventing mycobacterial transmission concludes the chapter. Throughout, there is an emphasis on the application of new methods to the detection and study of mycobacterial transmission.

## 2. Sources of Mycobacteria

### *2.1. Introduction*

The first factor to be considered when studying the transmission of mycobacteria is their source. Some mycobacteria are restricted to a narrow range of sources (i.e., infected humans and their *M. tuberculosis*-containing aerosols). By contrast, some mycobacterial species are found in a wide variety of habitats (i.e., *M. avium* in water, soil, and natural aerosols).

### 2.2. Infected Patients

Infected patients have been implicated in transmission of some mycobacteria (i.e., *M. tuberculosis*) (14–17). However, many mycobacterial species are not transmitted between patients (e.g., *M. avium*) (17).

The major source of *M. tuberculosis* is infected humans. The route of transmission is via aerosols generated by the infected patient coughing or sneezing. Consequently, aerosol infection is influenced by the patient's mycobacterial load and the frequency at which mycobacteria are aerosolized (14–16). Institutions where airborne infection has been demonstrated include hospitals, clinics, drug-treatment facilities, laboratories, morgues, shelters, jails, prisons, and long-term residential facilities—all places where *M. tuberculosis*-infected individuals can be found (11).

### 2.3. Infected Animals

Like *M. tuberculosis*-infected patients, animals (e.g., fish and cattle) infected with mycobacteria can be sources of infection for transmission to other animals as well as humans. Cattle infected with *M. bovis* can infect humans through infected milk or by generation of *M. bovis*-containing aerosols (18,19). An *M. bovis*-infected rhinoceros was proposed as the source of infection in zookeepers (20). *M. bovis* infection in wild badgers (i.e., *Meles meles*) has been proposed as the source of *M. bovis* infection in cattle in Great Britain (21–24).

### 2.4. Mycobacteria in Water

Many *Mycobacterium* species have been isolated from water. Because they have been repeatedly isolated from different sites by different investigators, it is likely that water is a normal habitat for these mycobacteria. Proof that *M. avium* in water is the source of infection in AIDS patients was recently obtained employing large restriction fragment fingerprinting of isolates from patients and their environment (25).

*Mycobacterium kansasii* (26,27), *M. marinum* (28), *M. xenopi* (27), *M. scrofulaceum* and members of the *M. avium* complex (i.e., *M. avium* and *M. intracellulare*) (29,30), and the rapidly growing mycobacteria *M. fortuitum, M. chelonae* and *M. abscessus,* have all been recovered from natural and municipal waters. Rapidly growing mycobacteria, particularly *M. fortuitum* and *M. abscessus,* have been recovered from water supplies and water solutions used in patient treatment (31–36). In particular, *M. abscessus* was recovered from a Gentian Violet solution used for skin marking (32).

### 2.5. *Mycobacteria in Aerosols*

There are at least three sources of mycobacteria in aerosols. Mycobacteria can be aerosolized by infected patients or animals coughing or sneezing (15,37). Mycobacteria in water can be aerosolized by natural and artificial mechanisms (38). Soil-associated mycobacteria can become part of the airborne microbial flora by agitation and wind as dust particles (see below).

The number of mycobacteria in aerosols, irrespective of the source, is directly influenced by the numbers in that primary source. Thus, mycobacterial numbers in aerosols generated from bodies of water are influenced by those conditions which influence their number in water (e.g., pH, temperature, and organic matter) (39,40). In addition, the numbers of aerosolized mycobacteria are influenced by ultraviolet light (41) and other factors such as humidity and visible light. Mycobacterial numbers in aerosols are expected to fall the longer mycobacteria are resident in aerosols or the farther they are from the original source (15).

In many studies, the presence of mycobacteria in aerosols has been inferred from their presence in a source (e.g., infected animal) and appearance in another, initially uninfected animal. In such a manner, Riley et al. (14) demonstrated aerosol transmission of *M. tuberculosis.* Because equipment for aerosol collection is available and most pathogenic mycobacteria can be cultivated on agar-based or liquid medium, it is possible to isolate and enumerate airborne mycobacteria. Employing the Andersen, six-stage viable sampler, Wendt et al. (42) was able to demonstrate the presence of members of the *M. avium* complex in aerosols. To date, there have been no other reports of direct isolation and enumeration of mycobacteria in aerosols.

### 2.6. *Mycobacteria in Soils*

There have been a number of systematic searches for mycobacteria in soils and a number of *Mycobacterium* species have been shown to be normal inhabitants of soil. These include *M. kansasii* (43), *M. malmoense* (44), *M. avium, M. intracellulare, M. scrofulaceum* (39,40,43–47), and rapidly growing mycobacteria (e.g., *M. smegmatis* and *M. fortuitum*) (43,45,48).

### 2.7. *Dust*

Dust particles can be of two types; tiny soil particles picked up and carried by the wind and nonsoil particulate matter (e.g., vegetable matter). The particles can be found in the air or on surfaces.

In studies carried out in Japan, it was shown that dust collected from the patients' floors had mycobacteria (49,50). Most other studies have inferred that mycobacteria were present on dust particles, based on isolation of the organisms in an environmental source and infected animal. The presence of *M. avium* of the same serotype in tuberculous lesions in pigs and in sputa of piggery workers led Reznikov and Robinson (51) to postulate that dusts generated in the piggery from *M. avium*-contaminated soil were the source of transmission for both pigs and man.

Although it is likely that the original source of mycobacteria in dust particles is soil, small dust particles can remain in the air and serve as a source for mycobacterial infection. Because of the cost of equipment and difficulties of collection, there have been only a few studies of mycobacteria in airborne dust particles. Using the Andersen cascade sampler (52), Falkinham et al. (53) were able to demonstrate that members of the *M. avium* complex were present in aerosols. Because the collection sites were far from water-droplet-generating bodies of water, the mycobacteria were presumed to be associated with dust particles (53).

### 2.8. Mycobacteria Equipment

Insufficient sterilization of endoscopes used during bronchoscopy can lead to transmission of infection to patients. Endoscope-mediated transmission of *M. tuberculosis* (54), *M. intracellulare* (55), and rapidly growing mycobacteria (56–61) have been demonstrated in patients. In a review of gastrointestinal endoscopy and bronchoscopy, it was shown that mycobacteria were common agents of infection transmission by endoscopy (58–61). This is undoubtedly due to the relative resistance of mycobacteria to disinfectants (58,59). Skin abscesses at the site of injections have been shown due to contaminated jet injectors (62–65). Mycobacteria can be found in humidifiers and nebulizers, like *Legionella pneumophila* (66), which could serve as sources for aerosol transmission. Mycobacterial infections found in patients undergoing dialysis (67,68) could be due to the presence of mycobacteria in water and their persistence in equipment. Contaminated prosthetic heart valves have also been implicated as sources of *M. chelonae* and *M. fortuitum* infection (69–71).

## 3. Mycobacterial Transmission Modes

### 3.1. General Considerations

A variety of routes have been proposed for transmission for mycobacteria. These include aerosol, water, soil, dust, food, and equipment. Care must be taken in interpretation, because in many instances, the only evidence for a transmission mode has been isolation of the same *Mycobacterium* species from a source and patient. Rarely have mycobacteria been isolated from a putative route (e.g., aerosols). Unambiguous proof of a particular transmission route requires that a *Mycobacterium* species of a particular type be isolated from the source, route, and target. Thus, studies of mycobacterial transmission require the capacity to fingerprint each isolate. Fingerprinting techniques which offer sufficient discrimination for use in epidemiological studies of mycobacteria include insertion sequence typing [e.g., IS6110 of *M. tuberculosis* (1) and IS1245 typing in *M. avium* (72–74)] and large restriction fragment analysis [e.g., *M. avium* (25,75,76)].

### *3.2. Transmission Modes*

#### AEROSOL TRANSMISSION

When considering aerosol transmission, two elements need consideration: aerosolization and aerosol transmission. Aerosolization is the process which entrains the mycobacterial cells into aerosols. Aerosolization occurs at the source of the mycobacteria and includes patients and bodies of water, soils, and dusts. It is the action at the source which generates an aerosol. Examples include a patient coughing (11,37), air bubbles bursting at the surface of water, resulting in the ejection of mycobacteria-laden droplets into the air (42), droplets generated by showers (77), and dust formation over soil generated by winds.

Once the mycobacterial cells are aerosolized, their ability to infect a patient is dependent on a number of factors. The residence time of a mycobacteria-laden droplet is dependent on the droplet size and atmospheric humidity and air movement (15). The longer the droplet remains in air, the longer the mycobacterial cells are exposed to the deleterious effects of visible and near and far ultraviolet light (UV). In addition to UV killing, mycobacteria are subject to killing by hydrogen peroxide, which is formed in organic matter-rich droplets by sunlight (78).

If the droplet is inhaled by an individual, droplet size will determine its site of deposition in the lung (15). Droplets and particles of less than 2 $\mu$m in diameter can impact on the alveoli surface (15). Although droplets of large size are generated by humans coughing or sneezing (11,15,37) or bubbles bursting at the surface of natural waters (38,42), a smaller distribution of droplet sizes is observed over time, suggesting that they evaporate (15). Thus, aerosolization of large droplets does not necessarily produce a noninfective aerosol.

#### WATER TRANSMISSION

Lack of evidence of person-to-person transmission of *M. avium* and the widespread presence of *M. avium* in natural and municipal water samples led to the hypothesis that the source of human infection of *M. avium* was water (17). Evidence for a waterborne source of *M. avium* infection in AIDS patients resulted from a comparison of isolates recovered from patients and their environment (i.e., drinking water). *M. avium* isolates from waters and patients were fingerprinted using large restriction fragment analysis (25). *M. avium* isolates from patients and their environment (e.g., drinking water) with identical fingerprints were identified (25).

Lymphadenitis in children has been historically caused by *Mycobacterium scrofulaceum* in water (17). Evidence that the distribution of serotypes of *M. scrofulaceum* isolates recovered from children was similar to the distribution of serotypes of *M. scrofulaceum* found in natural waters (79), reinforced that hypothesis. Interestingly, *M. avium* is more commonly found today as the agent of lymphadenitis (80). Further, it has been reported that there has been a fall in *M. intracellulare* and a rise in *M. avium* prevalence in Japan (81), and the recent absence of *M. scrofulaceum* in natural waters of the southeastern United States from which they have been historically recovered in high numbers (30) suggests a change in the prevalence and geographic distribution of *M. avium, M. intracellulare,* and *M. scrofulaceum.*

Strong evidence of waterborne transmission for other mycobacterial species employing DNA fingerprinting has not been published to date. However, a wide variety of mycobacteria have been recovered from water (82). These include *M. kansasii* (26,27), *M. marinum* (28), *M. xenopi* (27), and the rapidly growing mycobacteria, *M. fortuitum* and *M. chelonae* (31–36). In a number of cases (i.e., *M. kansasii, M. marinum, M. xenopi*) recovery of the mycobacterial species from patients was associated, in time, with recovery from water (26,27).

## SOIL TRANSMISSION

A number of mycobacterial species (e.g., *M. avium* and *M. intracellulare*) have been recovered from soils and sediments. However, there have been no studies in which fingerprints of patient isolates of mycobacteria have been compared to those of soil isolates, which would provide evidence that human infection was due to mycobacteria in soil. Such studies are also confounded by the fact that a possible identity of fingerprints of, for example, soil and animal isolates might be due to fecal contamination of soil by an infected animal, not soil-borne infection. That confounding factor was recognized in a study of soil and pig isolates of *M. avium,* where the authors found that both soil and pig isolates were of serotype 6 (46).

## DUST TRANSMISSION

Because of the presence of a number of mycobacterial species in soils and sediments (e.g., *M. kansasii, M. avium,* and *M. intracellulare*), it is logical that dust can be a source of mycobacterial infection. *M. avium, M. intracellulare,* and *M. scrofulaceum* have been isolated from aerosols containing dust particles

(39,40) and *M. avium* has been isolated from dusts collected from the floors of patients' rooms (49,50). Reznikov and Dawson (47) showed that the serotype of *M. avium* soil isolates was the same as the serotype of isolates recovered from patients. However, it was not known whether the aerosolized mycobacteria originated from soil or water.

*Mycobacterium avium* strains of the same serotype have been recovered from pigs, sputa of healthy piggery workers, and from the litter used as bedding material (51). That data led the authors to hypothesize that dust, which was generated in large amounts in the piggery, was the means of transmission (51). Thus, it is possible that dust is a source of mycobacterial infection.

Although *M. tuberculosis* is not thought to be an environmental *Mycobacterium,* except in cases of aerosol spread, dust collected from either a men's tuberculosis ward or the autopsy room was shown capable of transmission of fatal tuberculosis to guinea pigs (83).

## FOOD-BORNE TRANSMISSION

The classic case of food-borne transmission of mycobacteria is *Mycobacterium bovis* infection. Persons who drink milk from dairy cattle infected with *M. bovis* are at risk for tuberculosis. Recognition of this route of infection led to the institution of pasteurization of milk. Not only *M. bovis,* but other mycobacteria, including *M. intracellulare, M. scrofulaceum,* and *M. fortuitum* have been isolated from raw milk (84).

*Mycobacterium marinum* has been isolated from striped bass raised in an intensive culture system as a food source (85). Not only *M. marinum* but other *Mycobacterium* sp. (e.g., *M. simiae* and *M. scrofulaceum*) have been recovered from fish as well (86). The increased emphasis on intensive culture techniques (i.e., high populations in small volumes of water) for raising fish for food suggests that the incidence of food-borne mycobacterial infections can rise, unless the fish are monitored for evidence of mycobacterial disease, as are pigs.

## ANIMAL TRANSMISSION

Animals have been implicated as the source of mycobacterial infections of both other animals and humans. Although a variety of mycobacteria have been recovered from animals of all types (e.g., badgers, frogs, birds, and fish), most are nonpathogenic. However, *M. bovis* in cattle or other animals, *M. marinum* in fish and fish tanks, and *M. avium* in wild birds have been suggested as the source of human or other animal infection.

Although the prevalence of *M. bovis* infection in the United States is low, it can be a problem in those regions of the United States which have a high frequency of immigration of persons from areas where *M. bovis* infection in cattle is higher and milk is not routinely pasteurized. Evidence of this origin for *M. bovis* infection was presented in a study of tuberculosis patients in San Diego, California (19).

In England, studies have implicated the badger (*Meles meles*) as the source of bovine tuberculosis in cattle (21–24). Behavioral studies of badgers and cattle suggested that the route of infection of cattle was through the cattle eating grass contaminated with badger urine (21–24). Studies of feral pigs (87) and feral buffalo (88) have also demonstrated the presence of *M. bovis* and have thus implicated these wild animals as sources for transmission of *M. bovis* to humans.

*Mycobacterium avium* infection in pigs has been thought to be due to transmission by *M. avium*-infected birds [e.g., starlings (89)]. Although not a proven hypothesis, nonetheless there is the widespread employment of methods to prevent entry of birds in farms where either pigs or poultry are raised (89).

In a study carried out in an English wildfowl reserve, strains of *M. avium* recovered from ducks, geese, and swans were of a different serotype than those recovered from chickens (90). Based on this, Schaefer et al. felt that the infections had independent epidemiology (90). A similar study showed that *M. avium* strains isolated from domestic and wild animals also had different distributions of serotypes (91).

Although most of the literature concerning the epidemiology of *M. marinum* infection in man focuses on the presence of the organism in water, their presence, especially in aquaria water, might be due to infection in fish. Further, *M. marinum* infection of fish is an important cause of morbidity, mortality, and economic loss (85,92). The prevalence of mycobacterial infections in fish could be as high as 15% (86,92). Some of those mycobacteria infecting fish [*M. marinum, M. simiae,* and *M. scrofulaceum* (86)] can also infect man and other animals. Thus, mycobacterial infection in fish could be transmitted to humans or other animals via contact with water, aerosolization, or consumption of the fish as food.

## EQUIPMENT TRANSMISSION

Equipment has been implicated in transmission of mycobacteria between patients. Endoscopes used in bronchoscopy have been reported as the agent of *M. tuberculosis* transmission between patients (54,58). Rapidly growing mycobacteria have also been recovered from endoscopes (56–61). Dawson et al. (55) described the isolation of *M. intracellulare* from bronchoscopy specimens collected from three different patients within a 1-week period at the same medical facility. All three specimens yielded *M. intracellulare* of the same serotype (55). A review of the patient and laboratory records led to the conclusion that the second and third patient samples were contaminated with the *M. intracellulare* strain from the first patient. The authors further demonstrated the efficacy of 2% glutaraldehyde for sterilization of bronchoscopes (55). The widespread nature of this problem of mycobacterial transmission is evidenced by publication of a study comparing the mycobactericidal activities of germicides in disinfecting endoscopes (93).

Instruments used for injection of drugs or vaccines have also been associated with skin abscesses caused by rapidly growing mycobacteria (62–65).

Humidifiers and nebulizers can be sources for transmission of mycobacteria. Mycobacteria in the water contained within such equipment can be aerosolized, as has been described for *Legionella pneumophila* (66).

## 4. Selected Cases of Transmission of Mycobacteria

### *4.1 Aerosol Transmission of Mycobacterium tuberculosis*

Typically, transmission of *M. tuberculosis* occurs via the expiration of droplet nuclei from *M. tuberculosis*-infected patients. Nosocomial transmission can occur in hospitals or other medical-treatment facilities if patients are not recognized as having tuberculosis. In a case described by Catanzaro (16), a patient thought to have aspirated was subjected to fiber-optic bronchoscopy. Following bronchoscopy, the patient's breathing was assisted and expired air was ventilated into the room; the patient also received chest physical therapy (16). Following recognition that the patient was infected with *M. tuberculosis,* employees who were exposed to the patient and were tuberculin-negative were given a tuberculin-skin test (16). Fourteen (31%) of 45 exposed employees became tuberculin positive (16). The major risk factor for seroconversion was presence during the bronchoscopy. Ten of the 13 employees (77%) present during the bronchoscopy became tuberculin positive (16). The data permitted calculation of the number of *M. tuberculosis* infectious units present in the bronchoscopy site (i.e., 1 per 68.9 cubic feet) and the rate of production (i.e., 249 infectious units per hour) (16). The magnitude of these numbers [i.e., compare to those reported by Riley et al. (14)] is strong support for suggesting that alternatives to bronchoscopy should be considered and that rooms where bronchoscopy is performed should meet, or exceed, current ventilation standards (37).

Aerosol transmission of *M. tuberculosis* is a problem encountered in homeless shelters (94,95). A study conducted in a New York City homeless shelter reported the prevalence of tuberculosis infection (i.e., tuberculin skin test positive) was 79% (95). Individuals who had stayed longer in the shelter were more likely to be tuberculin positive (95). Tuberculosis transmission was not unexpected, because the shelter housed some 660 men whose beds were 18–20 in. apart (95).

Other examples of aerosol transmission have been reported in submarines (4), schools (5), and airplanes (6). A common element linking those outbreaks was that a *M. tuberculosis*-infected individual was in a closed area where other persons were exposed to *M. tuberculosis*-containing aerosols (4–6). An uninfected individual's proximity to the *M. tuberculosis*-infected person and length of exposure to the aerosol were associated with infection (4–6).

### 4.2 Nosocomial Transmission of M. tuberculosis from an Abscess

Transmission of *M. tuberculosis* is not always associated with the generation of respiratory aerosols from an individual with pulmonary tuberculosis. *M. tuberculosis*-containing aerosols can be generated in other ways. Specifically, an outbreak of tuberculosis involving hospital employees was traced to exposure of aerosols generated from a draining abscess (96). In this outbreak, 9 hospital employees developed tuberculosis and 59 became seropositive following exposure to a patient who had a large abscess of the hip and thigh which yielded *M. tuberculosis* (96). Aerosolization of *M. tuberculosis* from the wound was most likely due to the patient's agitation of the dressings which were soaked (96). Further, the wound was irrigated using a Water Pik oral hygiene appliance (96). Both activities would result in substantial aerosolization of *M. tuberculosis*-containing droplets. Those droplets could be the size able to enter human alveoli or could dry to that size (15). In fact, the investigators reported that 4% of the volume of the irrigating solution was aerosolized, with droplets in the 1–200-$\mu$m range (96). Finally, an important factor contributing to the dispersal of *M. tuberculosis* from the patient to persons in the hospital was that there existed a strong airflow from the patient's room into the corridor (96).

### 4.3 Airborne Transmission of Mycobacterium bovis

Recently, airborne *M. bovis* infection was reported. Seven of 23 (30%) zookeepers exposed to a rhinoceros infected with *M. bovis* demonstrated serological evidence of infection within 2 months (20). None came down with clinical illness. It was presumed that infection occurred via aerosols generated during the cleaning of the barn in which the rhinoceros was housed because six of the seven converting zookeepers was present during the cleaning operations (20). Human *M. bovis* infections have also been reported as an occupational hazard among abattoir workers in Australia (18).

### 4.4 Water Transmission of Mycobacterium marinum

Infections caused by *Mycobacterium marinum* are thought to be waterborne. Most commonly, human skin infections caused by *M. marinum,* are associated with previous injuries (e.g., cuts and scratches) and recovery of the organism from swimming pools and aquaria (28,82). *M. marinum* infections were also identified as being caused by exposure to the waters of the Chesapeake Bay (28). It appears that *M. marinum* skin infections are an occupational hazard of individuals exposed to *M. marinum*-containing natural or aquarium waters.

### *4.5 Water Transmission of Mycobacterium avium*

Lack of evidence of person-to-person transmission of *M. avium* (17) and the isolation of *M. avium* from natural (29,30) and municipal waters (30,97), including hospital water-supply systems (25,30,98) led to the hypothesis that one source of *M. avium* infection is water. Evidence that *M. avium* could be isolated from aerosols and droplets collected above natural waters which were of a size able to enter the alveoli (42) demonstrated the existence of a possible aerosol route of transmission. This route was logical because most infections in non-AIDS patients are pulmonary (17).

In AIDS patients, *M. avium* is most commonly disseminated (99,100) with infection being diagnosed by recovery of *M. avium* from blood (100). Evidence of either pulmonary or gastrointestinal *M. avium* infection correlates with disseminated infection (101,102). Although pulmonary infection can occur via an aerosol route, gastrointestinal infection or colonization can occur via the ingestion of water (101,102). Evidence for the existence of water as the source of *M. avium* infection in AIDS patients has recently been published (25). In the study, large restriction fragments (LRF) of *M. avium* isolates from AIDS patients and water in their environment were compared using pulsed-field gel electrophoresis. Three patients who resided in separate rural areas were infected with *M. avium* strains having identical LRF fingerprints (25). All three patients had been treated at a hospital at which an *M. avium* isolate with an identical fingerprint had been isolated repeatedly from water (25). Two patients, who had been treated at another hospital yet who shared no other common exposure, were infected with an *M. avium* strain whose fingerprints were identical to one another and an isolate from the hospital's water system (25). In light of the diversity of *M. avium* LRF types (75,76), identity of strains from the patients and their environment is strong evidence that water is at least one source of *M. avium* infection in AIDS patients.

### *4.6 Transmission of Mycobacterium ulcerans*

The geographic distribution of infection by *Mycobacterium ulcerans* resulting in skin ulceration (i.e., Buruli ulcer) has led to several hypotheses concerning its source and epidemiology. There has been no evidence of person-to-person transmission and most cases are characteristically found in persons living near large rivers (103). Evidence that persons drinking water from swamps had a higher incidence of *M. ulcerans* infection compared to persons who drank water from bore holes in the same area suggests that swamp waters are the source of *M. ulcerans* (103). Based on the observations that *M. ulcerans* disease occurs in tropical rain forests and the incidence is higher in years of high rainfalls, Hayman (104) postulated that heavy rains or disturbances led to the movement of *M. ulcerans* from its normal habitat into streams and rivers where persons can be exposed. Not only heavy rainfalls but also disturbances due to cultivation and erection of dams can lead to increased incidence of *M. ulcerans* disease (104).

### 4.7. Speculations on Transmission of Newly Emerging Mycobacterial Pathogens

#### MYCOBACTERIUM HAEMOPHILUM

In 1978, a new human mycobacterial pathogen, *Mycobacterium haemophilum* was described (105). Its name comes from its requirement for hemin (105). Since its discovery, *M. haemophilum* had been isolated from patients receiving immunosuppressive therapy following renal transplantation who had developed cutaneous lesions (105). The cutaneous nature of the infections correlated with the low temperature for optimal growth (i.e., 32°C) (105). Recently, another group of patients have been shown to be infected with *M. haemophilum.* (106,107) *M. haemophilum* has been isolated as the etiologic agent of disseminated cutaneous and deep tissue (e.g., lymphatics and lung) infections and bacteremia in AIDS patients (106,107).

The ability of *M. haemophilum* to grow over a wide range of pH values and the ability of ferric ammonium citrate to substitute for the hemin requirement (108) coupled with its low temperature optimum for growth (i.e., 32°C) suggests that its source is the environmental. Fortunately, the existence of polymorphisms in large restriction fragment patterns (109) offers an opportunity to test the hypothesis that the origin of *M. haemophilum* is environmental. An environmental source is likely, based on evidence that 12 of 16 *M. haemophilum* isolates, including 6—from a single hospital, shared the same LRF pattern (109). The inability of catalase to substitute for the hemin requirement (110) suggests that the iron requirement does not result in a catalase deficiency. The suggestion that the physiology of *M. haemophilum* resembles that of members of the *M. avium* complex (110) provides a guide for identifying the source of the organism. On that basis, logical sources for *M. haemophilum* include iron-rich waters whose temperatures do not rise appreciably. These could be in water-distribution systems (hot systems away from the source) in municipalities and hospitals. Systems which use iron pipes would be particularly attractive for sampling. It is to be expected that isolation of *M. haemophilum* from environmental samples will be difficult due to its unique requirements, slow growth, and hemin requirement.

#### MYCOBACTERIUM GENEVENSE

Another new mycobacterial pathogen is *Mycobacterium genevense.* This mycobacterial species has been identified as the etiological agent of disseminated disease in AIDS patients (111). Identification of *M. genevense* and its discrimination from other mycobacterial species has been based on 16S ribosomal (rRNA) sequences (111). The species is, to date, uncultivatable employing conventional techniques, although some limited consumption of $^{14}$C-palmitic acid results in the release of $^{14}CO_2$ (112). Recently, acid-fast microorganisms causing infections in pet birds were identified as *M. genevense* (113). That information suggests that the source of *M. genevense* infection in humans might be pet birds or their droppings. Quite possibly, the route of transmission via aerosolized fecal particles.

## 5. Experimental Study of Mycobacterial Transmission

### *5.1. Riley's Experiments on Airborne Tuberculosis Infection*

The "gold standard" of studies of airborne *M. tuberculosis* infection was carried out from 1950 to 1960 by Riley and his colleagues (14,41). Air from a special, isolated tuberculosis ward was exhausted into exposure chambers where guinea pigs were housed. Thus, rather than rely on recovery and enumeration of *M. tuberculosis* from aerosols, guinea pigs were used as the biological indicators. The guinea pigs were tuberculin tested to determine the incidence of infection and sacrificed to determine the number of discrete lung granuloma. The rationale of the latter measurement was that the number of discrete lung granuloma would be proportional to the number of infecting cells of *M. tuberculosis.* In one experiment, exhausted air from the tuberculosis ward was split and half was irradiated by a bank of ultraviolet lights. The incidence of positive tuberculin reactors and the number of lung granuloma were compared between guinea pigs exposed to irradiated and unirradiated exhausted air (14,41). Riley's studies demonstrated convincingly that *M. tuberculosis* could be transmitted by an airborne route. Further, the work also established that only one *M. tuberculosis* cell in 11,000 cubic feet of air was sufficient for infection. Finally, the studies proved that ultraviolet irradiation, under the conditions described by Riley et al. (14,41), was sufficient to prevent infection of guinea pigs by exhausted air.

### *5.2. Factors to Consider in Studying Routes of Mycobacterial Transmission*

The two studies reporting the results of investigations of the route of *M. avium* infection in AIDS patients (101,102) serve to illustrate one problem confounding studies of the route of mycobacterial transmission. The objective was to determine whether infection in the lungs or gastrointestinal tract was predictive of the disseminated *M. avium*-complex infection in AIDS patients. Isolation of *M. avium* from blood cultures was used as the indicator of dissemination. The problem lay in the fact that detection of the presence of *M. avium* in blood is highly sensitive because blood is normally sterile. In contrast, isolation of *M. avium* (or any *My-*

*cobacterium* sp. for that matter) from sputum or fecal material is less sensitive because of the need to decontaminate patient specimens (114). Consequently, it was impossible to detect low-level infection in sputum or feces (101,102) and provide a strong predictor of *M. avium* dissemination to the blood.

A second problem encountered in studies of mycobacterial transmission involves the inability to determine source when representatives of the same epidemiological type are isolated simultaneously from a putative environmental source and from an infected animal. This dilemma is best exemplified by those studies demonstrating isolation of mycobacteria from infected animals and soil in their environment. In most of the studies, it is not clear whether evidence of recovery of the same species from animals and their environment (i.e., soil) is due to infection of the animals by mycobacteria-containing soils or the animals are contaminating the soil (49,51,87,88).

Finally, no study of mycobacterial transmission should be initiated without two technical capabilities: mycobacterial isolation and typing. First, it should be possible to isolate the mycobacterial species from all possible sources, with the same level of sensitivity. If an aerosol route of transmission is hypothesized, aerosol samples must be collected and mycobacteria enumerated, isolated, and identified. Following isolation and identification, those and other mycobacteria of the same species must be typed by a method that permits sufficient discrimination. Depending on the species of *Mycobacterium,* rapid and discriminatory methods for typing strains have been identified and developed. Restriction fragment-length polymorphism of insertion sequences [e.g., IS6110 in *M. tuberculosis* (1) and IS1245 in *M. avium* (72–74)] have been used successfully for epidemiological studies of those species. Large restriction fragment analysis has been employed to show that water is a source of *M. avium* infection (25) to demonstrate polyclonal *M. avium* infection (75) and diversity of *M. avium* types infecting AIDS patients (75,76). Serotyping has also demonstrated polyclonal *M. avium* infection in AIDS patients (115) and DNA probes for mycobacterial species were used successfully to demonstrate simultaneous *M. avium* and *M. intracellulare* infection in an AIDS patient (116).

## 6. Factors Influencing Mycobacterial Transmission

### *6.1. Epidemiological Factors*

#### RISK FACTORS FOR OCCUPATIONAL INFECTION

It is well established that persons who work in nursing schools, sanatoria, and other medical facilities are at increased risk of *M. tuberculosis* infection (8). In 1928, when the prevalence of tuberculosis was much higher than today, Heimbach (117) showed that nurses in training were at risk for tuberculosis. Even today, physicians training in infectious disease are at increased risk for tuberculosis compared to physicians training in internal medicine (118). The variety of institutions where airborne *M. tuberculosis* infection has been demonstrated include hospitals, clinics, drug-treatment facilities, laboratories, morgues, homeless shelters, jails, prisons, and long-term residential facilities (11). Occupational or residential infection is even more likely because a substantial proportion of individuals in the institutions listed above are immunocompromised and thus at greater risk for infection (1–3). This means that it is likely that the total number of mycobacteria is likely to be higher today than before and that chances of infection are higher (11).

## RISK FACTORS FOR PATIENTS

Among nonimmunocompromised individuals, mycobacterial disease is most commonly pulmonary (17). Thus, it is not surprising that lung-weakening conditions (e.g., pneumoconiosis), lung abnormalities (e.g., pectus excavatum), and histories of chronic inhalation of dusts or other irritants (e.g., farmers lung and black lung) predispose for mycobacterial infections among the immunocompetent (17). This is especially important for infection caused by the weakly pathogenic mycobacteria, particularly *M. avium* and *M. intracellulare* (17).

Cystic fibrosis might be a risk factor for mycobacterial infection. Representatives of the *M. avium* complex were isolated from 7 of 64 cystic fibrosis patients (119). To date, it has not been established that the cystic fibrosis patients were truly infected or were simply colonized (119). However, the presence of members of the *M. avium* and other mycobacteria complex in the sputum of these patients might contribute to their morbidity.

Today, the most significant predisposing factor for mycobacterial infection is immunodeficiency. This can be a consequence of malignancy (120), immunosuppressive therapy following transplantation (121), or infection by the human immunodeficiency virus (1,3,100,121). In addition to increasing the susceptibility of individuals to mycobacterial infection, mycobacterial infections in human immunodeficiency–virus-infected patients are disseminated (1,3,100,121). Disseminated mycobacterial disease is seen most commonly in *M. avium*-infected patients (121,122) but is also quite common in *M. tuberculosis*-infected patients (123–125).

## EXPOSURE HISTORY

The dose-response relationship for numbers of aerosolized *M. tuberculosis* and the incidence of infection and disease in humans is well established (14). Not only did Riley and his colleagues prove that *M. tuberculosis* could be transmitted via an aerosol route, but they also established that only one *M. tuberculosis* cell in 11,000 cubic feet of air was sufficient for infection (14).

Dose-response relationships between numbers of nontuberculous mycobacteria and infection in humans and animals is unknown. This is particularly a problem in interpreting data reporting the prevalence and numbers of representatives of the *M. avium* complex in water, soil, or aerosol samples. Because methods have been developed for sensitive detection of *M. avium* in water [e.g., one per 25–250 ml water (30)], even water samples with low numbers of *M. avium* complex bacteria will be reported as containing *M. avium.* However, it is not known what value is significant for infection in AIDS patients. Consequently, it will be important to perform a study to determine the risk for development of disseminated *M. avium* infection for AIDS patients exposed to different levels of *M. avium* in water.

### *6.2. Mycobacterial Characteristics*

A number of characteristics of mycobacteria are likely to influence their transmission. Mycobacterial characteristics can influence transmission by affecting aerosolization, survival during transmission, or the ability to infect a susceptible host. Characteristics potentially able to affect transmission include cell surface hydrophobicity, cell surface charge, and colony type.

Studies of *M. avium, M. intracellulare,* and *M. scrofulaceum* have shown that cell surface hydrophobicity is a major determinant of aerosolization (38,126). In the experimental system, the number of cells ejected into the air in droplets from suspensions of low numbers of mycobacteria were measured. By dividing the number of mycobacteria per milliliter of droplet with the number per milliliter of the bulk suspension, a concentration factor could be calculated. The concentration factor of some *M. avium* strains was as high as 10,000 (38). Further investigations demonstrated that there was a relationship between cell surface hydrophobicity and aerosolization concentration. *M. avium* strains of greatest hydrophobicity were those with the highest concentration factor (126). Because the hydrophobicity of the interconvertible opaque and transparent colony variants of *M. avium* is different, (127), their concentration in ejected droplets differs (128). Cell surface charge did not appear to have any influence upon aerosolization (126,128).

The influence of mycobacterial characteristics on transmission by other means have not been investigated in as much detail. It is likely that cell surface hydrophobicity and charge can influence the distribution of mycobacteria in water systems. Charge and hydrophobicity can interact to promote the formation of biofilms (129) or adherence to filters used in water treatment (130).

The ability of a number of mycobacterial species to grown in natural waters (131) and the relative chlorine resistance of mycobacteria (132) are also important agents of mycobacterial transmission in water-distribution systems. Further, the ability of some mycobacterial species to grow at 45°C (i.e., *M. avium* and *M. xenopi*) are also thought to enable those mycobacteria to reside in hot-water-distribution systems. Thus, the presence of members of the *M. avium* complex in municipal water-distribution systems and hospital water-supply systems (25,98) is quite logical.

In addition to the possible role of catalase activity directly influencing the virulence and isoniazid susceptibility of *M. tuberculosis* (133), the presence of catalase and superoxide dismutase activities and carotenoid pigments in some *Mycobacterium* species (i.e., photochromogenic or scotochromogenic) is expected to influence survival in aerosols and water, where sunlight can produce toxic oxygen metabolites (e.g., hydrogen peroxide) in the presence of organic matter (78). Aerosols are enriched in hydrogen peroxide compared to the bulk water (134), and, thus, catalase activity of mycobacteria can be major factors influencing survival of mycobacteria in aerosols. Measurement of virulence of isoniazid-resistant, catalase-deficient strains of *M. tuberculosis* in the laboratory would not fully describe the effect of the catalase deficiency on the epidemiology of *M. tuberculosis.* To fully describe the role of catalase activity in the transmission dynamics of *M. tuberculosis,* animals would need to be exposed to aerosols containing either catalase-deficient, isoniazid-resistant strains or their catalase-containing, isoniazid-susceptible parents. To isolate the influence of aerosolization, animals would also need to be infected by injection by the same strains.

### *6.3. Physiochemical Factors*

In addition to those characteristics of the mycobacterial cells, a number of physiochemical factors are expected to influence mycobacterial transmission. These factors include salinity, temperature, humidity, and wind currents and velocity.

Aerosolization is directly influenced by water salinity (135). The concentration of *M. avium* cells in ejected droplets is increased in waters containing increased salt (135), presumably through an increase in hydrophobic interactions at higher salt concentrations. For mycobacteria transmitted by an aerosol or airborne route, humidity and wind velocity will be important determinants of transmission.

Temperature is expected to influence mycobacterial transmission through its influence on survival and growth. At high temperatures, cells will die. Although cell survival will be higher at low temperatures, growth and metabolism might be limited. As noted above, it is thought that one factor contributing to transmission of *M. avium* and *M. xenopi* in hospitals and water-distribution systems is their ability to grow at 45°C.

## 7. Experimental Methods for Detecting Mycobacterial Transmission

Direct demonstration of mycobacterial transmission requires the isolation or detection of a particular type of *Mycobacterium* in a source (e.g., patient or water), target (e.g., patient or animal), and the transmission vector (e.g., aerosol). The particular *Mycobacterium* species need not be isolated on bacteriologic medium, but, rather, detection can be performed by hybridization with a DNA probe (136) or by amplification of a particular DNA sequence employing the polymerase chain reaction (PCR) (137). Further, as noted above, the typing method must be able to provide the discrimination necessary to establish that isolates are from the same clone. For example, although serotyping has been widely employed in epidemiological studies of *M. avium, M. intracellulare,* and *M. scrofulaceum* (115), it lacks the necessary level of discrimination needed to identify the source of human *M. avium* complex infection.

Methods for the isolation of mycobacteria from environmental samples, namely water (29,30,38–40) and aerosols and dust (39,53) have been published. Likewise, there are published and well-established methods for the isolation of mycobacteria from patient samples, including blood and bone marrow (141), feces (114), and sputum (100). However, the existence of any one of those methods should not be taken that the best methods have been developed. The performance is far from that goal. Many of the methods for isolation of mycobacteria from sputum, feces, water, and soil rely on the use of germicides to kill bacteria or fungi, which grow faster than mycobacteria (29,30,42,138–140). However, those treatments also reduce the number of viable mycobacterial cells, although to a much lower extent (139,142,143). Thus, employment of these methods reduces the sensitivity of detection of mycobacteria. Even a selective medium for members of the *M. avium, M. intracellulare,* and *M. scrofulaceum* species does not support the growth of all members of those species (138).

Before collection of aerosol or dust samples is initiated, it is important to review the advantages and disadvantages of the two general types of samplers: impact and liquid capture samplers. Aerosol and dust samples can be collected without a decontamination step by employing an impact sampler (42,52,53). Contamination is usually minimal using an impact sampler which separates particle by size because fungal spores are larger and, consequently, impact onto the agar medium at stages different from those of particles which can enter the human alveoli (52). However, rapidly growing microorganisms can be inhibited by incorporating an antimicrobial agent in the agar medium which does not inhibit mycobacterial growth [i.e., Malachite Green (42) or cycloheximide for fungi (138)]. Collection of aerosol or dust samples by liquid capture (e.g., all glass impinger) requires that the liquid be sampled on bacteriological media. Liquid capture offers the advantage that detection of a particular *Mycobacterium* species or epidemiological type can be performed by a molecular typing method, such as by DNA probe (136) or PCR (137) without the necessity of isolation of individual colonies or cultures.

Data on comparisons of collection characteristics of aerosol samplers have shown that both liquid impingement and agar impaction are equally effective

(144,145). Those studies have, however, pointed out that the samplers are inaccurate estimators of true aerosol numbers when there are high numbers of microorganisms in aerosols or long collection times are employed (144,145). In a review of aerosol samplers, Nevalainen et al. (146) pointed out that comparison of field report data on microbial numbers and types in aerosols is confounded by the wide variety of methods and conditions under which different collections were performed. Another confounding factor affecting of collection of aerosols is human activity. Human activity was shown to significantly increase recovery of airborne fungal spores (146). That latter observation suggests that the number of mycobacteria in aerosols can be influenced by human activity as well, especially in rooms of small volume.

## 8. Preventing Mycobacterial Transmission

In light of the emergence of multidrug-resistant *M. tuberculosis* and the relative and general resistance of a number of nontuberculous mycobacteria (i.e., *M. avium*), prevention of mycobacterial transmission is being reemphasized.

### *8.1. Reducing Exposure to Mycobacteria*

Exposure to mycobacteria can be reduced by behavioral changes and protective measures. Avoidance of possible sources is one behavioral change; but if impossible (i.e., health care providers), protective measures, such as use of personal respirators, can be used (11). For those instances where water is a possible mycobacterial infection source (i.e., water in the case of AIDS patients), water can be boiled before consumption or use and showers with their attendant aerosols can be avoided.

A variety of building-associated interventions have been suggested and employed to reduce the chance of *M. tuberculosis* infection. They include ventilation (e.g., negative pressure ventilation and fresh air), directional airflow, air filtration [i.e., high-efficiency particulate air (HEPA) filtration], ultraviolet air disinfection, and patient isolation (11). In spite of their existence and trials, it has been suggested that they are going to ultimately prove to be ineffective (11). The problem lies in the fact that it takes only one *M. tuberculosis* cell to cause infection (14,147) and that one *M. tuberculosis* per 11,000 cubic feet of air was infectious in guinea pigs (14). It may prove beyond the ability of heating, ventilation, and air conditioning systems (HVAC) to practically reduce the numbers of *M. tuberculosis* in aerosols.

### *8.2. Reducing Mycobacteria in Sources*

The prevalence of sources yielding mycobacteria or the number of mycobacteria in possible sources can be reduced by a number of interventions. First, aerosols can be decontaminated by exposure to ultraviolet light (11,14,41). It should be pointed out that ultraviolet irradiation only kills cells, including mycobacteria, if irradiation occurs in the dark. Many organisms have a potent and effective mechanism for the removal of UV-light-induced damage in DNA (i.e., photoreactivation). Mycobacteria in water can be killed by boiling or another types of sterilization. Keep in mind, however, that mycobacteria are relatively resistant to disinfectants (148,149) and that disinfection can make the water undrinkable for a variety of reasons (e.g., toxicity or palatability).

It may also be possible that the routine employment of mycobactericides in waters and other mycobacterial sources can lead to selection of resistant mutants. Such mutants might not only be resistant to the mycobactericides but might also be resistant to antibiotics due to changes in cell permeation (150).

### *8.3. Reducing Mycobacterial Transmission Potential*

In the developed countries of the world, there is a role for intervention to prevent airborne *M. tuberculosis* infection. That is due to the fact that the chance of infection of certain locations (e.g., hospitals and homeless shelters) is high, whereas the chance of infection in other environments (e.g., home or work) is low (11). That is not the case in the developing world where the chance of infection in most human environments (e.g., work and home) is high. Measures to prevent infection in one environment would be negated by infection in another.

Ultraviolet-light sterilization of air containing *M. tuberculosis* was demonstrated by Riley et al. (14,41). Demonstration of the efficacy of UV sterilization of mycobacteria was confirmed by following the loss of colony-forming units of *M. bovis* BCG from a room with 12 air changes per hour (41). UV sterilization was achieved using a 17-W ultraviolet lamp in the upper position of a room, and the survival curve demonstrated that less than 10% of the original colonies survived beyond 10 min (41). In fact, because of the risk of airborne transmission of *M. tuberculosis* in shelters for the homeless, New York will implement UV sterilization in a number of homeless shelters (151).

Transmission of mycobacteria from water to air (e.g., *M. avium*) can be prevented by a number of interventions. Reduction of surface tension ought to reduce aerosolization of mycobacteria through reductions of hydrophobicity, because hydrophobicity is a determinant of aerosolization (126,135). It has been shown that the aerosol transmission of *Legionella pneumophila* can be reduced by reduction of foaming in water (152). Finally, it might be possible to increase the efficacy of removal of mycobacteria from waters and aerosols by employing hydrophobic traps. If the surface charge of the hydrophobic matrix is such that there is little or no charge difference between it and the mycobacterial cells (128), the hydrophobic mycobacterial cells (128,135) could possibly be removed quite effectively by entrapment without the necessity of using HEPA filters, which reduce airflow and have high-energy requirements.

## References


1. Daley CL, Small PM, Schecter GF, Schoolnik GK, McAdam RA, Jacobs WR Jr, Hopewell PC (1992) An outbreak of tuberculosis with accelerated progression among persons infected with the human immunodeficiency virus. New Engl J Med 326:231–235.
2. Dooley SW, Villarino ME, Lawrence M, Salinas L, Amil S, Rullan JV, Jarvis WR, Cauthen GM (1992) Nosocomial transmission of tuberculosis in a hospital unit for HIV-infected patients. J Am Med Assoc 267:2632–2634.
3. Selwyn PA, Sckell BM, Alcabes P, Friedland GH, Klein RS, Schoenbaum EE (1992) High risk of active tuberculosis in HIV-infected drug users with cutaneous anergy. J Am Med Assoc 268:504–509.
4. Houk VN, Baker JH, Sorensen K, Kent DC (1968) The epidemiology of tuberculosis infection in a closed environment. Arch Environ Hlth 16:26–35.
5. Braden CR, and an Investigative team (1995) Infectiousness of a university student with laryngeal and cavitary tuberculosis. Clin Infect Dis 21:565–570.
6. Kenyon TA, Valway SE, Ihle WW, Onorato IM, Castro KG (1996) Transmission of multidrug-resistant *Mycobacterium tuberculosis* during a long airplane flight. New Engl J Med 334:933–938.
7. Flora GS, Modilevsky T, Antoniskis D, Barnes PF (1990) Undiagnosed tuberculosis in patients with human immunodeficiency virus infection. Chest 98:1056–1059.
8. Sepkowitz KA (1994) Tuberculosis and the health care worker: A historical perspective. Ann Intern Med 120:71–79.
9. Mathur P, Sacks P, Auten G, Sall R, Levy C, Gordin F (1994) Delayed diagnosis of pulmonary tuberculosis in city hospitals. Arch Intern Med 154:306–310.
10. Glassroth J, Bailey WC, Hopewell PC, Schecter G, Warden JW (1990) Why tuberculosis is not prevented. Am Rev Respir Dis 141:1236–1240.
11. Nardell EA (1993) Environmental control of tuberculosis. Med Clin North Am 77:1315–1334.
12. Perriëns JH, Colebunders RL, Karahunga C, Willame J-C, Jeugmans J, Kaboto M, Mukadi Y, Pauwels P, Ryder RW, Priquot J, Piot P (1991) Increased mortality and tuberculosis treatment failure rate among human immunodeficiency virus (HIV) seropositive compared with HIV seronegative patients with pulmonary tuberculosis treated with "standard" chemotherapy in Kinshasa, Zaire. Am Rev Respir Dis 144:750–755.
13. Hopewell PC (1992) Impact of human immunodeficiency virus infection on the epidemiology, clinical features, management, and control of tuberculosis. Clin Infect Dis 15:540–547.

14. Riley RL, Mills CC, O'Grady F (1962) Infectiousness of air from a tuberculosis ward—ultraviolet irradiation of infected air: Comparative infectiousness of different patients. Am Rev Respir Dis 84:511–517.
15. Morrow PE (1980) Physics of airborne particles and their deposition in the lung. Ann NY Acad Sci 353:71–80.
16. Catanzaro A (1982) Nosocomial tuberculosis. Am Rev Respir Dis 125:559–562.
17. Wolinsky E (1979) Nontuberculous mycobacteria and associated diseases. Am Rev Respir Dis 119:107–159.
18. Robinson P, Morris D, Antic R (1988) *Mycobacterium bovis* as an occupational hazard in abattoir workers. Aust NZ J Med 18:701–703.
19. Dankner WM, Waecker NJ, Essey MA, Moser K, Thompson M, Davis CE (1993) *Mycobacterium bovis* infections in San Diego: A clinicoepidemiologic study of 73 patients and a historical review of a forgotten pathogen. Medicine 72:11–37.
20. Dalovisio JR, Stetter M, Mikota-Wells S (1992) Rhinoceros' rhinorrhea: Cause of an outbreak of infection due to airborne *Mycobacterium bovis* in zookeepers. J Infect Dis 15:598–600.
21. Clifton-Hadley RS, Wilesmith JW, Stuart FA (1993) *Mycobacterium bovis* in the European badger (*Meles meles*): Epidemiological findings in tuberculous badgers from a naturally infected population. Epidemiol Infect 111:9–19.
22. White PCL, Brown JA, Harris S (1993) Badgers (*Meles meles*), cattle and bovine tuberculosis (*Mycobacterium bovis*): A hypothesis to explain the influence of habitat on the risk of disease transmission in southwest England. Proc R Soc London B 253:277–284.
23. White PCL, Harris S (1995) Bovine tuberculosis in badger (*Meles meles*) populations in southwest England: The use of a spatial stochastic simulation model to understand the dynamics of the disease. Phil Trans R Soc London B 349:391–413.
24. White PCL, Harris S (1995) Bovine tuberculosis in badger (*Meles meles*) populations in southwest England: An assessment of past present and possible future control strategies using simulation modeling. Phil Trans R Soc London B 349:415–432.
25. von Reyn CF, Maslow JN, Barber TW, Falkinham JO III, Arbeit RD (1994) Persistent colonisation of potable water as a source of *Mycobacterium avium* infection in AIDS. Lancet 343:1137–1141.
26. Kaustova JZ, Olsovsky Z, Kubin M, Zatloukal O, Pelikan M, Hradil Y (1981) Endemic occurrence of *Mycobacterium kansasii* in water supply systems. J Hyg Epidem Immunol 25:24–30.
27. Wright EP, Collins CH, Yates MD (1985) *Mycobacterium xenopi* and *Mycobacterium kansasii* in a hospital water supply. J Hosp Infect 6:175–178.
28. Zeligman I (1972) *Mycobacterium marinum* granuloma. A disease acquired in the tributaries of the Chesapeake Bay. Arch Dermatol 106:26–31.
29. Falkinham JO III, Parker BC, Gruft H (1980) Epidemiology of infection by nontuberculous mycobacteria. I. Geographic distribution in the eastern United States. Am Rev Respir Dis 120:89–94.

30. von Reyn CF, Waddell RD, Eaton T, Arbeit RD, Maslow JN, Barber TW, Brindle RJ, Gilks CF, Lumio J, Lähdevirta J, Ranki A, Dawson D, Falkinham JO III (1993) Isolation of *Mycobacterium avium* complex from water in the United States, Finland, Zaire, and Kenya. J Clin Microbiol 31:3227–3230.

31. Kuritsky JN, Bullen MG, Broome CV, Silcox VA, Good RC, Wallace RJ Jr (1983) Sternal wound infections and endocarditis due to organisms of the *Mycobacterium fortuitum* complex. Ann Intern Med 98:938–939.

32. Safranek TJ, Jarvis WR, Carson LA, Cusick LB, Bland LA, Swenson JM, Silcox VA (1987) *Mycobacterium chelonae* wound infections after plastic surgery employing contaminated gentian violet skin-marking solution. New Engl J Med 317:197–201.

33. Wallace RJ Jr, Musser JM, Hull SI, Silcox VA, Steele LC, Forrester GD, Labidi A, Selander RK (1989) Diversity and sources of rapidly growing mycobacteria associated with infections following cardiac surgery. J Infect Dis 159:708–716.

34. Burns DN, Wallace RJ Jr, Schultz ME, Zhang Y, Zubairi SQ, Pang Y, Gibert CL, Brown BA, Noel ES, Gordin FM (1991) Nosocomial outbreak of respiratory tract colonization with *Mycobacterium fortuitum:* Demonstration of the usefulness of pulsed-field gel electrophoresis in an epidemiologic investigation. Am Rev Respir Dis 144:1153–1159.

35. Hector JSR, Pang Y, Mazurek GH, Zhang Y, Brown BA, Wallace RJ Jr (1992) Large restriction fragment patterns of genomic *Mycobacterium fortuitum* DNA as strain-specific markers and their use in epidemiologic investigation of four nosocomial outbreaks. J Clin Microbiol 30:1250–1255.

36. Wallace RJ Jr, Zhang Y, Brown BA, Fraser V, Mazurek GH, Maloney S (1993) DNA large restriction fragment patterns of sporadic and epidemic nosocomial strains of *Mycobacterium chelonae* and *Mycobacterium abscessus.* J Clin Microbiol 31:2697–2701.

37. Nardell EA (1990) Dodging droplet nuclei. Am Rev Respir Dis 142:501–503.

38. Parker BC, Ford MA, Gruft H, Falkinham JO III (1983) Epidemiology of infection by nontuberculous mycobacteria. IV. Preferential aerosolization of *Mycobacterium intracellulare* from natural waters. Am Rev Respir Dis 128:652–656.

39. Brooks RW, Parker BC, Gruft H, Falkinham JO III (1984) Epidemiology of infection by nontuberculous mycobacteria. I. Numbers in eastern United States soils and correlation with soils characteristics. Am Rev Respir Dis 130:630–633.

40. Kirschner RA Jr, Parker BC, Falkinham JO III (1992) Epidemiology of infection by nontuberculous mycobacteria. X. *Mycobacterium avium, Mycobacterium intracellulare,* and *Mycobacterium scrofulaceum* in acid, brown-water swamps of the southeastern United States and their association with environmental variables. Am Rev Respir Dis 145:271–275.

41. Riley RL, Knight M, Middlebrook G (1976) Ultraviolet susceptibility of BCG and virulent tubercle bacilli. Am Rev Respir Dis 113:417–422.

42. Wendt SL, George KL, Parker BC, Gruft H, Falkinham JO III (1980) Epidemiology

of infection by nontuberculous mycobacteria. I. Isolation of potentially pathogenic mycobacteria from aerosols. Am Rev Respir Dis 122:259–263.

43. Jones RJ, Jenkins DE (1965) Mycobacteria isolated from soil. Can J Microbiol 11:127–133.

44. Saito H, Tomioka H, Sato K, Tasaka H, Dekio S (1994) *Mycobacterium malmoense* isolated from soil. Microbiol Immunol 38:313–315.

45. Wolinsky E, Rynearson TK (1968) Mycobacteria in soil and their relation to disease-associated strains. Am Rev Respir Dis 97:1032–1037.

46. Costallat LF, Pestana de Castro AF, Rodrigues AC, Rodrigues FM (1977) Examination of soils in the Campinas rural area for microorganisms of the *Mycobacterium avium-intracellulare-scrofulaceum* complex. Austr Vetin J 53:349–350.

47. Reznikov M, Dawson DJ (1980) Mycobacteria of the *intracellulare–scrofulaceum* group in soils from the Adelaide area. Pathology 12:525–528.

48. Paull A (1973) An environmental study of the opportunistic mycobacteria. Med Lab Technol 30:11–19.

49. Tsukamura M, Mizuno S, Murata H, Memoto H, Yugi H (1974) A comparative study of mycobacteria from patients' room dusts and from sputa of tuberculous patients. Jpn J Microbiol 18:271–277.

50. Ichiyama S, Shimokata K, Tsukamura M (1988) The isolation of *Mycobacterium avium* complex from soil, water, and dusts. Microbiol Immunol 32:733–739.

51. Reznikov M, Robinson E (1970) Serologically identical Battey mycobacteria from sputa of healthy piggery workers and lesions of pigs. Austr Vetin J 46:606–607.

52. Andersen AA (1958) New sampler for the collection, sizing, and enumeration of viable airborne particles. J Bacteriol 76:471–484.

53. Falkinham JO III, George KL, Ford MA, Parker BC (1990) Collection and characteristics of mycobacteria in aerosols. In: Morey PR, Feeley JC Sr, Otten JA, eds. Biological Contaminants in Indoor Environments, pp. 71–79. Philadelphia: American Society for Testing and Materials.

54. Leers WD (1980) Disinfecting endoscopes: how not to transmit *Mycobacterium tuberculosis* by bronchoscopy. J Can Med Assoc 123:275–283.

55. Dawson DJ, Armstrong JF, Blacklock ZM (1982) Mycobacterial cross-contamination of bronchoscopy specimens. Am Rev Respir Dis 126:1095–1097.

56. Nye K, Chadha DK, Hodgkin P, Bradley C, Hancox J, Wise R (1990) *Mycobacterium chelonae* isolation from broncho-alveolar lavage fluid and its practical implications. J Hosp Infect 16:257–261.

57. Fraser VJ, Jones M, Murray PR, Medoff G, Zhang Y, Wallace RW Jr (1992) Contamination of flexible fiberoptic bronchoscopes with *Mycobacterium chelonae* linked to an automated bronchoscope disinfection machine. Am Rev Respir Dis 145:853–855.

58. Gubler JGH, Salfinger M, von Gravenitz A (1992) Pseudoepidemic of

nontuberculous mycobacteria due to contaminated bronchoscope cleaning machine. Chest 101:1245–1249.

59. Brown NM, Hellyar EA, Harvey JE, Reeves DS (1993) Mycobacterial contamination of fibreoptic bronchoscopes. Thorax 48:1283–1285.
60. Takigawa K, Fujita J, Negayama K, Terada S, Yamaji Y, Kawanishi K, Takahara J (1995) Eradication of contaminating *Mycobacterium chelonae* from bronchofibrescopes and an automated bronchoscope disinfection machine. Respir Med 89:423–427.
61. Spach DH, Silverstein FE, Stamm WE (1993) Transmission of infection by gastrointestinal endoscopy and bronchoscopy. Ann Intern Med 118:117–128.
62. Inman PM, Beck A, Stanford AE (1969) Outbreak of injection abscesses due to *Mycobacterium abscessus.* Arch Dermatol 100:141–147.
63. Borghans JGA, Stanford JL (1973) *Mycobacterium chelonae* in abscesses after injection of diphtheria-pertussis-tetanus-polio vaccine. Am Rev Respir Dis 107:1–8.
64. Gremillion DH, Mursch SB, Lerner CJ (1983) Injection site abscesses caused by *Mycobacterium chelonae.* Infect Control 4:25–28.
65. Wenger JD, Spika JS, Smithwick RW (1990) Outbreak of *Mycobacterium chelonae* infection associated with use of jet injectors. J Am Med Assoc 264:373–376.
66. Woo AH, Goetz A, Yu VL (1992) Transmission of *Legionella* by respiratory equipment and aerosol generating devices. Chest 102:1586–1590.
67. Bolan G, Rheingold AL, Carson LA (1985) Infections with *Mycobacterium chelonae* in patients receiving dialysis and using processed hemodialyzers. J Infect Dis 152:1013–1019.
68. Lowry PW, Beck-Sague CM, Bland LA (1990) *Mycobacterium chelonae* infection among patients receiving high-flux dialysis in a hemodialysis clinic in California. J Infect Dis 161:85–90.
69. Altmann G, Horowitz A, Kaplinsky N, Frankl O (1975) Prosthetic-valve endocarditis due to *Mycobacterium chelonae.* J Clin Microbiol 1:531–533.
70. Repath F, Seabury JH, Sanders CV, Domer J (1976) Prosthetic valve endocarditis due to *Mycobacterium chelonae.* South Med J 69:1244–1246.
71. Narasimhan SL, Austin TW (1978) Prosthetic valve endocarditis due to *Mycobacterium fortuitum.* Can Med Assoc J 119:154–155.
72. Guerrero C, Bernasconi C, Burki D, Bodmer T, Telenti A (1995) A novel insertion element from *Mycobacterium avium,* IS1245, is a specific target for analysis of strain relatedness. J Clin Microbiol 33:304–307.
73. Roiz MP, Palenque E, Guerrero C, Garcia MJ (1995) Use of restriction fragment length polymorphism as a genetic marker for typing *Mycobacterium avium* strains. J Clin Microbiol 33:1389–1391.
74. Picardeau M, Vincent V (1996) Typing of *Mycobacterium avium* isolates by PCR. J Clin Microbiol 34:389–392.
75. Arbeit RD, Slutsky A, Barber TW, Maslow JN, Niemczyk S, Falkinham JO III,

O'Connor GT, von Reyn CF (1993) Genetic diversity among strains of *Mycobacterium avium* causing monoclonal and polyclonal bacteremia in patients with AIDS. J Infect Dis 167:1384–1390.

76. Mazurek GH, Hartman S, Zhang Y, Brown BA, Hector JSR, Murphy D, Wallace RJ Jr (1993) Large DNA restriction fragment polymorphism in the *Mycobacterium avium, M. intracellulare* complex: A potential epidemiologic tool. J Clin Microbiol 31:390–394.
77. Collins CH, Yates MD (1984) Infection and colonisation by *Mycobacterium kansasii* and *Mycobacterium xenopi:* Aerosols as a possible source? J Infect 8:178–179.
78. Cooper WJ, Zika RG (1983) Photochemical formation of hydrogen peroxide in surface and ground waters exposed to sunlight. Science 220:711–712.
79. Hoffner SE, Källenius G, Petrini B, Brennan PJ, Tsang AY (1990) Serovars of *Mycobacterium avium* complex isolated from patients in Sweden. J Clin Microbiol 28:1105–1107.
80. Wolinsky E (1995) Mycobacterial lymphadenitis in children: A prospective study of 105 nontuberculous cases with long-term follow-up. Clin Infect Dis 20:954–963.
81. Miyachi T, Shimokata K, Dawson D, Tsukamura M (1988) Changes of the biotype of *Mycobacterium avium–Mycobacterium intracellulare* complex causing lung disease in Japan. Tubercle 69:133–137.
82. Collins CH, Grange JM, Yates MD (1984) Mycobacteria in water. J Appl Bacteriol 57:193–211.
83. Hetherington HW, McPhedran FM, Landis HRM, Opie EL (1935) Further studies of tuberculosis among medical and other university students, occurrence and development of lesions during the medical course. Arch Intern Med 55:709–734.
84. Dunn BL, Hodgson DJ (1982) 'Atypical' mycobacteria in milk. J Appl Bacteriol 52:373–376.
85. Hedrick RP, McDowell T, Groff J (1987) Mycobacteriosis in cultured striped bass from California. J Wildlife Dis 23:391–395.
86. Lansdell W, Dixon B, Smith N, Benjamin L (1993) Isolation of several *Mycobacterium* species from fish. J Aquat Anim Hlth 5:73–76.
87. Corner LA, Barrett RH, Lepper AWD, Lewis V, Pearson CW (1981) A survey of mycobacteriosis of feral pigs in the northern territory. Austr Vetin J 57:537–542.
88. Hein WR, Tomasovic AA (1981) An abattoir survey of tuberculosis in feral buffaloes. Austr Vetin J 57:543–547.
89. Bickford AA, Ellis GH, Moses HE (1966) Epizootiology of tuberculosis in starlings. J Am Vetin Med Assoc 149:312–317.
90. Schaefer WB, Beer JV, Wood NA, Jenkins PA, Marks J (1973) A bacteriological study of endemic tuberculosis in birds. J Hyg Camb 71:549–557.
91. Saxegaard F (1981) Serological investigations of *Mycobacterium avium* and *M. avium*-like bacteria isolated from domestic and wild animals. Acta Vetin Scand 22:153–161.

92. Parisot TJ, Wood JN (1970) Fish mycobacteriosis (tuberculosis). U.S. Fish and Wildlife Service, Fish Disease Leaflet 7.

93. Mbithi JN, Springthorpe VS, Sattar SA, Pacquette M (1993) Bactericidal, virucidal, and mycobactericidal activities of reused alkaline glutaraldehyde in an endoscopy unit. J Clin Microbiol 31:2988–2995.

94. Torres RA, Mani S, Altholz J, Brickner PW (1990) Human immunodeficiency virus infection among homeless me in a New York City shelter. Arch Intern Med 150:2030–2036.

95. Paul EA, Lebowitz SM, Moore RE, Hoven CW, Bennett BA, Chen A (1993) Nemesis revisited: Tuberculosis infection in a New York City men's shelter. Am J Publ Hlth 83:1743–1745.

96. Hutton MD, Stead WW, Cauthen GM, Bloch AB, Ewing WM (1990) Nosocomial transmission of tuberculosis associated with a draining abscess. J Infect Dis 161:286–295.

97. duMoulin GC, Stottmeier KD (1986) Waterborne mycobacteria: an increasing threat to health. Am Soc Microbiol News 52:525–529.

98. duMoulin GC, Stottmeier KD, Pelletier PA, Tsang AY, Hedley-Whyte J (1988) Concentration of *Mycobacterium avium* by hospital hot water systems. J Am Med Assoc 260:1599–1601.

99. Young LS, Inderlied CB, Berlin OG, Gottlieb MS (1986) Mycobacterial infections in AIDS patients, with an emphasis on the *Mycobacterium avium* complex. Rev Infect Dis 8:1024–1033.

100. Inderlied CB, Kemper CA, Bermudez LEM (1993) The *Mycobacterium avium* complex. Clin Microbiol Revs 6:266–310.

101. Chin DP, Hopewell PC, Yajko DM, Vittinghoff E, Horsburgh CR Jr, Hadley WK, Stone EN, Nassos PS, Ostroff SM, Jacobson MA, Matkin CC, Reingold AL (1994) *Mycobacterium avium* complex in the respiratory or gastrointestinal tract and the risk of *M. avium* complex bacteremia in patients with human immunodeficiency virus infection. J Infect Dis 169:289–295.

102. Jacobson MA, Hopewell PC, Yajko DM, Hadley WK, Lazarus E, Mohanty PK, Modin GW, Feigal DW, Cusick PS, Sande MA (1991) Natural history of disseminated *Mycobacterium avium* complex infection in AIDS. J Infect Dis 164:994–998.

103. Barker DJP (1973) Epidemiology of *Mycobacterium ulcerans* infection. Trans R Soc Trop Med Hyg 67:43–47.

104. Hayman J (1991) Postulated epidemiology of *Mycobacterium ulcerans* infection. Int J Epidemiol 20:1093–1098.

105. Sompolinsky D, Lagziel A, Naveh D, Yankilevitz T (1978) *Mycobacterium haemophilum* sp. nov., a new pathogen of humans. Int J System Bacteriol 28:67–75.

106. Kiehn TE, White M, Pursell KJ, Boone N, Tsivitis M, Brown AE, Polsky B, Armstrong D (1993) A cluster of four cases of *Mycobacterium haemophilum* infection. Eur J Clin Microbiol Infect Dis 12:114–118.

107. Straus WL, Ostroff SM, Jernigan DB, Kiehn TE, Sordillo EM, Armstrong D, Boone N, Schneider N, Kilburn JO, Silcox VA, LaBonbardi V, Good RC (1994) Clinical and epidemiologic characteristics of *Mycobacterium haemophilum,* an emerging pathogen in immunocompromised patients. Ann Intern Med 120:118–125.

108. Dawson DJ, Jennis F (1980) Mycobacteria with a growth requirement for ferric ammonium citrate, identified as *Mycobacterium haemophilum.* J Clin Microbiol 11:190–192.

109. Yakrus MA, Straus WL (1994) DNA polymorphisms detected in *Mycobacterium haemophilum* by pulsed-field gel electrophoresis. J Clin Microbiol 32:1083–1084.

110. Sompolinsky D, Lagziel A, Rosenberg J (1979) Further studies of a new pathogenic mycobacterium (*Mycobacterium haemophilum* sp. nov.). Can J Microbiol 25:217–226.

111. Böttger EC, Teske A, Kirschner P, Bost S, Chang HR, Beer V, Hirshel B (1992) Disseminated "*Mycobacterium genevense*" infection in patients with AIDS. Lancet 340:76–80.

112. Coyle MB, Carlson LDC, Wallis CK, Leonard RB, Raisys VA, Kilburn JO, Samadpour M, Böttger EC (1992) Laboratory aspects of "*Mycobacterium genevense,*" a proposed species isolated from AIDS patients. J Clin Microbiol 30:3206–3212.

113. Hoop RK, Böttger EC, Ossent P, Salfinger M (1993) Mycobacteriosis due to *Mycobacterium genevense* in six pet birds. J Clin Microbiol 31:990–993.

114. Yajko DM, Nassos PS, Sanders CA, Gonzalez PC, Reingold AL, Horsburgh CR Jr, Hopewell PC, Chin DP, Hadley WK (1993) Comparison of four decontamination methods for recovery of *Mycobacterium avium* complex from stools. J Clin Microbiol 31:302–306.

115. Dawson DJ (1990) Infection with *Mycobacterium avium* complex in Australian patients with AIDS. Med J Austral 153:466–468.

116. Conville PS, Keiser JF, Witebsky FG (1989) Mycobacteremia caused by simultaneous infection with *Mycobacterium avium* and *Mycobacterium intracellulare* detected by analysis of a BACTEC 13A bottle with the Gen-Probe kit. Diagn Microbiol Infect Dis 12:217–219.

117. Heimbeck J (1928) Immunity to tuberculosis. Arch Intern Med 41:336–342.

118. Malasky C, Jordan T, Potulski F, Reichman LB (1990) Occupational tuberculosis infections among pulmonary physicians in training. Am Rev Respir Dis 142:505–507.

119. Aiken ML, Burke W, McDonald W, Wallis C, Ramsey B, Nolan C (1993) Nontuberculosis mycobacterial disease in adult cystic fibrosis patients. Chest 103:1096–1099.

120. Winter SM, Bernard EM, Gold JWM, Armstrong D (1985) Humoral response to disseminated infection by *Mycobacterium avium–Mycobacterium intracellulare* in acquired immunodeficiency syndrome and hairy cell leukemia. J Infect Dis 151:523–527.

121. Young LS (1988) *Mycobacterium avium* complex infection. J Infect Dis 157:863–867.

122. Guthertz LS, Damsker B, Bottone EJ, Ford EG, Midura TF, Janda JM (1989) *Mycobacterium avium* and *Mycobacterium intracellulare* infections in patients with and without AIDS. J Infect Dis 160:1037–1041.

123. Barber TW, Craven DE, McCabe WR (1990) Bacteremia due to *Mycobacterium tuberculosis* in patients with human immunodeficiency virus infection. Medicine 69:375–383.

124. Braun MM, Byers RH, Heywood WL, Ciesielski CA, Bloch AB, Berkelman RL, Snider DE (1990) Acquired immunodeficiency syndrome and extrapulmonary tuberculosis in the United States. Arch Intern Med 150:1913–1916.

125. Theuer CP, Hopewell PC, Elias D, Schecter GF, Rutherford GW, Chaisson RE (1990) Human immunodeficiency virus infection in tuberculosis patients. J Infect Dis 162:8–12.

126. George KL, Falkinham JO III (1989) Aerosolization of mycobacteria. In: Comtois P. ed. Aerobiology, Health, and Environment, pp. 211–220. Montreal: Centre Recherches Ecologiques de Montreal.

127. Stormer RS, Falkinham JO III (1989) Differences in antimicrobial susceptibility of pigmented and unpigmented colonial variants of *Mycobacterium avium.* J Clin Microbiol 27:2459–2465.

128. George KL, Pringle AT, Falkinham JO III (1986) The cell surface of *Mycobacterium avium-intracellulare* and *M. scrofulaceum:* Effect of specific chemical modifications on cell surface charge. Microbios 45:199–207.

129. Marshall KC (1992) Biofilms: An overview of bacterial adhesion, activity, and control at surfaces. Am Soc Microbiol News 58:202–207.

130. Ridgway HF, Rigby MG, Argo DG (1984) Adhesion of a *Mycobacterium* sp. to cellulose diacetate membranes used in reverse osmosis. Appl Environ Microbiol 47:61–67.

131. George KL, Parker BC, Gruft H, Falkinham JO III (1980) Epidemiology of infection by nontuberculous mycobacteria. II. Growth and survival in natural waters. Am Rev Respir Dis 122:89–94.

132. Haas CN, Meyer MA, Paller MS (1983) The ecology of acid-fast organisms in water supply, treatment, and distribution systems. Am Water Works Assoc J 75:139–144.

133. Jackett PS, Aber VR, Lowrie DB (1978) Virulence and resistance to superoxide, low pH and hydrogen peroxide among strains of *Mycobacterium tuberculosis.* J Gen Microbiol 104:37–45.

134. Blanchard DC (1989) Bacteria and other materials in drops from bursting bubbles. In: Monahan EC, Van Patten MA, eds. Climate and Health Implications of Bubble-Mediated Sea–Air Exchange, pp. 1–16. Groton, CT: Connecticut Sea Grant Institute.

135. Falkinham JO III (1989) Factors influencing the aerosolization of mycobacteria. In: Monahan EC, Van Patten MA, eds. Climate and Health Implications of Bubble-

Mediated Sea–Air Exchange, pp. 17–26. Groton, CT: Connecticut Sea Grant Institute.

136. Falkinham JO III (1994) Nucleic acid probes. In: Gerhardt P, ed. Methods for General and Molecular Bacteriology, pp. 701–710. Washington, DC: American Society for Microbiology.
137. Atlas RM, Bej AK (1994) Polymerase chain reaction. In: Gerhardt P, ed. Methods for General and Molecular Bacteriology, pp. 418–435. Washington, DC: American Society for Microbiology.
138. George KL, Falkinham JO III (1986) Selective medium for the isolation and enumeration of *Mycobacterium avium-intracellulare* and *M. scrofulaceum.* Can J Microbiol 32:10–14.
139. Carson LA, Cusick LB, Bland LA, Favero MS (1988) Efficacy of chemical dosing methods for isolating nontuberculous mycobacteria from water supplies of dialysis centers. Appl Environ Microbiol 54:1756–1760.
140. Kamala R, Paramasivan CN, Herbert D, Venkatesan P, Prabhakar R (1994) Evaluation of procedures for isolation of nontuberculous mycobacteria from soil and water. Appl Environ Microbiol 60:1021–1024.
141. Askgaard D, Fuursted K, Gottshau A, Bennedsen J (1992) Detection of mycobacteria from blood and bone marrow: a decade of experience. APMIS 100:609–614.
142. Brooks RW, George KL, Parker BC, Falkinham JO III (1984) Recovery and survival of nontuberculous mycobacteria under various growth and decontamination conditions. Can J Microbiol 30:1112–1117.
143. Songer JG (1981) Methods for selective isolation of mycobacteria from the environment. Can J Microbiol 27:1–7.
144. Buttner MP, Stetzenbach LD (1991) Evaluation of four aerobiological sampling methods for the retrieval of aerosolized *Pseudomonas syringae.* Appl Environ Microbiol 57:1268–1270.
145. Buttner MP, Stetzenbach LD (1993) Monitoring airborne fungal spores in an experimental indoor environment to evaluate sampling methods and the effects of human activity on air sampling. Appl Environ Microbiol 59:219–226.
146. Nevalainen A, Pastuszka J, Liebhaber F, Willeke K (1992) Performance of bioaerosol samplers: collection characteristics and sampler design considerations. Atmos Environ 26A:531–540.
147. Wells WF, Ratcliff HL, Crumb C (1948) On the mechanism of droplet nuclei infection. II. Quantitative experimental airborne infection in rabbits. Am J Hyg 47:11–28.
148. Best M, Sattar SA, Springthrope VS, Kennedy ME (1990) Efficacies of selected disinfectants against *Mycobacterium tuberculosis.* J Clin Microbiol 28:2234–2239.
149. Rutala WA, Cole EC, Wannamaker NS, Weber DJ (1991) Inactivation of *Mycobacterium tuberculosis* and *Mycobacterium bovis* by 14 hospital disinfectants. Am J Med 91 (Suppl 3B):267S–271S.
150. Rastogi N, Frehel C, Ryter A, Ohayon H, Lesourd M, David HL (1981) Multiple

drug resistance in *Mycobacterium avium:* Is the wall architecture responsible for the exclusion of antimicrobial agents? Antimicrob Agents Chemother 20:666–677.

151. Wald ML (1994) Testing due of ultraviolet to fight TB. New York Times, 4 Jan.

152. Colbourne JS, Dennis PJ, Lee JV, Bailey MR (1987) Legionnaires' disease: reduction in risks associated with foaming in evaporative cooling towers. Lancet March 21. 1:684, 8534.

# 7

# Nosocomial *Mycobacterium Tuberculosis* Outbreaks: Risk Factors, Prevention Intervention Efficacy, and Guidelines

*William R. Jarvis*

## 1. Introduction

*Mycobacterium tuberculosis* can be transmitted to patients (nosocomial or hospital-acquired infection) or to health care workers (occupational-acquisition) in health care facilities. Before the 1980s, reports of nosocomial *M. tuberculosis* transmission were rare (1–3). The lack of reports of nosocomial patient to patient *M. tuberculosis* transmission in health care facilities can be related to the fact that only approximately 5–10% of immunocompetent persons exposed to a patient with infectious tuberculosis (TB) will develop TB in their lifetime; if an immunocompetent patient were exposed to an infectious TB patient in the hospital, development of infection and then active disease would occur months to years after the hospital exposure and nosocomial acquisition would never be considered. Thus, unless there is a cluster of patients with newly diagnosed TB or patients infected with a *M. tuberculosis* strain with an unusual antimicrobial resistance pattern, nosocomial *M. tuberculosis* transmission might not be suspected or detected. Thus, although rarely reported, nosocomial transmission of *M. tuberculosis* can occur and remain undetected. In contrast, occupational acquisition has been documented as a risk to health care workers in the United States since the early 1930s. From 1935 through 1939, Israel followed a cohort of 637 nursing students during their hospital training (4). At entry into the study, 360 (57%) nurses were tuberculin skin test (TST) negative and 277 (43%) were TST positive. All of the TST-negative nurses had TST conversions during their training. Of the 637 nursing students, 68 (11%) developed active tuberculosis. Subsequent studies in the

1930s and 1940s also documented high (79–85%) TST conversion rates among nurses (5,6). During the 1940s and 1950s, studies were conducted showing that the risk of active TB among medical school students was over three times higher than the general population (7). Risk factors for disease among the medical students included exposure to unrecognized TB patients, the men's TB ward, or presence in the autopsy room.

Despite these studies suggesting that nosocomial transmission of *M. tuberculosis* could occur and that occupational acquisition of TB by health care workers might be common, efforts to introduce preventive interventions were slow. Many hospitals established "tuberculosis wards," sanitaria were common, and at more and more hospitals, routine patient entry chest radiographs were initiated to identify the previously unrecognized TB patient. At the same time, improvements in suspicion of TB, diagnostic and therapeutic modalities, and preventive therapy resulted in a dramatic reduction in the incidence of TB in the United States from the mid-1940s until the mid-1980s (8).

## 2. Nosocomial *M. tuberculosis* in Human Immunodeficiency Virus-Infected Patients

Tuberculosis is a common infection in human immunodeficiency virus (HIV)-infected patients (9,10). The majority of these episodes of TB are thought to be a reactivation of latent infection. In the mid-1980s, the incidence of TB began to rise in the United States; it is unclear how much of this rise in TB was due to HIV-infected persons who are at increased risk of both reactivation of latent infection or new primary infection (8). Nevertheless, between 1985 and 1992, over 50,000 excessive episodes of TB occurred above what would have been expected in the United States had the downward trend in TB been maintained; much of this increase was in HIV-infected persons (8). The interaction of the HIV and TB outbreaks have resulted in a major public health challenge. In the late 1980s, nosocomial outbreaks of drug-susceptible and/or multidrug-resistant (MDR) *M. tuberculosis* began to occur in U.S. hospitals (11). In 1989, Dooley et al. investigated an outbreak of TB among patients in an HIV ward of a hospital (12). They found that HIV-ward patients with exposure to an infectious TB patients were 11 times more likely to develop TB than were HIV-ward patients without such an exposure (8/48 versus 2/192 patients or 9.7 versus 0.8 infections per 10,000 person-days). In addition, nurses working on the HIV or internal medicine wards were significantly more likely to have a positive TST than were clerical staff working on other wards at that hospital (9/19 or 45/90 versus 35/188, $p = .0005$). This investigation documented that HIV-infected patients sharing a room with an HIV-infected patient with infectious TB were at increased risk of nosocomial acquisition of *M. tuberculosis,* that the incubation period was most consistent with primary infection, that the risk of developing active disease given exposure to an infectious TB patient was high (35.7 per 100 person-years), and that the duration from *M. tuberculosis* infection to disease (incubation period) is shortened in HIV-infected patients. This and another outbreak documented the risk of nosocomial transmission of *M. tuberculosis* among hospitalized HIV-infected patients (13).

## 3. Nosocomial Multidrug-Resistant *M. tuberculosis* Outbreaks

In the late 1980s and early 1990s, numerous nosocomial TB outbreaks occurred in U.S. hospitals caused by strains of *M. tuberculosis* resistant to two or more antituberculous agents, most commonly isoniazid and rifampin. In 1990, Edlin et al. investigated an outbreak of multidrug-resistance tuberculosis (MDR-TB) among acquired immunodeficiency syndrome (AIDS) patients in a New York City hospital (14). From 1989 through 1990, 18 AIDS patients at this hospital had infections with *M. tuberculosis* strains resistant to isoniazid and streptomycin compared to only three patients with such infections in the preceding 3 years. When these 18 MDR-TB AIDS patients were compared to 30 AIDS patients with TB caused by drug-susceptible strains of *M. tuberculosis,* the MDR-TB patients were more likely to be homosexual men, to have had AIDS for a longer period, or to have been hospitalized at the outbreak hospital within the 6 months preceding diagnosis of their TB. Furthermore, the MDR-TB patients were significantly more likely than AIDS patients infected with drug-susceptible strains of *M. tuberculosis* to have been hospitalized during their exposure period in the same ward at the same time as another patient with infectious MDR-TB. When the MDR-TB AIDS patients were compared to a group of AIDS patients infected with drug-susceptible strains of *M. tuberculosis* and similar durations of hospitalization, the MDR-TB patients were more likely to occupy rooms closer to the room of an infectious MDR-TB patient. Restriction fragment-length polymorphism (RFLP) analysis of the strains of 16 MDR-TB AIDS patients showed that 13 had an identical pattern. Thus, both epidemiologic and laboratory data supported nosocomial acquisition of *M. tuberculosis* by these MDR-TB AIDS patients. The attack rate was 21/346 (6.1%) among all AIDS patients and 18/189 (9.5%) among AIDS patients hospitalized in rooms two or less rooms away from another infectious MDR-TB patient's room. The estimated incubation period (from exposure to the development of active disease) was estimated at 1½–6 months. An environmental evaluation showed that only 1 of 16 patient's rooms had negative pressure. A TST survey conducted at the time of the outbreak investigation documented TST conversions (from documented negative to positive) in 11/60 (18.3%) health care workers; those with a follow-up TST > 2 years before their baseline negative TST had a TST conversion rate of 9/31 (29%), whereas those with a follow-up TST 2 years or less of their negative baseline had a TST conversion rate of 2/29 (6.9%) (15). One health care worker with active MDR-TB had an *M. tuberculosis* strain with an identical RFLP pattern to that of the MDR-TB AIDS patient's strains.

At about the same time, another MDR-TB outbreak, caused by a strain of *M. tuberculosis* resistant to isoniazid and rifampin, was occurring at a hospital in Miami, Florida (16,17). From January 1988 through January 1990, 25 MDR-TB patients were identified among HIV-infected patients admitted to the HIV ward at one hospital. When these MDR-TB patients were compared to HIV-ward patients with TB caused by drug-susceptible strains, the MDR-TB patients were more likely to have had an opportunistic infection before being diagnosed with TB, to have been exposed to a sputum smear acid-fast bacillus (AFB)-positive MDR-TB patient during their hospitalization preceding the diagnosis of TB, to have failed to respond to antituberculosis therapy, or to have died. MDR-TB patients remained AFB sputum smear positive for a significantly greater proportion of their hospitalization than did HIV-infected patients with TB caused by drug-susceptible strains (375/860 versus 197/1445 person-days, $p < .001$). Exposure to AFB sputum smear-positive, culture-positive patients was significantly more likely to result in transmission of *M. tuberculosis* than was exposure to AFB sputum smear-negative, culture-positive patients. Exposures to infectious MDR-TB patients occurred both in the HIV ward and the HIV outpatient clinic. Patients who attended the clinic to receive aerosolized pentamidine were at greater risk of developing MDR-TB than were those attending the HIV clinic but not receiving aerosolized pentamidine. All of the available 13 MDR-TB patient *M. tuberculosis* isolates had one of two RFLP patterns. Health care workers in the HIV ward and clinic were significantly more likely than health care workers in a comparison ward to have a TST conversion during the study period (13/39 versus 0/15, $p < .001$). There was a strong correlation between risk of health care worker *M. tuberculosis* infection and the number of days that an AFB sputum smear-positive MDR-TB patient was hospitalized on the HIV ward. Six of the 23 AFB isolation rooms tested were found to have positive pressure. In the HIV clinic, the aerosolized pentamidine administration rooms had positive pressure compared to the treatment room, which was positive relative to the discharge waiting room; air also was recirculated back into the clinic.

From 1990 through 1992, the Centers for Disease Control and Prevention conducted eight nosocomial MDR-TB outbreak investigations (11,14,16,18–24) (Table 7.1). These outbreaks all occurred between 1988 and 1992 and involved 7–70 (mean = 31.6) patients at each hospital. Six of eight outbreaks occurred in New York State with five of these in New York City hospitals. In all the outbreaks, the *M. tuberculosis* strain transmitted was resistant to isoniazid; in seven outbreaks, the strain also was resistant to rifampin. Depending on the outbreak, the strains also had resistance to streptomycin, ethambutol, ethionamide, kanamycin,

Table 7.1. Characteristics of Centers for Disease Control and prevention investigated nosocomial multidrug-resistant *Mycobacterium tuberculosis* outbreak, 1989—1992

| Hospital | Outbreak period | No. of case-patietns | No. (%) case-patients HIV-positive | Outbreak ward | Mortality number (%) | Infecting strain: Resistance pattern[a] | Infecting strain: RFLP[b] type | No. (#) health care worker TST[c] conversions | No. (#) of isolation rooms negative pressure | Refs. |
|---|---|---|---|---|---|---|---|---|---|---|
| 1 | Jan. 88–Jan. 89 | 25 | 25 (100) | HIV | 21 (84) | INH, RIF | 2 strains | 13/39 (33) | 17/23 (74) | 16 |
| 2 | Jan. 88–Jan. 90 | 18 | 18 (100) | Medicine | 10 (56) | INH, Strep | 1 strain | 11/60 (18) | 1/16 (6) | 14 |
| 3 | Sept. 89–Mar. 91 | 17 | 16 (94) | Medicine | 14 (82) | INH, RIF, Strep. | 1 strain | 88/352 (25)[d] | None | 25 |
| 4 | Jan. 90–Mar. 91 | 23 | 21 (91) | HIV | 19 (83) | INH, RIF | 1 strain | 6/12 (50) | None | 19 |
| 5 | May 91–Oct. 91 | 8 | 1 (2.5) | Medicine | 1 (12.5) | INH, RIF, EMB, ETA, CRM, SM, RBT | 1 strain | 46/696 (6.5) | None | 18 |
| 6 | Jan. 90–Dec. 91 | 16 | 14 (88) | Medicine | 14 (88) | INH, RIF, Strep. | 1 strain | Unknown | None | 22 |
| 7 | June 90–Apr. 92 | 13 | 13 (100) | Infectious disease | 11 (85) | INH, RIF | 1 strain | 5/10 (50) | None | 21 |

[a]INH = isoniazid; RIF = rifampin; EMB = ethambutol; ETA = etionamide; KM = kanamycin, SM = streptomycin; RBT = rifabutin.
[b]RFLP = Restriction fragment length polymoprhism.
[c]TST = Tuberculin skin test.
[d]11 HCWs with active TB.

and/or rifibutin. In two outbreaks, some of the infecting strains were resistant to seven antituberculous agents (18,25). The proportion of patients in these outbreaks who had HIV infections ranged from 14% to 100%; the majority of these outbreaks occurred either in HIV wards or in wards where the majority of patients had HIV infection. Mortality ranged from 43% to 89% (median = 82.5%). The interval from diagnosis of TB until death ranged from 4 to 16 (median = 4) weeks.

## 4. Risk Factors for Nosocomial MDR-TB

In each investigation, various studies were conducted to identify risk factors for nosocomial acquisition of MDR-TB. Factors identified included having HIV or AIDS, prior hospitalization at the outbreak hospital, the close proximity of AFB sputum smear-positive (i.e., infectious) MDR-TB patients and patients with HIV infection or AIDS, having exposure to another infectious MDR-TB patient either because of close proximity of the rooms or exposures outside of the patient's room, delayed recognition of TB in the patient because of nonclassical chest radiograph, clinical findings or low index of suspicion, delayed recognition of infecting multidrug-resistant strains because of delayed laboratory identification and/or communication to the clinicians, and delayed institution of effective antituberculosis therapy (Table 7.2). In addition, there were a number of inadequate infection control practices, including delayed initiation of AFB isolation; isolation rooms without six or more air changes per hour, negative pressure, or air exhausted to the outside; failure to isolate patients until they were no longer infectious, lapses in AFB isolation such as allowing patients in isolation to leave their rooms while they were infectious for nonmedical reasons (to attend group social events, to walk the halls, go to common bathrooms, visit the lounge or television areas, etc.); inadequate duration of follow-up after AFB isolation; and inadequate precautions during aerosol-generating procedures such as sputum induction or aerosolization of pentamidine. Finally, delay in the institution of effective therapy contributed to prolonged infectiousness and increased the risk of transmission of *M. tuberculosis.*

In these outbreaks, the failure to rapidly identify and appropriately isolate infectious MDR-TB patients combined with the prolonged infectiousness of these patients, secondary to delays in diagnosis and treatment, led to exposure of other patients and health care workers. In each outbreak, the major risk factor for acquisition of MDR-TB or for a TST conversion was exposure to an infectious MDR-TB patient. Delayed identification of infectious MDR-TB patients might have been contributed to by a low index of suspicion by the clinicians for TB in HIV-infected patients with pulmonary symptoms and the fact that many of the MDR-TB patients did not present with classical signs, symptoms, or chest radi-

Table 7.2 Risk factors for nosocomial transmission of MDR-TB

**Patient factors**

- HIV-Infection
- AIDS
- Low CD-4 *t*-lymphocyte count
- Prior hospital admission
  - Admission to a room near (<3 rooms) from an infections MDR-TB patient
  - Exposure to an AFB sputum smear positive MDR-TB patient

**Clinical factors**

- Delayed diagnosis
  - Nonclassical signs/symptoms
  - Nonclassical chest radiograph
  - Low index of suspicion of TB
  - Delays in laboratory results (identification and susceptibility)
- Delayed patient isolation
  - Low index of suspicion of TB
  - Delay recognition of MDR-TB
  - Inadequate number of isolation rooms
- Delayed institution of effective therapy
  - Delayed recognition of MDR-TB
  - Delayed laboratory susceptibility data

**Infection control factors**

- Inadequate isolation
  - Isolation rooms
    - • positive pressure
    - • <6 air changes per hour
    - • air recirculated
    - • doors open
- Infectious patients leave rooms
  - • for nonmedical reasons (social, smoking, TV)
  - • premature discontinuation of isolation
- Inadequate precautions for aerosol-generating procedures
- Inadequate microbiologic methods
  - AFB smears not done
  - Slow turnaround of identification and susceptibility testing
  - Rapid methods not performed for:
    - culture
    - identification
    - antimicrobial susceptibility testing
  - Delayed communication of results from referral laboratories
  - Delayed communication of results to clinicians
  - Failure to monitor *M. tuberculosis* antimicrobial susceptibility results
  - Failure to maintain or analyze results/records of *M. tuberculosis* identification or susceptibility testing

ographs (i.e., cavitary or miliary patterns) for TB (21). In some instances, neither TSTs nor AFB smears of sputum were performed on potentially infectious patients. Most MDR-TB patients had abnormal chest radiographs; however, they usually had interstitial patterns rather than classical miliary or cavitary patterns. When cultures were obtained, the results of cultures and/or antimycobacterial susceptibility testing were not available for a median of 7 weeks because the methods used were not the most rapid; in some instances, the results were not available for 6 months. These delays in diagnosis resulted in delays in recognition of MDR-TB, delays in addition of effective antituberculosis therapy, and delays in the institution of appropriate isolation, and thus prolonged periods of exposure to other patients and health care workers.

## 5. Genetic Analysis of MDR-TB Outbreak *M. tuberculosis Isolates*

A critical element of each of the MDR-TB outbreak investigations was the molecular typing of the MDR-TB infecting strains. In the early 1990s, a new method to type *M. tuberculosis* isolates, RFLP, had been developed (26). Having documented epidemiologically that *M. tuberculosis* was being transmitted in these health care facilities, it was essential that molecular typing data support this conclusion. Thus, in each MDR-TB outbreak, available isolates from patients and health care workers with active TB disease were obtained and subjected to RFLP analysis. Although occasionally, a similar strain was identified at more than one facility, particularly in New York City, where patients might have either had contact in or outside of the hospital, in most instances one or more unique strains were documented to be transmitted within each facility (Table 7.1). In one instance, use of a polymerase chain reaction-based RFLP method was necessary because the source-case isolate was no longer viable (27,28). The combination of epidemiologic and molecular typing data conclusively proved that multidrug-resistant strains of *M. tuberculosis* were being transmitted within these facilities.

## 6. Risk of TB Infection in Health Care Workers

In each outbreak, occupational acquisition of TB was suspected. However, conclusively documenting this was difficult because, in most instances, the health care workers had infection but not disease. Furthermore, at a number of the MDR-TB outbreak hospitals, the health care worker TST program was inadequate; often, the health care workers had not had a TST within the past 1–2 years (14,15). At one MDR-TB outbreak hospital, 18% of health care workers had a positive TST in whom a TST was applied during the investigation; one health care worker developed MDR-TB and the infecting isolate had the same RFLP pattern as that recovered from MDR-TB outbreak patients (14,15). In the investigation by Beck-

Sague et al., outbreak ward health care workers with or without TST conversions were similar in age, race, sex, duration of employment in the outbreak ward, occupation, and shift worked (16). When health care workers in the outbreak ward and clinic were compared to those in control wards where TB patients were not admitted, outbreak ward health care workers were at significantly greater risk of a TST conversion. Risk of TST conversion was associated with exposure to infectious MDR-TB patients rather than to drug-susceptible TB patients and to MDR-TB patients who were AFB sputum smear positive. In the three MDR-TB outbreak hospitals in which TST data were available, between 22% and 50% of health care workers had TST conversions (14,16,19,24). At the time of these investigations, at least 16 health care workers had developed active MDR-TB; 7 were HIV-positive, 7 were HIV-negative, and the HIV status of 2 was unknown. At that time, 5/16 (31.2%) health care workers with active MDR-TB had died; 4/5 were known to be HIV infected. In many if not most instances, health care workers were not wearing the CDC-recommended respiratory protection, that is, particulate (dust–mist, dust–fume–mist, or high-efficiency particulate air filter (HEPA) respirators; often, the health care workers either wore no respiratory protection at all or improperly wore their masks or respirators.

## 7. Infection Control Programs in the United States

With the onset of these nosocomial MDR-TB outbreaks, considerable concern was raised about the status of TB infection control programs at U.S. hospitals. Several surveys were conducted to assess these programs. In 1992, the American Hospital Association (AHA) and CDC conducted a survey of all U.S. municipal, veterans administration, and university hospitals and a 20% random sample of private hospitals ($N$ = 1076 hospitals) (29). Of the 763 (71%) respondants, MDR-TB patients were admitted to 178 (25%) hospitals in 39 states. The number of AFB isolation rooms meeting CDC recommendations ranged from 0 to 60 (median = 7); 219 (29%) hospitals reported having no AFB isolation rooms meeting recommended criteria. Health care worker TST programs varied widely. Of the respondants, 15 (2%) hospitals reported nosocomial transmission of TB to patients and 91 (13%) reported TB transmission to health care workers. In March 1993, the Society of Healthcare Epidemiology of America (SHEA) and CDC conducted a survey of the SHEA membership, most of whom are affiliated with medical school teaching hospitals (30,31). From 1989 through 1992, the number of SHEA member facilities admitting MDR-TB patients increased from 10/166 (10%) to 49/166 (30%). During this period, the median TST positivity rate at the time of hire increased from 0.54% to 0.81%, whereas the median TST conversion rate remained stable at 0.35% in 1992 and 0.33% in 1989, respectively. Of 181 facilities from which data were reported, 113 (62%) had AFB isolation rooms meeting CDC recommendations. During the study period, the proportion of facilities in which surgical submicron masks or dust–mist or dust–fume–mist respirators were used increased from 9/196 (5%) to 85 (43%). In another survey conducted by the Association for Practitioners in Infection Control and Epidemiology (APIC) and CDC covering the same time period, the majority of respondants ($N = 1494$) were at smaller community hospitals and the proportion of hospitals admitting patients with TB and MDR-TB increased from 46.4% to 56.6% and 0.8% to 4.5%, respectively (32). During the study period, the health care worker TST positivity rate at hire rose from 0.95% to 1.14% and the TST conversion rate increased from 0.4% to 0.5%. In 1992, 66% of the hospitals reported that they had rooms which were compliant with CDC recommendations for AFB isolation rooms. At 64% of the hospitals, health care workers still used surgical masks for respiratory protection; from 1989 through 1992, the number of hospitals in which particulate respirators were used increased from 0.4% to 13.8%. In 1993, Moran et al. conducted a survey of infection control practices in emergency departments at a sample of the hospitals responding to the AHA/CDC survey (33). Of the 446 facilities surveyed, 298 returned completed questionnaires. The proportion of emergency departments in which TB patients were seen daily, weekly, monthly, or less frequently was 12.6%, 17.2%, 23.3%, and 46.9%, respectively. Emergency departments in which TB isolation rooms meeting CDC recommendations were available in triage/waiting rooms and the emergency department itself was 1.7% and 19.6%, respectively. One or more health care workers had TST conversions in 16.1% of the surveyed emergency departments in 1991 and 26.9% in 1992. Surveys have also documented that the microbiologic methods used in many laboratories are not the most rapid and that communication between the laboratories and the clinicians often is inadequate (14,16,19,34,35). Rapid methods were used for AFB microscopy in 47%, primary culture in 72%, *M. tuberculosis* identification in 38%, and drug-susceptibility testing in 13% of the laboratories surveyed in the AHA/CDC survey (29). Approximately 46% of hospitals surveyed and an estimated 30% of laboratories at all U.S. hospitals with 100 or more beds performed the minimal number of mycobacterial cultures deemed necessary to maintain competence (35,36).

The outbreak and survey data show that the emergence of MDR-TB as a major public health problem can be traced to the incomplete implementation of the CDC TB Guidelines (37,38). Many hospitals do not have recommended isolation facilities, health care worker TST programs, respiratory protection devices, or laboratory methods. Although the APIC/CDC survey shows that as many as 50% of community hospitals do not admit TB patients routinely and few admit MDR-TB patients, the SHEA/CDC and AHA/CDC surveys show that larger hospitals and those affiliated with medical schools in all parts of the U.S. routinely admit TB and MDR-TB patients (29–32). There also is evidence that there have been improvements in the degree of implementation of CDC TB Guideline recommendations from the time of the AHA/CDC survey until the SHEA/CDC and APIC/CDC surveys.

## 8. Efficacy of the CDC TB Guideline Recommendations

The nosocomial TB and MDR-TB outbreaks raised concern in the infectious disease, infection control, and industrial hygiene communities about the effectiveness of the CDC TB Guideline recommendations (37). To assess the efficacy of the control measures, follow-up investigations were conducted at three of the hospitals where MDR-TB outbreaks had occurred. In each hospital, a wide variety of infection control measures similar to the CDC 1990 TB Guidelines were implemented (Table 7.3 and 7.4). Because same ward exposure to an infectious MDR-TB patient was identified as the most significant risk factor in the initial MDR-TB outbreak investigations, this was assessed as a measure of nosocomial or occupational acquisition of TB in the follow-up investigations. In the first investigation at an MDR-TB outbreak in a New York City hospital, Maloney et al. documented that the proportion of patients with multidrug-resistant strains of *M. tuberculosis* decreased, the proportion of MDR-TB patients with same ward exposures decreased, and the TST conversion rates of health care workers assigned to the outbreak wards were lower in the intervention period compared to the outbreak period (Table 7.3) (39). In another follow-up investigation, Wenger et al. showed that after implementation of control measures in the MDR-TB outbreak ward, no episodes of MDR-TB could be traced to contact with infectious MDR-TB patients and health care worker TST conversions were terminated (40). Finally, Stroud et al., in a follow-up investigation at another New York City hospital, documented that after implementation of recommended TB infection control measures, the MDR-TB attack rate for AIDS patients decreased from 19/216 (8.8%) to 5/193 (2.6%) (41). Each of these studies showed that with more complete implementation of administrative and engineering controls and use of respiratory protective devices, MDR-TB outbreaks could be terminated and further transmission of *M. tuberculosis* from patients to other patients or health care workers could be either terminated or reduced to background rates seen in wards where TB patients were not admitted routinely (42). These data document that the CDC TB Guidelines work if they are fully implemented.

## 9. CDC TB Guidelines

In 1990, shortly after the investigation of the first nosocomial MDR-TB outbreak in the United States, the CDC published "Guidelines for Preventing the Transmission of Tuberculosis in Health Care Settings, with Special Focus on HIV-Related Issues" (37). These Guidelines provided the basic elements of infection

Table 7.3 Evaluation of preventive intervention on patient to patient MDR-TB transmission at three MDR-TB outbreak hospitals

| | | | No. of Case-Patient | | | Proportion of patients with MDR-TB[a] | | Mortality | | Nosocomial MDR-TB exposure | |
|---|---|---|---|---|---|---|---|---|---|---|---|
| Hospital | Initial period | Intervention period | Outbreak period | Intervention period | Total | Outbreak period | Intervention period[a] | Outbreak period | Intervention period | Outbreak period | Intervention period |
| 1 | Jan. 90– May 90 | June 90– June 92 | 15 | 11 | 26 | 26/180 | 28/498[b] (TB) | — | — | 12 | 0 |
| 2 | Jan. 89– Mar. 90 | Apr. 90– Sept. 92 | 16 | 22 | 38 | 19/216 | 9/277[c] (A) | 6 | 8 | 11 | 5 |
| 4 | Jan. 90– June 91 | July 91– Aug. 92 | 30 | 10 | 40 | 30/95 | 10/70[c] (O) | 25 | 4 | 20 | 1[d] |

[a]All TB patients (TB), all outbreak ward patients (O), or all AIDS patients (A).

[b]$p < .01$.

[c]Four patients were exposed on HIV ward in initial period.** p = .02.

[d]$p = 0.003$.

Table 7.4 Evaluation of preventive interventions on patient to health care worker MDR-TB transmission at Three MDR-TB outbreak hospitals

| | | | Susceptible HCWs | | No. of HCWs with Tuberculin skin test conversion | | No. of HCWs with active TB | | |
|---|---|---|---|---|---|---|---|---|---|
| Hospital | Initial period | Intervention period | Initial period | Intervention period | Initial period | Intervention period | Initial period | Intervention period | Total |
| 1 | Jan. 90–May 90 | June 90–June 92 | 25 | 27 | 7 | 3[a,b] | 1 | 0 | 1 |
| 2 | Jan. 89–Mar. 90 | Apr. 90–Sept. 92 | 60 | 29 | 11 | 5 | 4 | 0 | 4 |
| 4 | Jan. 89–June 91 | Jul 91–Aug. 92 | 90 | 78 | 15 | 4[c] | 1 | 1 | 2 |

[a]All had exposure to unknown infectious MDR-TB patient.

[b] $p < .01$.

[c] $p < .02$.

control programs to prevent the transmission of *M. tuberculosis* in health care facilities. Emphasis was placed on early identification of infectious TB patients, preventing the generation of infectious droplet nuclei, preventing spread through source controls, reducing microbial contamination of the air, disinfection and sterilization, and surveillance for TB in health care workers. Recommendations for AFB isolation rooms included having six or more air changes per hour, having the room with negative pressure in relation to other rooms or corridors, and exhausting the air from the room directly to the outside. The Guideline also recommended use of particulate respirators for the protection of health care workers.

As a result of the MDR-TB outbreaks and in response to requests for changes and/or clarifications in the CDC 1990 TB Guidelines, the CDC revised these Guidelines, published the revision in the *Federal Register* for public comment, made final revisions based on these comments, and then published the "Guidelines for Preventing the Transmission of *Mycobacterium tuberculosis* in Health-Care Facilities, 1994" (38). The 1994 CDC TB Guidelines include a relatively concise recommendations section followed by four extensive supplements. The recommendations include sections on assignment of responsibility, risk assessment, development of the TB infection control program, and periodic risk assessment; identifying, evaluating, and initiating treatment for patients who might have active TB; management of patients who might have active TB in ambulatory-care settings and emergency departments; management of hospitalized patients who have confirmed or suspected TB; engineering controls; respiratory protection; cough-inducing and aerosol-generating procedures; education and training of health care workers (HCWs); HCW counseling, screening and evaluation; problem evaluation; coordination with the Public Health Department; and additional considerations for selected areas in health care facilities and other health care settings. The supplements cover determining the infectiousness of a TB patient; diagnosis and treatment of latent TB infection and active TB; engineering controls [i.e., ventilation and ultraviolet germicidal irradiation (UVGI)]; respiratory protection; and decontamination—cleaning, disinfecting, and sterilizing of patient-care equipment. Finally, a glossary of terms, references, and a detailed index are provided.

## 10. Hierarchy of Controls

The 1994 CDC TB Guidelines emphasize the importance of understanding the hierarchy of TB control measures: administrative controls for identifying infectious TB patients rapidly and reducing exposure to such patients, engineering controls to reduce the production and spread of airborne droplet nuclei, and respiratory protective devices. Although each of these measures was discussed in the 1990 CDC TB Guidelines, the 1994 CDC TB Guidelines highlight the importance of these measures. The first and most important TB control measure is the use of administrative controls to reduce the risk of exposure to persons with infectious TB. This includes developing and implementing effective written policies and protocols to ensure a high index of suspicion for TB by clinicians, appropriate workup of patients suspected to have TB, rapid laboratory techniques (fluorescence stains, radiometric methods for species identification and susceptibility testing, genetic probes), and communication of results rapidly to the clinician and prompt placement of the infectious TB patients in an area where exposure to other patients or unprotected health care workers is minimized (34,36). Emphasis is placed on education of health care workers about the epidemiology of TB, the importance of suspecting TB particularly in patients with HIV infection, the importance of health care worker TST programs, and the need to know the susceptibility of prevalent *M. tuberculosis* strains in the hospital and community, the need to rapidly and appropriately isolate TB patients, and the importance of initiating an effective treatment regimen in infectious TB patients.

The next level of the hierarchy is the use of engineering controls to prevent the spread and reduce the concentration of infectious droplet nuclei. This includes (1) direct source control using local exhaust ventilation, (2) controlling the direction of airflow to prevent contamination of air in adjacent areas, (3) dilution and removal of contaminated air by general ventilation, and (4) air cleaning via air filtration or ultraviolet germicidal irradiation.

The third and last level of the hierarchy is the use of personal respiratory protective devices by health care workers in selected areas where risk of occupational acquisition of TB is suspected to be higher (i.e., AFB isolation rooms and where cough-inducing or aerosol-producing procedures are performed). Neither the outbreak investigations, follow-up studies at outbreak hospitals, or surveys have documented the independent importance of these different measures. However, a high index of suspicion leading to rapid diagnosis and prompt placement of infectious TB patients in isolation is critical. The majority of nosocomial or occupational acquisition of TB results from exposure to an undetected infectious TB patient.

## 11. Risk Assessment

The next important addition in the 1994 CDC TB guidelines is risk assessment. The purpose of the risk assessment is to identify areas in the facility in which the risk of patient-to-patient or patient-to-health care worker transmission of tuberculosis is minimal, very low, low, intermediate or high. This was an attempt to individualize the TB control program so that one approach would not be expected of everyone and individual institutions could be given flexibility both between and within institutions. In the risk assessment, each facility and individual areas of the institution are assessed for risk of nosocomial or occupational transmission of *M. tuberculosis.* Based on this assessment, the frequency of health care worker TST, repeat risk assessment, and ventilation evaluation is determined or supplemental engineering interventions are suggested. If a cluster of patients or health care workers develop *M. tuberculosis* infection or disease, more frequent monitoring is required. Several hypothetical examples are provided to assist in the risk assessment approach.

## 12. TB Control Program

Next, the elements of the TB control program are outlined and discussed (Table 7.5). In this section, emphasis is placed on use of laboratory methods to facilitate rapid diagnosis of TB patients, recommendation that smear results be available in ≤24 h of specimen collection, the importance of considering TB in HIV-infected patients even if another pathogen is identified, the importance of rapid evaluation and triage in outpatient settings, the importance of directly observed therapy (DOT), that AFB isolation rooms (i.e., private room, negative pressure, ≥6 air changes per hour, air exhausted directly outside) can have recirculation of air to the same patients room or general ventilation if the air is passed through a HEPA filter in the ventilation system, that all infectious TB patients (drug susceptible or MDR) should remain in isolation until clinically improved *and* they have negative AFB sputum smears on 3 consecutive days (because the decision to remove from isolation is usually made before susceptibility results are available), that consideration should be given to keeping MDR-TB patients in AFB isolation throughout their hospitalization, and that discharge should be coordinated with the public health department and private clinicians.

In the 1990 CDC TB Guidelines, the respiratory protective devices recommended were particulate respirators which include dust–mist, dust–fume–mist, or HEPA filter respirators. In the 1994 CDC TB Guidelines, criteria for a new TB respirator are made. These criteria include (1) the ability to filter particles 1 $\mu$m in size (in the unloaded state) with a filter efficiency of ≥95% (i.e., filter leakage ≤5%), given flow rates of up to 50 L/min, (2) the ability to be qualitatively or quantatively fit tested in a reliable way, (3) the ability to be adequately fit checked before each use in a reliable way, and (4) the ability to fit health care workers with different facial sizes and characteristics, which can usually be met by the availability of at least three sizes of respirator. The criteria were based on the characteristics thought to be most desirable in a respirator and on the in-use experience at MDR-TB outbreak hospitals where patient-to-health care worker MDR-TB transmission was terminated by using submicron masks or dust–mist respirators together with implementation of the 1990 CDC TB Guidelines recommendations (39–43).

When the 1990 CDC TB Guidelines first recommended particulate respirators

Table 7.5 Characteristics of an effective tuberculosis (TB) Infection control program[a]

I. Assignment of responsibility
   A. Assign responsibility for the TB infection control program to qualified person(s)
   B. Ensure that persons with expertise in infection control, occupational health, and engineering are identified and included.

II. Risk assessment, TB infection control plan, and periodic reassessment
   A. Initial risk assessments
      1. Obtain information concerning TB in the community.
      2. Evaluate data concerning TB patients in the facility.
      3. Evaluate data concerning purified protein derivative (PPD) tuberculin skin test conversions among health care workers (HCWs) in the facility.
      4. Rule out evidence of person-to-person transmission.
   B. Written TB infection control program
      1. Select initial risk protocol(s).
      2. Develop written TB infection control protocols.
   C. Repeat risk assessment at appropriate intervals.
      1. Review current community and facility surveillance data and PDD tuberculin skin test results.
      2. Review records of TB patients.
      3. Observe HCW infection control practices.
      4. Evaluate maintenance of engineering controls.

III. Identification, evaluation, and treatment of patients who have TB
   A. Screen patients for signs and symptoms of active TB
      1. On initial encounter in emergency department or ambulatory-care setting.
      2. Before or at the time of admission.
   B. Perform radiologic and bacteriologic evaluation of patients who have signs and symptoms suggestive of TB.
   C. Promptly initiate treatment.

IV. Managing outpatients who have possible infectious TB
   A. Promptly initiate TB precautions.
   B. Place patients in separate waiting areas or TB isolation rooms.
   C. Give patients a surgical mask, a box of tissues, and instructions regarding the use of these items.

V. Managing inpatients who have possible infectious TB
   A. Promptly isolate patients who have suspected or known infectious TB.
   B. Monitor the response to treatment.
   C. Follow appropriate criteria for discontinuing isolation.

VI. Engineering recommendations
   A. Design local exhaust and general ventilation in collaboration with persons who have expertise in ventilation engineering.
   B. Use of single-pass air system or air recirculation after high-efficiency particulate air (HEPA) filtration in areas where infectious TB patients receive care.
   C. Use additional measures, if needed, in areas where TB patients may receive care.
   D. Design TB isolation rooms in health-care facilities to achieve $\geq$ 6 air changes per hour (ACH) for existing facilities and $\geq$ 12 ACH for new or renovated facilities.
   E. Regularly monitor and maintain engineering controls.

Table 7.5 (*continued*)

- F. TB isolation rooms that are being used should be monitored daily to ensure they maintain negative pressure relative to the hallway and all surrounding area.
- G. Exhaust TB isolation room air to outside or, if absolutely unavoidable, recirculate after HEPA filtration.

VII. Respiratory protection
- A. Respiratory protective devices should meet recommended performance criteria.
- B. Respiratory protection should be used by persons entering rooms in which patients with known or suspected infectious TB are being isolated, by HCWs when performing cough-inducing or aerosol-generating procedures on such patients, and by persons in other settings where administrative and engineering controls are not likely to protect them from inhaling infectious airborne droplet nuclei.
- C. A respiratory protection program is required at all facilities in which respiratory protection is used.

VIII. Cough-Inducing procedures
- A. Do not perform such procedures on TB patients unless absolutely necessary.
- B. Perform such procedures in areas that have local exhaust ventilation devices (e.g., booths or special enclosures) or, if this is not feasible, in a room that meets the ventilation requirements for TB isolation.
- C. After completion of procedures, TB patients should remain in the booth or special enclosure until their coughing subsides.

IX. HCW TB training and education
- A. All HCWs should receive periodic TB education appropriate for their work responsibilities and duties.
- B. Training should include the epidemiology of TB in the facility.
- C. TB education should emphasize concepts of the pathogenesis of and occupational risk for TB.
- D. Training should describe work practices that reduce the likelihood of transmitting *M. tuberculosis.*

X. HCW counseling and screening
- A. Counsel all HCWs regarding TB and TB infection.
- B. Counsel all HCWs about the increased risk to immunocompromised persons for developing active TB.
- C. Perform PPD skin tests on HCWs at the beginning of their employment, and repeat PPD tests at periodic intervals.
- D. Evaluate symptomatic HCWs for active TB.

XI. Evaluate HCW PPD test conversions and possible nosocomial transmission of *M. tuberculosis*

XII. Coordinate efforts with public health department(s)

[a]A program such as this is appropriate for health-care facilities in which there is a high risk for transmission of *M. tuberculosis.*

[b]Adapted for the 1994 CDC TB Guidelines (13).

for use to protect health care workers from occupational-acquisition of *M. tuberculosis* and it became clear that the use of respirators was for protection of health care workers from patient infections not vice versa, this area came under the jurisdiction of the Occupational Safety and Health Administration (OSHA) in the United States (44). Thereafter, no matter which respirator was recommended, the law required a respirator training program, education, and fit testing. OSHA requires that when a health care worker wears a respiratory protective device, that they use a National Institute for Occupational Safety and Health (NIOSH)-certified respirator. In 1992, OSHA asked NIOSH for a respirator recommendation to protect HCWs from TB. Because of a lack of data about the concentration of *M. tuberculosis* in droplet nuclei, the minimal infectious dose, the actual droplet nuclei particle size and size distribution, together with the possibility of a lethal outcome, NIOSH recommended a powered air-purifying respirator (PAPR) with a HEPA filter for moderate exposures and a positive pressure air-line respirator with a tight-fitting half-mask respirator for high-risk exposures (45). By law, NIOSH may not consider either cost or practicality in making it's recommendation. During 1992 and 1993, various OSHA regions made respirator recommendations ranging from dust–mist to PAPRs.

To understand the complex nature of arriving at an appropriate respirator recommendation for health care workers to prevent occupational acquisition of *M. tuberculosis,* it is important to understand the relationship between OSHA and NIOSH. OSHA requires that any respiratory protective device used to protect health care workers must be NIOSH-certified (44). Before 1996, NIOSH certifies respirator filtration in two ways (46). HEPA filter respirators are challenged with 0.03-$\mu$m particles; 99.97% of these particles must be excluded ($\leq$ 0.03% penetration). In contrast, dust–mist or dust–fume–mist respirators are challenged with 144 mg of silica dust; 99% of particles must be excluded ($<$1.5 mg or $<$1% penetration). That respirator certification process did not adequately test the efficacy of dust–mist or dust–fume–mist respirators against low-concentration aerosols in the size range of *M. tuberculosis* droplet nuclei (estimated at 1–5 $\mu$m). Neither test used a biologic or particle size similar to *M. tuberculosis.* Data show that there is wide variability in the penetration of dust–mist or dust–fume–mist respirators when challenged by either 1-$\mu$m particles or *M. chelonae* aerosols (47). On the other hand, some dust–mist or dust–fume–mist respirators prevent $\geq$95% of 1-$\mu$m particles from penetrating, whereas others allow up to 40% penetration. Thus, in contrast to HEPA filter respirators, which all prevent penetration of all but 0.03% or less of 0.3 micron size particles, the efficacy of dust–mist and dust–fume–mist respirators varies widely. For this reason, the 1994 CDC TB Guidelines indicate that although some dust–mist or dust–fume–mist respirators would meet the new criteria, only HEPA-filtered respirators meet these criteria and currently are certified by NIOSH to do so (38,43,46).

In January 1996, NIOSH initiated a new respirator certification process (48).

When approved, the new process will use a 0.03-$\mu$m test particle and have nine classes of filters (three levels of filter efficiency, each with three categories of resistance to filter efficiency degradation). The three levels of filter efficiency are 99.95%, 99%, and 95%. All of these classes of respirators will surpass the filter criteria recommended in the 1994 CDC TB Guidelines (specifically the ability to filter 95% of 1-$\mu$m particles) and will be available for use to protect health care workers from occupational exposure to *M. tuberculosis.* Follow-up data at several of the MDR-TB outbreak hospitals show that use of submicron surgical masks or dust–mist respirators by health care workers together with implementation of recommendations similar to those in the 1990 CDC TB Guidelines terminated patient-to-health care worker *M. tuberculosis* transmission on outbreak wards (39,40,42). Furthermore, survey data have shown that at health care facilities with more than six TB patients or with more than 200 beds that respirators with submicron filter capability reduce the risk of health care worker TST conversion (30,31). In addition, studies show that some dust–mist or dust–fume–mist respirators will filter >95% of particles with a mean size of 0.8 $\mu$m, smaller than the estimated particle size of TB droplet nuclei (47). The new NIOSH certification process will permit the use of respirators similar to those shown to be effective in outbreak settings and reduce the controversy surrounding the use of respirators in health care facilities to protect workers from *M. tuberculosis.*

Other factors to consider when selecting a respirator included face-seal leakage, the ability to fit test, and to fit check. HCWs with facial hair (beards, etc.) will not be able to get adequate face-seal with particulate respirators; positive-air-pressure respirators might be an alternative for these workers and for those performing very high-risk procedures on the TB patient.

## 13. HCW Education/Problem Evaluation

Another new section in the 1994 CDC TB Guidelines is on education and training of health care workers. All health care workers with possible *M. tuberculosis* exposure should be taught the epidemiology of TB, the potential for occupational exposure, the principles of TB infection control, the importance of routine and periodic health care worker TST, the principles of preventive therapy, the importance of seeking evaluation if the health care worker develops symptoms of TB, the higher risk of immunocompromised, particularly HIV-infected health care workers to develop disease given infection, counseling of health care workers, and options for voluntary work reassignment. The importance is emphasized of the two-step Mantoux TST of all health care workers at the time of hire and then retesting based on the risk assessment. Furthermore, a section on evaluating problems provides guidance on investigation of TST conversions or active disease in health care workers or possible patient-to-patient TB transmission.

## 14. Supplements

Finally, in the five supplements, extensive details are provided on the rationale, methods, and guidance for implementation of various elements of the TB control program. In the supplement on diagnosis and treatment, a table on TST interpretation, treatment options, and drug dosages are provided. In the supplement on engineering controls, extensive discussion of the basis and current science on the ventilatory recommendations is provided. In the supplement on respiratory protection, an extensive discussion of the factors to consider when choosing a respirator is provided, including the key components of a respirator training program and the advantages and disadvantage of the various respirator types.

## 15. Summary

Nosocomial transmission of *M. tuberculosis* is a long-recognized risk to health care workers. The recent collision of the TB and HIV outbreaks has resulted in a resurgence of TB in the United States and increased risk of nosocomial or occupational transmission. Most recently, the emergence of multidrug-resistant strains of *M. tuberculosis* has resulted in patients who are infectious for prolonged periods. These MDR-TB patients have shown that incomplete implementation of previously recommended measures, in both nosocomial and occupational transmission of *M. tuberculosis.* Most of the episodes of patient-to-patient or patient-to-health care worker *M. tuberculosis* transmission resulted from exposure to an unknown infectious TB patient and/or the failure to adequately implement the CDC TB Guidelines. Surveys of U.S. health care facilities show that although there has been improvement between 1989 and 1993, many still have not fully implemented the current CDC TB Guideline recommendations. Follow-up studies have documented that complete implementation of these Guidelines will terminate TB outbreaks and prevent further transmission to either patients or health care workers. The 1994 CDC TB Guidelines are the result of extensive debate and discussion including experts in infectious diseases, infection control, occupational health, industrial hygiene, laboratory medicine, and others. To the greatest extent possible, the scientific basis of recommendations is given. Unfortunately, much of the data needed to base all recommendations on available scientific studies is absent. Hopefully, future studies will provide the data needed to confirm the utility of the individual recommendations or provide data on which to base their modification. Now that the 1994 CDC TB Guidelines are finalized, the infection control community together with those in occupational medicine, health care facility administrators, worker unions, and national organizations must join with federal agencies (OSHA, CDC) to fully implement these guidelines and reduce the risk of further nosocomial TB transmission in health care facilities.

**References**


1. Pope AS (1942) An outbreak of tuberculosis in infants due to hospital infection. J Pediatr 40:297–300.
2. Ehrenkranz JN, Kicklighter LJ (1972) Tuberculosis outbreak in a general hospital: Evidence for airborne spread of infection. Ann Intern Med 77:377–382.
3. Alpert ME, Levison ME (1965) An epidemic of tuberculosis in a medical school. New Engl J Med 272:718–721.
4. Israel HL, Hetherington HW, Ord JG (1941) A study of tuberculosis among students of nursing. J Am Med Assoc 117:839–844.
5. Brahdy L (1941) Immunity and positive tuberculin reaction. Am J Publ Health 31:1040–1043.
6. Badger TL, Ayvazian LF (1949) Tuberculosis in nurses: Clinical observations on its pathogenesis as seen in a fifteen year follow-up of 745 nurses. Am Rev Tuberc 60:305–327.
7. Abruzzi WA, Hummel RJ. Tuberculosis: Incidence among American medical students, prevention and the use of BCG. New Engl J Med 248:722–729.
8. Centers for Disease Control (1994) Expanded tuberculosis surveillance and tuberculosis morbidity—United States, 1993. Morbid Mortal Wkly Rep 43:361–366.
9. Barnes PF, Bloch AB, Davidson PT, Snider DE Jr (1991) Tuberculosis in patients with human immunodeficiency virus infection. New Engl J Med 324:1644–1650.
10. Selwyn PA, Hartel D, Lewis VA, Schoenbaum EE, Vermund SH, Klein RS, Walker AT, and Freidland GH (1989) A prospective study of the risk of tuberculosis among intravenous drug users with human immunodeficiency virus infection. New Engl J Med 320:545–550.
11. Centers for Disease Control (1991) Nosocomial transmission of multidrug-resistant tuberculosis among HIV-infected person—Florida and New York, 1988–1991. Mortal Wkly Rep 40:585–591.
12. Dooley SW, Villarino ME, Lawrence M, Salinas L, Amil S, Rullan JV, Jarvis WR, Bloch AB, Cauthen GM (1992) Tuberculosis in a hospital unit for patients infected with the human immunodeficiency virus (HIV): Evidence of nosocomial transmission. J Am Med Assoc 267:2632–2634.
13. Di Perri G, Cruciani M, Danzi MC, Luzzati R, DeChecchi G, Malena M, Pizzighella S, Mazzi R, Solbiati M, and Concia E (1989) Nosocomial epidemic of active tuberculosis among HIV-infected patients. Lancet 2:1502–1504.
14. Edlin BR, Tokars JI, Grieco MH, Crawford JT, Williams J, Sordillo EM, Ong KR, Kilburn JO, Dooley SW, Castro KG, Jarvis WR, Holmberg SD (1992) An outbreak of multidrug-resistent tuberculosis among hospitalized patients with the acquired immunodeficiency syndrome: Epidemiologic studies and restriction fragment length polymorphism analysis. New Engl J Med 326:1514–1522.
15. Tokars JI, Jarvis WR, Edlin BR, Dooley S, Grieco M, Gilligan ME, Schneider N, Montonez M, and Williams J (1992) Tuberculin skin testing of hospital employees during an outbreak of multidrug-resistant tuberculosis in human immunodeficiency virus (HIV)-infected patients. Infect control Hosp Epidemiol 13:509–510.

16. Beck-Sague CM, Dooley SW, Hutton MD, Otten J, Breeden A, Crawford JT, Pitchenik AE, Woodley C, Cauthen G, Jarvis WR (1992) Outbreak of multidrug-resistant tuberculosis among persons with HIV infection in an urban hospital: Transmission to staff and patients and control measures. J Am Med Assoc 268:1280–1286.
17. Fischl MA, Uttamchandani RB, Daikos GL, Poblete RB, Moreno JN, Reyes RR, Boota AM, Thompson LM, Cleary TJ and Lai S (1992) An outbreak of tuberculosis caused by multiple-drug-resistant tubercle bacilli among patients with HIV infection. Ann Intern Med 117:177–183.
18. Ikeda RM, Birkhead GS, DiFerdinando GT Jr, et al. (1995) Nosocomial tuberculosis: An outbreak of a strain resistant to seven drugs. Infect Control Hosp Epidemiol 16:152–159.
19. Pearson ML, Jereb JA, Frieden TR, Crawford JT, Davis B, Dooley SW, Jarvis WR (1992) Nosocomial transmission of multidrug-resistant *Mycobacterium tuberculosis:* A risk to hospitalized patients and health-care workers. Ann Intern Med 117:191–196.
20. Valway SE, Greifinger RB, Papania M, Kilburn JO, Woodley C, Diferdinando GT, and Dooley SW (1994) Multidrug-resistant tuberculosis in the New York State prison system 1990–1991. J Infect Dis 170:151–156.
21. Coronado VG, Valway S, Finelli L, Dato V, Pineda MR, Yanaguchi E, Woodley CL, Crandell MS, and Jarvis, WR (1993) Nosocomial transmission of multidrug-resistant *Mycobacterium tuberculosis* among intravenous drug users with human immunodeficiency virus infection (Abstract S50). Abstracts of the Third Annual Meeting of the Society for Hospital Epidemiology of America.
22. Coronado VG, Beck-Sague CM, Hutton MD, Davis BJ, Nicholas P, Villareal C, Woodley CL, Kilburn JO, Crawford JT, Frieden TR, Sinkowitz RL, Jarvis WR (1993) Transmission of multidrug-resistant *Mycobacterium tuberculosis* among persons with human immunodeficiency virus infection in an urban hospital: Epidemiologic and restriction fragment length polymorphism analysis. J Infect Dis 168:1052–1055.
23. Jarvis WR (1993) Nosocomial transmission of multidrug-resistant *Mycobacterium tuberculosis.* Res Microbiol 144:117–122.
24. Coronado VG, Beck-Sague CM, Pearson ML, Dooley SW, Valway SE, Pineda MR and Jarvis WR (1993) Multidrug-resistant *Mycobacterium tuberculosis* among patients with HIV infection. Infect Dis Clin Pract 2:297–302.
25. Jereb JA, Klevens RM, Privett TD, Smith PJ, Crawford JT, Sharp VL, Davis BJ, Jarvis WR, and Dooley SW (1995) Tuberculosis in health care workers at a hospital with an outbreak of multidrug-resistant *Mycobacterium tuberculosis.* Arch Intern Med 155:854–859.
26. Cave MD, Eisenbach KD, McDernott PF, Bates JH, Crawford JT (1991) IS6110: Conservation of sequence in the *Mycobacterium tuberculosis* complex and its utilization in DNA finger printing. Molec Cell Probes 5:73–80.
27. Haas WH, Butler WR, Woodley CL, Crawford JT (1993) Mixed-linker polymerase chain reaction: A new method for rapid fingerprinting of the *Mycobacterium tuberculosis* complex. J Clin Microbiol 31:1293–1298.
28. Zaza S, Blumberg HM, Beck-Sague C, Parrish C, Pineda M, Woodley C, Crawford JT Jr, McGowan JE Jr, Jarvis WR (1995) Occupational Risk of Infection with *Mycobacterium tuberculosis:* Patients and healthcare workers as a source of infection. J Infect Dis 172:1542–1549.

29. Rudnick JR, Kroc K, Manangan L, Banerjee S, Pugliese G, Jarvis W (1992) How prepared are U.S. hospitals to control nosocomial transmission of tuberculosis? Presented at the First World Congress on Tuberculosis.
30. Fridkin SK, Manangan L, Bolyard E, SHEA, Jarvis WR (1995) SHEA–CDC TB survey, Part I: Status of TB infection control programs at member hospitals, 1989–1992. Infect Contr Hosp Epidemiol 16:129–134.
31. Fridkin SK, Manangan L, Bolyard E, SHEA, Jarvis WR (1992) SHEA–CDC survey—Part II: Efficacy of TB infection control programs at member hospitals, 1992. Infect Control Hosp Epidemiol 16:135–140.
32. Sinkowitz R, Fridkin S, Manangan L, Wenger P, APIC, Jarvis WR (1995) Status of tuberculosis infection control programs at U.S. hospitals, 1989–1992. Annual Meeting of the Association for Professional in Infection Control and Epidemiology.
33. Moran GJ, Fuchs MA, Jarvis WR, Talan DA (1995) Tuberculosis infection control practices at emergency departments in the United States, 1993–1994. J Am Med Assoc Ann Emerg Med 26:283–289.
34. Huebner RE, Good RC, Tokars JI (1993) Current practices in mycobacteriology: Results of a survey of state public health laboratories. J Clin Microbiol 31(4):771–775.
35. Tenover FC, Crawford JT, Huebner RE, Geiter LJ, Horsburgh CR, Good RC (1993) Guest Commentary. The Resurgence of tuberculosis: Is your laboratory ready? J Clin Microbiol 31(4):767–770.
36. Tokars JI, Rudnick JR, Kroc K, Manangan L, Pugliese G, Huebner RE, Chan J, and Jarvis WR (1996) U.S. hospital mycobacteriology laboratories: Status and comparison with state public health department laboratories. J Clin Microbiol 34(3):680–685.
37. CDC (1990) Guidelines for Preventing the Transmission of Tuberculosis in Tuberculosis in Health-Care Settings, with Special focus on HIV-Related Issues. Morbid Mortal Wkly Rep 39:No. RR-17.
38. CDC (1994) Guidelines for Preventing the Transmission of *Mycobacterium tuberculosis* in Health-Care Facilities, 1994. Morbid Mortal Wkly Rep 43:No. RR1-13.
39. Maloney SA, Pearson ML, Gordon MT, DelCastillo R, Boyle JF, and Jarvis WR (1995) Efficacy of control measures in preventing nosocomial transmission of multidrug-resistant tuberculosis to patients and health care workers. Ann Intern Med 122:90–95.
40. Wenger PN, Otten J, Breeden A, Orfas D, Beck-Sague CM, and Jarvis WR (1995) Control of nosocomial transmission of multidrug-resistant *Mycobacterium tuberculosis* among healthcare workers and HIV-infected patients. Lancet 345:235–240.
41. Stroud LA, Tokars JI, Grieco MH, Crawford JT, Culver DH, Edlin BR, Sordillo EM, Woodley CL, Gilligan ME, Schneider N, and Jarvis WR (1995) Evaluation of infection control measures in preventing the nosocomial transmission of multidrug-resistant *Mycobacterium tuberculosis* in a New York City hospital. Infect Control Hosp Epidemiol 16:141–147.
42. Jarvis WR (1995) Nosocomial transmission of multidrug-resistant *Mycobacterium tuberculosis.* Am J Infect Control 23:146–151.

43. Jarvis WR, Bolyard EA, Bozzi CJ, Burwen DR, Dooley SW, Martin LS, Mullan RJ and Simone PM (1995) Respirators, recommendations, and regulations: The controversy surrounding protection of health care workers from tuberculosis. Ann Intern Med 122:142–146.

44. OSHA 29DFR-OSH (1972) 29 CFR 1910.134—occupational safety and health standards, personal protective equipment, respiratory protection. Code of Federal Regulation. Fed Reg 37:1910–1934.

45. Leidel NA, Mullan RJ (1992) NIOSH recommended guidelines for personal respiratory protection of workers in health-care facilities potentially exposed to tuberculosis. Atlanta, GA: U.S. Department of Health and Human Services, Public Health Service, Centers for Disease Control, National Institute for Occupational Safety and Health.

46. NIOSH 30CFR-NIOSH (1972) 30 CFR Part II—respiratory protective devices; tests for permissibility, fees. Code of Federal Regulations. Fed Reg 37:6243–6271.

47. Chen SK, Vesley D, Brosseau LM, and Vincent JH (1994) Evaluation of single-use masks and respirators for protection of health care workers against mycobacterial aerosols. Am J Infect Control 22:65–74.

48. US Department of Health and Human Services (1994) 42 CFR Part 84: Respiratory protective devices; proposed rule. Fed Reg 59:26849–26889.

# 8

# Immunology of Tuberculosis

*Venkata M. Reddy and Burton R. Andersen*

## 1. Introduction

The foundation of tuberculosis immunity was laid by Robert Koch with the discovery of the "tuberculin test." At the 10th International Congress of Medicine held at Berlin in 1890, he announced the discovery of a curative agent for tuberculosis. This "agent" called tuberculin is, in fact, a broth culture filtrate of tubercle bacilli. The announcement was greeted with enthusiasm and some skepticism by the medical profession. It was soon found that tuberculin was not a curative agent, but it was of diagnostic value in both human and veterinary medicine. Later, it became the basis for the study of immunity in humans and experimental animals, and much attention has been focused on the mechanism involved in the tuberculin reaction.

A century after the discovery of the tubercle bacillus and tuberculin, the nature of host resistance against tuberculosis is still not completely understood. The slow progress is due to several factors, including the organism's complex structure, its slow generation time, its intracellular nature, and chronicity of the disease. It is likely that several resistance mechanisms operate concurrently, and the degree to which one mechanism appears to contribute to the state of resistance will be influenced by the experimental conditions which are used to evaluate a particular immune mechanism. In addition, the immunological status, genetic makeup, and the environment of the host are other parameters which influence the immune process.

In a chronic disease such as tuberculosis, the clinical status and prognosis of the disease depends on the balance between host factors and the pathogenicity of the parasite. *Mycobacterium tuberculosis,* being a facultative intracellular pathogen, causes disease by virtue of its ability to resist intracellular killing by macrophages, multiply within macrophages, and destroy them liberating the organisms to be taken up by fresh macrophages.

Acquired immunity in tuberculosis is mediated by immunocompetent cells which influence macrophages to develop an increased capacity to suppress and ultimately kill the bacteria. With the development of cell-mediated immunity (CMI) against tubercle bacilli, the patient concurrently exhibits delayed hypersensitivity (DH), an allergic reaction to the antigens of tubercle bacilli. The role of DH in immunity and/or pathogenesis of tuberculosis is much debated. The immune response to tuberculosis appears to act like a double-edged sword by contributing to the pathology of the diseases as well as leading to the protection of the host. The outcome of the interaction between the host and the parasite is determined by the attributes of the host as well as the parasite. In order to understand the problem in its proper perspective, it is essential to take into account the immunological responses of the host and the ways and means by which tubercle bacilli overcome the threat posed by the immune responses.

## 2. Mycobacterial Antigens

The antigens of *M. tuberculosis* consists of proteins, polysaccharides, and lipid complexes. Many of the unique immunopathological responses against mycobacteria are attributed to the high lipid content.

### *2.1. Old Tuberculin and PPD*

Old tuberculin (OT) is the oldest mycobacterial antigen preparation and was first produced by Koch (1890–1892). It is a concentrated crude extract of tubercle bacilli. Using this preparation, Koch produced an exaggerated ulcerating lesion in guinea pigs which were already sensitized to tubercle bacilli. This has been called the Koch phenomenon. Florence Seibert, in 1932, prepared chemical fractions of old tuberculin. These fractions were partially purified protein derivatives but were still complex mixtures of proteins and polysaccharides. The heterogenity of the preparation contributed to its nonspecificity and cross-reactivity. Later, Seibert described protein fractions A, B, C, and D, and polysaccharide fractions I and II. The protein fractions still contained some polysaccharides. Proteins A and B were found to be effective in eliciting the tuberculin reaction. In 1941, she described the preparation of "tuberculin purified protein derivative" (PPD) from OT by repeated precipitation with 50% ammonium sulfate. The purified protein derivative rapidly became the standard preparation for tuberculin testing, and for investigating CMI response (1,2).

Many efforts have been made to purify the culture filtrate and bacillary extracts using improved immunochemical techniques. Using goat antisera, 11 major antigens have been recognized by immunoelectrophoresis, and a reference system was developed (3). Marked cross-reactivity among various species of mycobacteria was well established by using techniques like fused rocket immunoelectrophoresis. Advanced biochemical and immunological techniques have revealed the complexity of the *M. tuberculosis* antigens. The antigens investigated include cell-wall-derived and cytoplasmic constituents. It has been observed that the antigenic composition varied with growth conditions, age of the culture, temperature of incubation, and extraction procedures.

Based on immunodiffusion analysis, mycobacterial antigens can be divided into four groups. Group I consists of at least four antigens present in all mycobacteria and nocardiae. Group II has at least three antigens which are shared by all species of slow growers, and Group III has four antigens shared by most fast growers. Group IV antigens are limited to individual species and at least some of these are present in every strain of a given species (4). Techniques such as crossed immunoelectrophoresis and double rocket immunoelectrophoresis have helped in understanding the complexity of mycobacterial antigens. Because various species of mycobacteria share antigens, the analysis of a large number of precipitin bands among species becomes a tremendous task when the lines of identity are used to establish relationships. Using hyperimmune serum, as many as 50 or more precipitins have been identified, and to relate a number of species with homologous and heterologous antisera is a very difficult task. To make the task easier, the concept of percent quality sharing (PQS) was introduced for analysis of rocket immunoelectrophoresis. Species and strains were related to various reference antigen–antibody systems by dividing the number of precipitin bands produced with a heterologous antigen preparation reacting with reference antigens (5,6).

### *2.2. Protein Antigens*

The major source of protein antigens of *M. tuberculosis* are the cytoplasmic proteins liberated following autolysis. Secretory proteins, especially the early secretory proteins, are thought to play an important role in stimulation of the host's protective immune response (7). Culture filtrate antigens contain both secretory and cytoplasmic proteins. A large number of protein antigens of *M. tuberculosis* have been identified using modern immunochemical techniques. Monoclonal antibodies have been raised for several of the cytoplasmic and secretory proteins, and genes for some of them have been cloned (8). Most of the cytoplasmic and cell-wall peptides with molecular weights of 500–10,000 were found to be immunoreactive, causing either cellular immune responses or serological reactions. These small peptides can be associated with polysaccharides. The cell membrane is the other source of protein antigens obtained following disruption of bacilli. A major problem encountered during the preparation of many of the protein antigens is denaturation, which occurs during the course of purification. Another problem is the presence of a polysaccharide component of the cell wall. Cell wall-associated protein antigens can be covalently linked to peptidoglycon (9). These antigens might be responsible for immunoreactivity of the purified mycobacterial cell walls. Ribosomes and ribosomal proteins of mycobacteria have also been shown to be immunologically active by stimulating delayed hypersensitivity (10,11), and immune response to these antigens has been shown to protect against infection (12).

### 2.3. Polysaccharide Antigens

Earlier workers like Seibert, Afronti, and Stacey have studied polysaccharide antigens extensively, and there is still interest in the role of these cell-wall mycobacterial antigens. Polysaccharide preparations of mycobacteria are made up of arabinogalactans, arabinomannans, mannans, and glucans. These polysaccharides alone or in combination with peptides and lipids are responsible for various immune responses (1). The mycobacterial cell wall contains peptidoglycan, and crude preparations of peptidoglycans are immunologically active. The arabinogalactans and arabinomannans with oligosaccharide side chains constitute specific antigenic determinants of mycobacterial species. Polysaccharides devoid of proteins fail to induce delayed skin hypersensitive reactions; however, they are able to induce immediate skin reactions and circulating antibodies. Lipoarabinomannan (LAM) containing arabinose, mannose, glycerol, and a polyphosphate has been found to be an immunodominant antigen in *M. tuberculosis* (13).

## 3. The Immune Response

The first line of defense in mammalian hosts is phagocytosis of the invading organisms by polymorphonuclear leukocytes and mononuclear macrophages. The invading organisms are ingested and compartmentalized in the phagosomes. They are killed by oxidative and other bactericidal mechanisms and are degraded by the secretion of lysosomal enzymes. Facultative intracellular parasites, however, tend to resist these mechanisms and continue to multiply and kill the macrophages. To increase the efficiency of macrophages in controlling facultative intracellular organisms, the host has devised ways to activate these cells.

Normally, when the first line of defense fails to curtail the invading parasite, the second line of defense takes over. This involves specific immune response and humoral and/or cellular response against antigenic components of the invader. Humoral antibodies are particularly effective in combating extracellular organisms, whereas specific cell-mediated immunity has been reported to be the major mechanism of destroying intracellular organisms (14–16).

In acquired cellular resistance against infection, immunologically committed lymphocytes act as directive and motivating factors that bring phagocytes to the focus of infection and stimulate them metabolically, thus creating necessary conditions that interfere with the survival of the parasite. The antigenic stimulus provokes the production of a large number of pyrininophilic blasts in the thymus-dependent areas of the spleen and regional lymph nodes, and these cells are delivered into the lymphatics and blood. The central lymph from appropriately immunized animals contains cells that can confer sensitivity to tuberculin upon injection into normal recipients (14–16).

The following events take place when a sensitized animal is injected with specific antigen:

A few specifically sensitized lymphocytes interact with the antigen and release lymphokines.

The lymphokines cause the accumulation of blood-borne macrophages or their precursors at the site of antigen localization, followed by activation of the macrophages.

Consequently, the macrophages acquire enhanced microbicidal capacity characteristic of acquired cellular resistance.

## 4. Immunocompetent Cells

From the foregoing description, it is clear that T lymphocytes, B lymphocytes, and macrophages are the cells that take part in immune response against an invading organisms. T lymphocytes are mainly responsible for CMI, whereas B lymphocytes alone or in concert with T lymphocytes are necessary for humoral immunity. Human T and B lymphocytes are indistinguishable by light microscopy; however, they can be distinguished by cell surface molecules identified with suitable antisera or other reagents. T and B lymphocytes also differ in their reaction to reagents such as mitogens.

## 5. T Lymphocytes and Immunity to Tuberculosis

Acquired immunity to tuberculosis is cell mediated, and the major partner in the generation of the cellular immune response is the thymus-dependent (T) lymphocyte. Adoptive immunity studies in mice have shown that T cells confer immunity to naive animals. T lymphocytes are a heterogenic population of cells present in the circulation and in major lymphoid organs. Cell surface antigens have been used as phenotypic markers of immunocompetent cell subtypes. These antigens, referred to as cluster differentiating (CD) antigens, are glycoprotein molecules resembling surface immunoglobulins. The surface glycoproteins play a very important role in recognition of antigens and signal transduction into the

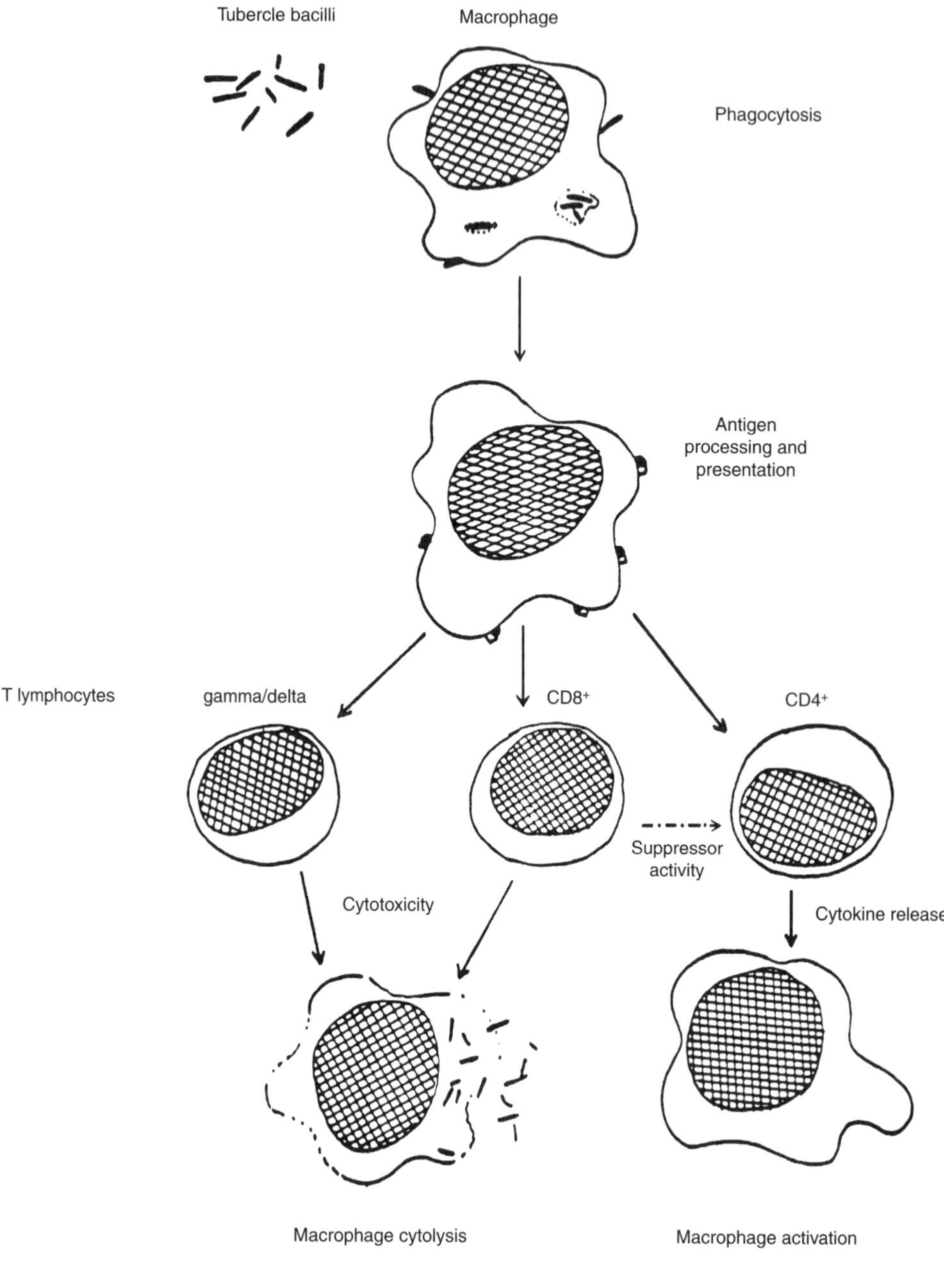


Figure 8.1. Protective cellular immune response in tuberculosis.

cell interior. Based on the presence of different CD antigens, T lymphocytes can be classified into different subsets, which also reflects on different functions. All mature T lymphocytes express CD2 and CD3 on their surface. A 57-kDa glycoprotein, CD4 is present in a subpopulation of T lymphocytes called "helper" cells. Another subpopulation called "suppressor" and "cytotoxic" cells express a 32-kDa protein (CD8). Helper and suppressor/cytotoxic T lymphocytes are also referred to as T4 and T8 cells. Functional classification is not very rigid, as CD4+ cytotoxic cells do exist; thus, their functional characteristics overlap to a certain extent. The major immune functions of T lymphocytes is to recognize antigen, initiate the immune response, direct and modulate the function of other cells, and maintain memory.

The antigen binding site is the T-cell receptor (TCR), which consists of an aggregation of several polypeptide chains. Most T cells possess on their surface a 90-kDa heterodimeric glycopeptide consisting of alpha and beta chains linked by disulfide bonds. Another TCR subset has a heterodimer consisting of gamma and delta chains. The alpha/beta and gamma/delta receptors are associated with the CD3 complex on the surface. The TCR and CD3 complex is essential for the effective function of T cells. The TCR is analogous to surface immunoglobulin, consisting of terminal variable chains and distal constant chains. Unlike immunoglobulins, T cells do not bind to free antigen in order to initiate a response. The T cells recognize antigens in conjunction with major histocompatibility (MHC) molecules expressed on most cells. The majority of the cells express class I MHC molecules; however, only a limited number of cells express class II MHC molecules. These Class II expressing cells include macrophages, dendritic cells, Langerhan cells of skin, Kupffer cells of liver, and activated B lymphocytes.

T lymphocytes do not recognize native antigens; the antigens need to be processed before being presented to the T cells. Macrophages process the antigens and present the antigens in context with MHC molecules on the surface. Processing involves altering the tertiary structure and fragmentation of the protein. In the case of antigen-presenting cells (APC), endocytozed antigens are processed in an acidic environment, followed by fusion of acidified antigen vesicles with other vesicles containing MHC class II molecules, and are exported to the cell surface. In case of non-APCs, the cytosolic-degraded antigens are entrapped within vesicles containing class I MHC molecules and are expressed together on the surface. T-cell-mediated immune response is restricted to antigens associated with specific autologous allelic form of MHC molecules, because T cells recognize antigens only in association with MHC molecules. CD4+ T cells recognize foreign antigen in association with class II MHC molecules, whereas CD8+ T cells recognize antigen in association with Class I molecules. Recognition of antigen by CD4+ T cells results in stimulation of the effector functions by producing several lymphokines; however, CD8+ T cells upon recognition of antigen cause lysis of target cells through direct contact. The general belief is that the

CD4+ T lymphocytes mediate immunity against intracellular bacteria by activating microbicidal capacity of macrophages via lymphokines, whereas CD8+ T lymphocytes mediate immunity against viral infection by lysis of target cells. In murine tuberculosis, the production CD4+ and CD8+ cytotoxic cells has been demonstrated which can lyse mycobacteria infected macrophages, also both of the subpopulations produce gamma interferon, the most important lymphokine that activates macrophages.

## 6. Cytokines in Tuberculosis

A complex network of cytokines serve as messengers of the immune system and are produced by a variety of cells of the body. They mediate immunological and nonimmunological functions of the body and are of crucial importance in maintaining homeostasis. However, it should be emphasized that cytokines not only mediate protection but also contribute to pathogenesis.

Interferon-gamma (INF-gamma) plays central role in activation of macrophages against intracellular organisms. In the murine macrophage system, IFN-gamma causes tuberculostasis, whereas in human macrophages, it fails to show any activity. Mycobacterial antigens are known to stimulate release of (TNF) from macrophages. Whether TNF confers protection against tuberculosis is not conclusive; however, TNF shows synergy with IFN-gamma in murine macrophages. Because TNF is an endogenous pyrogen and induces granuloma formation, it can contribute to the pathogenesis of the disease. Infection with mycobacteria and mycobacterial antigens cause increased release of interleukin-1 (IL-1) from macrophages, and IL-1 suppresses lymphocyte blastogenesis. Other cytokines like IL-4, IL-6, and IL-8 are also produced during mycobacterial infections; however, to what extent these cytokines take part in protection and/or pathogenesis of tuberculosis is still unknown (16).

Cytokine profiles differ in each species of animal. In the murine system CD4+ T cells can be divided into TH-1 and TH-2 types with different patterns of cytokine secretion. Following Bacille–Calmette–Guérin (BCG) vaccination, mice producing a TH-1 pattern of cytokines developed a more efficient protective immunity than the mice producing a TH-2 pattern of cytokines. This pattern of cytokine response is genetically dependent (17). TH-1 cells producing IFN-gamma and IL-2 confer resistance to intracellular infections, whereas susceptible mice with a TH-2–type response produce cytokines IL-4 and IL-5 (18). A similar kind of response has also been observed in humans, wherein tuberculosis patients exhibited a TH-2 pattern of response, whereas tuberculin-positive healthy individuals showed a TH-1 type of cytokine production (19). The TH-1 type of cytokine response produced locally in the tuberculous pleuritis (20) possibly results in a protective immune response.

## 7. Cytotoxic Cells in Tuberculosis

During the course of mycobacterial infection, in addition to the development of cytokine-mediated T cells, mycobacterial antigens stimulate antigen-specific MHC I restricted CD8+ cytotoxic T cells, and the kinetics of appearance of these two cell populations seem to differ (21). Mycobacterial antigens also stimulate nonspecific, non-MHC restricted natural killer (NK) cells (22). Cytotoxic cells have been shown to lyse the macrophages infected with mycobacteria (23). Cytotoxic cells might play a role in protection against tuberculosis by lysis of infected macrophages, resulting in the liberation of organisms to be taken up by activated macrophages with increased capacity to inhibit. Liberation of the organisms into external environment in the granulomatous lesions might be deleterious to the organisms. However, cytotoxicity might also contribute to the pathogenesis and dissemination of the organisms.

## 8. Macrophages and Tuberculosis Immunity

Macrophages bear the major responsibility in the generation of the protective immune response against tuberculosis. In doing so, macrophages also directly or indirectly participate in the pathogenesis of the disease. Upon encounter with mycobacteria, macrophages engulf the organisms. The first stage of engulfment is adherence of the organism to the macrophage, followed by phagocytosis. It has been shown that phagocytosis of *M. tuberculosis* is through Fc and complement receptors on the macrophages (24). However, *M. tuberculosis* can enter and multiply in nonprofessional phagocytic cells like fibroblasts, HeLa cells, human amnion cells, and epithelial cells, probably through binding to fibronectin (25).

The most important function of macrophages in the generation of protective immunity is to kill the organisms which have been phagocytozed. Following interaction of bacteria with the macrophage membrane and phagocytosis, the antibacterial function of the macrophage is activated. Macrophages possess several mechanisms of combating microorganisms, such as (a) generation of reactive oxygen intermediate (ROI), (b) generation of reactive nitrogen intermediates (RNI), (c) acidification of phagosomes, (d) secretion of lysosomal enzymes, (e) limitation of intracellular iron, and (f) production of defensins (26). The defense mechanisms responsible for mycobacterial killing is currently unclear.

Binding to the cell membrane and phagocytosis trigger membrane-bound NADH-oxidase, causing production of ROI such as $O_2^-$, $H_2O_2$, $OH^-$, and $O_2$ (1). The role of ROI in protection against mycobacterial infection is not conclusive. Mycobacteria also produce certain enzymes like superoxide dismutase (SOD) and catalase to overcome antimicrobial action of ROI. Activated macrophages and other cells produce RNI, and these RNI have been shown to inhibit growth of *M.*

*tuberculosis*. Phagocytosis causes entrapment of the organisms in phagosomes. Macrophages initiate fusion of lysosomes with phagosomes dumping the lysosomal enzymes into phagosomes causing degradation of microbes. Mycobacteria by various mechanisms can prevent phagosome–lysosome fusion. Neutrophils and alveolar macrophages secrete into phagosomes certain proteins with antibacterial potential called defensins; however, it is not clear whether they play a role in killing *M. tuberculosis*.

Acquired cellular resistance in tuberculosis can be defined as a state in which macrophages have been activated and possess an increased capacity to destroy tubercle bacilli. Even though unstimulated naive macrophages are capable of some antimicrobial activity, only activated macrophages are capable of suppressing and killing intracellular tubercle bacilli. Activation of host macrophages is effected through substances resulting from specific interactions between sensitized lymphocytes and the organism. Activation results in macrophage proliferation and increased digestive and microbicidal capacity, leading to fewer intracellular bacilli. Macrophages prior to activation will not kill the bacilli; however, the multiplying bacilli will kill the macrophages. Macrophages entering lesions become activated rapidly enough to kill the bacillus before being killed (27). These activated macrophages contain increased enzyme levels, such as $\beta$-galactosidase, acid phosphatase, $\beta$-glucuronidase, succinic dehydrogenase, and cytochrome oxidase. It has been found that the microbicidal capacity is proportional to $\beta$-galactosidase activity of the macrophages (27–29). Epitheloid cells with the highest enzyme activity generally contain fewer bacilli than epitheloid cells with lower enzyme activity (28). The mechanism of activation might differ between murine and human macrophages. IFN-gamma has been shown to enhance intracellular killing in mouse macrophages (30); however, the effect on human macrophages has been variable. Douvas et al. (31) and Rook et al. (32) have not observed increased killing of *M. tuberculosis* by IFN-gamma—stimulated macrophages. Activated murine macrophages produce increased levels of NO and this agent may be responsible for killing *M. tuberculosis*. In human macrophages, in addition to IFN-gamma and TNF-alpha, 1,25,-dihydroxycholecalciferol (calcitriol) plays an important role in the activation of macrophages.

*Mycobacterium tuberculosis* circumvents the antibacterial mechanism of the macrophages by various mechanisms. *M. tuberculosis* strains are capable of scavenging oxygen-derived free radicals by producing enzymes like catalase and superoxide dismutase. Bacterial components like lipoarabinomannan when released are known to interfere with the production of cytokines, especially IFN-gamma (33). At another level, *M. tuberculosis* or its products might interfere with phagosomes–lysosomes fusion (34).

## 9. Polymorphonuclear Neutrophils in Tuberculosis

Following infection with tubercle bacilli, neutrophils arrive first at the site and liberate mononuclear chemotactic factors (35), resulting in the accumulation of mononuclear cells, the precursors of macrophages. Neutrophils from tuberculosis patients exhibit increased adherence (36), increased phagocytosis (37), and increased reduction of nitroblue tetrazolium (38). In addition to participation in granuloma formation, neutrophils from tuberculosis are known to kill *M. tuberculosis* (39). The killing of *M. tuberculosis* by neutrophils is a nonoxidative process (40); however, the killing mechanism remains to be identified.

## 10. Delayed Hypersensitivity

Most intracellular parasites induce delayed hypersensitivity to one or more of the microbial antigens in the infected host. Delayed hypersensitivity is one of the manifestations of CMI and is thought to be protective against a variety of infectious agents. Concomitantly, it can also contribute to pathogenesis, as seen in some of the serious respiratory diseases. Delayed hypersensitivity is mediated by specially derived, sensitized lymphocytes rather than immunoglobulins and it also depends on collaboration between lymphocytes and macrophages. In the induction phase, macrophages present the antigen to T lymphocytes, involving physical interaction of two cell types, resulting in activated T lymphocytes responsible for production of lymphokines. Lymphokines such as chemotactic factor, migration inhibition factor (MIF), and perhaps others cause accumulation of blood-borne macrophages at tissue sites containing antigen and also cause their activation and increased microbicidal capacity (41).

Since the recognition of tuberculin hypersensitivity, efforts have been directed toward isolation of the antigenic components responsible for tuberculin hypersensitivity. Different proteins complexed with carbohydrates and nucleic acids were isolated first. Later, attention was directed toward crude proteins and purified proteins. Some workers concentrated on the mycobacterial protoplasmic components. One such component which caught much attention was the ribosomal protein which induced DH in BCG-sensitized animals. However, purified ribosomal protein failed to elicit DH (10).

The primary function of pulmonary macrophages is to act as scavengers of foreign material (42). Many of the antigens entering the lungs are disposed of without reaching the immunological system. However, some materials are able to make effective contact with the immunological apparatus causing hypersensitivity reactions. Considering the variety of antigens one inhales, the lungs are poor initiators of hypersensitive reactions. One of the reasons for limited pulmonary sensitization might be the difficulty of alveolar macrophages in presenting antigens to lymphocytes in a manner that evokes an immune response. Despite low hypersensitive reactions, it has been shown that cellular immune response takes place in the lungs (42,43). Two populations of monocytes should be distinguished when assessing the roles of these cells in CMI; resident alveolar macrophages and "migrant" macrophages newly arrived from the blood (42). The resident alveolar macrophages do not participate in CMI; however, the lungs develop CMI after the arrival of specifically sensitized T lymphocytes and blood-borne monocytes. It has been shown that migrant macrophages which enter the lung as a result of immunologic mobilization are capable of responding to MIF, as well as to other lymphokines (44). Further, some antigens from *M. tuberculosis* and a variety of pathogenic fungi are quite capable of penetrating beyond the lungs, a circumstance that brings them in close contact with the immunological apparatus. Sensitized lymphocytes which are quickly produced bring about an inflammatory reaction at the site of antigen (45). Activated alveolar macrophages show increased ability to ingest as well as kill both specific and nonspecific organisms.

The antigens of tubercle bacilli are highly concentrated in the local lesions, causing sensitization of lymphocytes and release of cytokines; consequently, the specific stimulus for cell activation is greater in the local lesion than anywhere else in the host (28,46). In inducing DH, cells localized to the regional lymph nodes are sensitized, and these sensitized cells are present in greater numbers in the tissue where the sensitization occurred (47). Macrophages continually enter the tuberculous lesions, can divide once or twice, and become activated. They first become immature epitheloid cells and then mature in the site containing the bacilli (46). With the accumulation of activated macrophages, the lesion begins to heal. Delayed hypersensitivity seems to influence the process by increasing the number of mononuclear cells entering the tuberculous lesion, their local rate of division, the rate of death, and the rate of activation. If the lesions are to regress and heal, the number of mononuclear cells entering and getting activated must exceed the rate of death (27,29).

Elicitation of DH demands the availability of mononuclear cells, specifically sensitized lymphocytes and a normal vascular system. Blocking and sequestration of lymphocytes in the lymphoid organs can often lead to misleadingly negative results. Interaction between antigen and lymphocytes, resultant production of mediators, and the influx of mononuclear cells to the reaction site must achieve completion within 24–48 h. Failure in any of the above conditions can lead to a negative result (48).

## 11. Beneficial and Detrimental Effects of Delayed Hypersensitivity

Whereas macrophages and other immunocompetent cells of the defense mechanism determine the course of tuberculosis, the dose of antigen determines whether cellular hypersensitivity is detrimental or beneficial. If tuberculinlike products of the bacilli are in high concentration, they cause localization and extensive activation of macrophages which can result in tissue necrosis (caseous necrosis). In addition, part of the tissue necrosis arises from factors accompanying the allergic inflammation and perhaps toxic products released from dead and dying cells. Thus, DH is responsible for much of the tissue damage in tuberculosis.

On the other hand, if the tuberculinlike products of the bacilli are in low concentration, they cause the accumulation and multiplication of activated macrophages so that the local defense is augmented but tissue injury is avoided. Thus, DH contributes to the development of cellular immunity (28).

In the necrotic and liquified areas, RNase and DNase are especially active; phosphatase and glucourodinase are also detectable. It therefore seems likely that nucleases and other lysosomal enzymes contribute to the cell autolysis and liquifaction of caseous lesions. Liquifaction is responsible for endobronchial spread of bacilli and for exogenous infection of other individuals (46). It has been observed that a single intrapulmonary injection of the lipid–protein mixture produces a cavity in sensitized animals, but lipid or the protein alone fail to do so (49). In the process of cavity formation, the antigenicity of the protein fraction was enhanced by adjuvant capacity of the cell-wall lipid.

## 12. Delayed Hypersensitivity and Protective Immunity

One of the greatest paradoxes in the immune response to *M. tuberculosis* is the apparent lack of correlation between tuberculin hypersensitivity and immunological protection. For example, patients who have had tuberculosis once and retain a strongly positive DH reaction might experience reactivation of the disease. These cases certainly have all the manifestations of tuberculin hypersensitivity, but their CMI to tubercle bacilli fails to protect against recurrence of tuberculosis. On the other hand, persons who have received BCG vaccine seem to have immunity to tuberculosis that often remains even after the tuberculin reaction disappears (50).

It is not clear whether the inflammatory reaction in tuberculosis is responsible for the increased resistance noted in infected or vaccinated animals (51). The probability of increased resistance mediated by tuberculin hypersensitivity stems from the fact that the intensity of the reaction depends on the amount of antigen administered. Inflammation is a protective mechanism of essential importance in resistance against bacterial infection, because it can wall off and prevent the spread of bacteria. An inflammatory exudate can kill bacteria or at least inhibit the growth of the tubercle bacilli. Hypersensitivity permits the body to respond with a more rapid and abundant inflammation at any site where tubercle bacilli are lodged (51,52).

Because desensitized guinea pigs show immunity in the absence of hypersensitivity (53), it could be argued that acquired immunity to tuberculosis and tuberculin hypersensitivity are two different phenomena and there is no quantitative correlation between the two. Hypersensitivity alone is incapable of giving protection, whereas acquired immunity can remain intact even after hypersensitivity has waned. Acquired cellular immunity can be specifically stimulated and recalled but acts nonspecifically (54). Nonspecific resistance to tuberculosis is effected by tuberculin hypersensitivity, because the two phenomena coexist after vaccination in animals. There is a direct relationship between the degree of immunity and hypersensitivity in vaccinated animals. Both are recalled with booster injections of vaccine and both can be adoptively transferred with cells from vaccinated animals. It has also been observed that viable attenuated mycobacterial cells are more effective immunogens than killed organisms. Even though viable attenuated bacteria are better immunogens, in vitro multiplication is not a prerequisite for higher immunogenicity. Higher efficacy of the viable mycobacteria depends on a heat-labile active factor, and the lower immunizing capacity of killed mycobacteria is due to the weakly immunogenic cell-wall components. Tuberculin hypersensitivity is not involved in the specific immune response and there are both specific and nonspecific responses involved when viable attenuated organisms are used as immunogens. When killed organisms are used, only a nonspecific response is activated. In view of the above, it could be inferred that tuberculin hypersensitivity and acquired immunity are separate immune responses elicited by different components of mycobacterial cells (55). However, it is also possible that delayed hypersensitivity and acquired immunity are different expressions of the same underlying process being manifested by CMI. The two are so closely associated during the development of immunity after primary infection with tubercle bacilli that they might be indivisible (48).

It has been postulated that the ability to respond specifically to a wide variety of antigens is under genetic control. The genes that control the immune response can predispose an individual to resistance or susceptibility to certain diseases; that is, the presence or absence of a particular immune response gene to one of the numerous antigenic components of *M. tuberculosis* might determine the nature of that particular individual's immune response to *M. tuberculosis* (48). In the murine model, the *Bcg* gene has been determined to regulate susceptibility to mycobacterial infections. The *Bcg* gene is located on the chromosome-1, which has structural homology to a conserved region on human chromosome-2 (56).

The controversy over DH versus protective immunity can be extended to the protective immunity produced by BCG. One of the patterns of responses to mycobacterial infection is the development of Koch's phenomenon, an important part of which is the process responsible for destruction of cells which contain mycobacteria or have mycobacterial antigens on their surface. This phenomenon can help in the containment of the disease. The second type of response is the Listeria type of response, which causes activation of macrophages by specifically primed lymphocytes. These activated macrophages can provide a better protective immunity from infection than does Koch's phenomenon. Mycobacteria differ in their ability to induce the two patterns of responses. Some species induce Listeria type and others Koch's type. BCG vaccination in the former case boosts immunity, in the latter case makes one susceptible to tuberculosis (57).

## 13. Correlates of Delayed Hypersensitivity and Immunity

Sensitized lymphocytes, when cultured in the presence of specific antigen, undergo defined morphological and biochemical alterations associated with the secretion of a variety of nonantibody mediators which attract, stimulate, and modify the behavior of the host and foreign cells. Morphological change include increased DNA synthesis and transformation into lymphoblasts. Blast transformation is merely a device for expanding the clone of cells of a given antigenic specificity. Lymphocyte activation is associated with concurrent secretion of lymphokines.

Many workers have observed a close correlation among DH, MIF production, and lymphoblast transformation (58,59). These tests were considered to be analogous to each other. It seems plausible that MIF-like mediators could be liberated with DH reaction as one of the components of DH complex. It is not surprising that the correlation between DH and MIF production, although strong, is imperfect. The relation between blast transformation and DH is remote (60,61). There is no evidence of prior transformation of sensitized lymphocytes while participating in DH. Blast transformation reaches a peak at 5–7 days, long before the DH reaction has subsided, and lymphocyte transformation closely corresponds to acquired immunity (48).

## 14. Local Versus Systemic Immune Response

Cell-mediated immunity in tuberculosis seems to be compartmentalized into local and systemic, and the local immune response is more intense than the systemic response. Guinea pigs immunized via the respiratory tract showed greater local respiratory immunity than the animals immunized subcutaneously. On the contrary, animals immunized parentarily developed greater systemic immunity. However, both systemic and respiratory immune response resulted with a high dose of local or systemic immunization (62,63). A marked activation takes place in the local lesions where the antigens are in high concentration and the macrophages are sensitive to them (29). In patients with tuberculous pleuritis, greater levels of INF-gamma and IL-2 were seen in the pleural fluids than in serum, and pleural fluid lymphocytes responded to mycobacterial antigen with increased production of cytokines than did the peripheral blood lymphocytes (20).

## 15. Humoral Response and Immunodiagnosis of Tuberculosis

Most *M. tuberculosis*-infected patients produce antibodies to various antigenic components of the organism during the course of the disease. The role of humoral antibodies in protection against the disease appears to be very limited. The antibody isotype and titer depends on the organs involved, the extent of disease, and treatment status. Diagnostic value of the humoral response has been studied by many investigators. Immunodiagnosis of tuberculosis has great value in smear-negative pulmonary cases and extrapulmonary disease. Demonstration and isolation of tubercle bacilli from extrapulmonary tuberculosis such as tuberculous meningitis, tuberculous pleurisy, tuberculosis peritonitis, and renal tuberculosis is difficult and time-consuming. Detection of either specific antibodies in the sera or antigen in the clinical samples is an attractive approach in the diagnosis of tuberculosis; however, this approach has not been successful. There has recently been some success in using PCR assays for the identification of *M. tuberculosis* infection.

Serological tests using crude and partially purified antigens of *M. tuberculosis* have been used for many years with limited success. Simple tests consisting of precipitation, bacterial agglutination, and agglutination of particles coated with antigens gave way to more sophisticated tests like ELISA and RIA. Initially, crude antigens like culture extracts and partially purified antigens were used, followed by more specific recombinant antigens. Monoclonal antibody-based serodiagnostic tests are found to be more sensitive and more acceptable serological tests. Using specific monoclonal antibodies and assays like the solid-phase antibody competition test (SACT), sensitivities of 74% and specificities of 90% have been achieved. Tuberculosis serology is a complex process involving a chronic disease wherein patients are exposed to various antigens of the organism for a long time; the isotype and idiotype of the antibodies probably depend, in addition to the other things, on the status of the disease. The genetic makeup of the population undoubtedly plays an important role in the antibody response. Geographic location also determines the extent of exposure of the population to cross-reacting organisms which can interfere with the specificity of the test. Combination of two or more tests may improve sensitivity and specificity and, therefore, hold some potential for use at least in diagnosing extrapulmonary tuberculosis (64,65).

An interesting alternate approach for the diagnosis of extrapulmonary tuberculosis is to demonstrate antigens in clinical samples such as cerebrospinal fluid, pleural fluid, peritoneal fluids, urine, and so on. Detection of mycobacterial DNA using PCR assays is an exciting new area of intense research discussed elsewhere in this book. Detection of the *M. tuberculosis* products like tuberculosteric acid by gas chromatography, and lipoarabinomannan using monoclonal antibody (ML34) have been evaluated for the diagnosis of tuberculosis; however, both tests show frequent false positivity (64).

## 16. Immunomodulation

Patients with active tuberculosis show a general suppression of cellular immune responses. Treatment with a combination of effective chemotherapeutic agents for 6–9 months usually results in the elimination of detectable organisms associated with improvement in the clinical condition. However, in certain percentage of patients, relapse of the disease after varying periods of quiescence is observed. Immunotherapy of tuberculosis patients with *M. vaccae* following chemotherapy showed more improvement of clinical and immunological parameters than those without immunotherapy (66), indicating enhancement of recovery following immunopotentiation. Cytokines like IL-2 and IFN-gamma have been used in the treatment of leprosy, with a significant reduction in the bacillary burden in the lesions (67,68). Recombinant IL-2 has also been shown to limit the replication of other mycobacteria in mice (69), indicating that immunomodulation has definite place as an adjunct in treatment of tuberculosis. However, the cytokines have a very short half-life and also are very toxic; the development of efficient delivery systems will probably boost the therapeutic potential of immunomodulators.

## 17. Immunological Mechanism and Nature of Protective Immunity in Tuberculosis

An inhaled bacillus is ingested by pulmonary alveolar macrophages, which either destroy the organism or allow it to multiply within, in which case a primary lesion is established. About 2 weeks after the primary tubercle begins, cellular immunity and DH develops. The products of the bacillus become stimulatory to macrophages, which differentiate into epitheloid cells with an increased capacity to destroy bacilli, and are rich in lysosomes (46). Macrophages process the mycobacterial antigens and present on the surface in association with MHC molecules to be recognized by the lymphocytes. Depending on the nature of the antigen and its association with either class I or class II MHC molecules, particular subsets of T cells are selectively stimulated. Endogenous or intracytoplasmic antigens are known to be processed and presented along with class I molecules, whereas the antigens from a the phagocytic vesicles are presented with class II molecules. Thus the location of the antigen inside the macrophage determines the nature of T cell response. Mycobacteria generally are localized in the phagosomes; however, occasionally, the organisms could be seen in the cytoplasm (70). Consequently stimulation of both CD4+ and CD8+ T lymphocytes is observed in tuberculosis. Evidence has been presented for the role of class I restricted T cells in protective immune response against tuberculosis (71). For an effective immune response to ensue, macrophages, in addition to presenting the antigen with appropriate MHC molecules, should produce secondary signals in the form of cytokines. Mycobacterial products can interfere with the production of necessary cytokines like IL-1 directly or indirectly through stimulation of prostaglandins.

Lymphocytes, upon reaction with mycobacterial antigens, are transformed into blasts, multiply, and produce antibodies (B lymphocytes) or secrete cytokines, which influence the function of macrophages. Presentation of processed mycobacterial antigens along with class I MHC molecules results in stimulation of CD8+ T lymphocytes, whereas presentation with class II molecules results in stimulation of CD4+ T lymphocytes. CD8+ lymphocytes are cytotoxic cells causing release of intracellular organisms from macrophages to be taken up by fresh, incoming, activated macrophages with increased capacity to kill the bacteria. Activated CD4+ lymphocytes influence antibody production by B lymphocytes, secrete cytokines which activate macrophages, and induce inflammatory response. Gamma/delta T cells, which can also be stimulated by mycobacterial antigens, are cytotoxic and secrete cytokines and, therefore, can also play a role in the immune process.

The objective of protective immune response is to eliminate the pathogen with the least amount of toxicity to the host. Sequential interaction between immunocompetent cells and several cytokines culminates in the generation of macrophages with increased capacity to kill phagocytosed organisms.

## 18. Vaccination Against Tuberculosis:

Even though development of protective measures was initiated by Koch, it did not make any impact until the discovery of BCG vaccine by Calmette and Guérin. They attenuated a *M. bovis* strain by repeated subcultures on medium containing bile and glycerol, resulting in the BCG strain. The efficacy of BCG vaccination has been controversial ever since its introduction. Vaccine trials conducted in different parts of the world have shown efficacy ranging from 0% to 80%. BCG is more effective against childhood tuberculosis than adult tuberculosis. The ineffectiveness of BCG vaccine has been attributed to various factors like the genetic makeup of the population, variation in BCG strains and preexposure to environmental mycobacteria. Perhaps the lack of a clear understanding of the protective immune mechanism and the antigens involved are the chief reasons for the failure to develop an effective vaccine against tuberculosis.

Because the antigenic profile of *M. tuberculosis* is very complex, it is difficult to identify the protective antigens. Although a large number of mycobacterial antigens which stimulate T lymphocytes have been purified, characterized, and cloned, no clear proof exists of a particular antigen being protective. From the accumulated data, it seems likely that the protective immune response involves stimulation with more than one antigen. The type of T cells stimulated by an antigen depends on its presentation in context with either class I or II MHC molecules. For effective immunity to ensue, antigens should be delivered using appropriate carrier systems so that antigens are presented utilizing both class II and I pathways. Microbial carriers like vaccinia virus, BCG, *S. typhimurium* are being actively considered. Other carrier systems like liposomes, ISCOMs, and biodegradable polylactic glycolic acid (PLGA) microspheres are alternative choices. The nonviable carrier systems have the advantage of accommodating a blend of antigens and adjuvants or immunomodulators. However, the prospects of developing such an effective vaccine against tuberculosis depends on the identification of protective antigens and the carrier system. Further advances in vaccine technology and our knowledge on the immunology of tuberculosis and other related diseases might lead to the development of a better vaccine than BCG.

## 19. Conclusions

In the past few decades, tremendous progress has been made in our understanding of the immunology of tuberculosis. From the knowledge gained so far, it is clear that several cell types involving more than one mechanism take part in the generation of the immune response. However, the relative role of each population of immunocompetent cells and the cytokines responsible for protection and pathogenesis of the disease is not clear. In addition to understanding the intricate mechanisms involved in in the immunopathogenesis of tuberculosis, identification, and characterization of the antigens stimulating a protective immune response, the development of effective carrier systems should lead to an effective vaccine.

## References

1. Goren MB (1982) Immunoreactive substances of mycobacteria. Am Rev Respir Dis 125(Suppl 3):50.
2. Daniel TM, Janicki BW (1978) Mycobacterial antigens: A review of their isolation, chemistry and immunological properties. Microbiol Rev 42:84.
3. Janicki BW, Chaparas SD, Daniel TM, Kubica GP, Wright GL Jr, Yee GS (1971) A reference system for antigens of *Mycobacterium tuberculosis.* Am Rev Respir Dis 104:602.
4. Stanford JL (1983) Immunologically important constituents of mycobacteria: Antigens. In: Ratledge C, ed. The Biology of Mycobacteria, vol. II, p. 92. New York: Academic Press.
5. Chaparas SD (1982) The immunology of mycobacterial infections. CRC Crit Rev Microbiol 9:139.

6. Chaparas SD, Brown TM, Hyman IS (1978) Antigenic relationships of various mycobacterial species with *Mycobacterium tuberculosis.* Am Rev Respir Dis 117:1091.

7. Andersen P (1994) Effective vaccination of mice against *Mycobacterium tuberculosis* infection with a soluble mixture of secreted mycobacterial proteins. Infect Immun 62:2536.

8. Young D, Garbe T, Lathigra R, Abou-Zeid C (1990) Protein antigens: Structure, function and regulation. In: McFadden J, ed. Molecular Biology of Mycobacteria, p. 1. London: Surrey University Press.

9. Brennan PJ (1989) Structure of mycobacteria: Recent developments in defining cell wall carbohydrates and proteins. Rev Infect Dis 11:s420.

10. Baker RW, Hill WE, Larson CL (1973) Ribosomes of acid-fast bacilli: Immunogenicity, serology, and *in vitro* correlates of delayed hypersensitivity. Infect Immun 8:236.

11. Ortiz-Ortiz L, Solarolo EB, Bojalil LF (1971) Delayed hypersensitivity to ribosomal protein from BCG. J Immunol 107:1022.

12. Youmans AS, Youmans GP (1969) Factors affecting immunogenic activity of mycobacterial ribosomal and ribonucleic acid preparations. J Bacteriol 99:42.

13. Hunter SW, Gaylord H, Brennen PJ (1986) Structure and antigenicity of the phosphorylated lipopolysaccharide antigens from the leprosy and tubercle bacilli. J Biol Chem 261:12345.

14. Campbell PA (1976) Immunocompetent cells in resistance to bacterial infections. Bacteriol. Rev 40:284.

15. Mackaness GB (1971) Resistance to intracellular infection. J Infect Dis 123:439.

16. Kaufmann SHE (1993) Immunity to intracellular bacteria. Annu Rev Immunol 11:129.

17. Huygen K, Abramowicz D, Vandenbussche P, Jacobs F, De Bruyan J, Kentos A, Drowart A, van Vooren J, Goldman M (1992) Spleen cell cytokine secretion in *Mycobacterium bovis* BCG-infected mice. Infect Immun 60:2880.

18. Noelle R, Snow EC (1992) T-helper cells. Curr Opin Immun 4:333.

19. Sanchez FO, Rodriguez JI, Agudelo G, Garcia LF (1994) Immune responsiveness and lymphokine production in patients with tuberculosis and healthy controls. Infect Immun 62:5673.

20. Barnes PE, Lu S, Abrams JS, Wang E, Yamamura M, Modlin RL (1993) Cytokine production at the site of disease in human tuberculosis. Infect Immun 61:3482.

21. Orme IM, Miller ES, Roberts AD, Furney SK, Griffin JP, Dobos KM, Chi D, Rivoire B, Brennen PJ (1992) T lymphocyte mediating protection and cellular cytolysis during the course of *Mycobacterium tuberculosis* infection. J Immunol 148:189.

22. Ab BK, Kiessling R, Van Embden JDA, Thole JER, Kumararatne DS, Pisa P, Wondimu A, Ottenhoff HM (1990) Induction of antigen-specific CD4+ HLA-DR restricted cytotoxic T lymphocytes as well as nonspecific nonrestricted killer cells by the recombinant mycobacterial 65 kDa heat-shock protein. Eur J Immunol 20:369.

23. Ottenhoff THM, Ab BK, Van Embden JDA, Thole JER, Kiessling R (1988) The recombinant 65 kD heat shock protein of *Mycobacterium bovis* Calmette–Guerin/*M. tuberculosis* is a target molecule for CD+ cytotoxic T lymphocytes that lyse human monocytes. J Exp Med 168:1947.
24. Schlesinger LS, Bellinger-Kawahara CG, Payne NR, Horwitz MA (1990) Phagocytosis of *Mycobacterium tuberculosis* is mediated by human monocyte complement receptors and complement component C3. J Immunol 144:2771.
25. Abou-Zeid C, Ratliff TL, Wiker HG, Harboe M, Bennedsen J, Rook GAW (1988) Characterization of fibronectin-binding antigens released by *Mycobacterium tuberculosis* and *Mycobacterium bovis* BCG. Infect Immun 56:3046.
26. Kaufmann, SHE (1993) Immunity to intracellular bacteria. Annu Rev Immunol 11:129.
27. Ando M, Dannenberg AM, Shima K (1972) Macrophage accumulation, division, maturation, and digestive and microbicidal capacities in tuberculosis. J Immunol 109:8.
28. Dannenberg AM (1968) Cellular hypersensitivity and cellular immunity in the pathogenesis of tuberculosis: Specificity, systemic and local nature and associated macrophage enzymes. Bacteriol Rev 32:85.
29. Dannenberg AM, Ando M, Shima K (1972) Macrophage accumulation, division, maturation, digestive and microbicidal capacities in tuberculosis lesions: II. The turnover of macrophages and its relation to their activation and antimicrobicidal immunity in primary BCG lesions and those of reinfection. J Immunol 109:1109.
30. Flesch I, Kaufmann SHE (1987) Mycobacterial growth inhibition by interferon-gamma activated bone marrow macrophages and differential susceptibility among strains of *Mycobacterium tuberculosis.* J Immunol 138:4408.
31. Douvas GS, Looker DL, Vatter AE, Crowle AJ (1985) Gamma interferon activates human macrophages to become tumoricidal and leishmanicidal but enhances replication of macrophage-associated mycobacteria. Infect Immun 50:1.
32. Rook GAW, Steele J, Ainsworth M, Chapion BR (1986) Activation of macrophages to inhibit proliferation of *Mycobacterium tuberculosis:* Comparison of the effects of recombinant gamma-interferon on human monocytes and murine peritoneal macrophages. Immunology 59:333.
33. Chan J, Fan XD, Hunter SW, Brennan PJ, Bloom BR (1991) Lipoarabinomannan, a possible virulence factor involved in persistence of *Mycobacterium tuberculosis* within macrophages. Infect Immun 59:1755.
34. Hart PD, Young MR, Gordon AH, Sullivan KH (1987) Inhibition of phagosome–lysosome fusion in macrophages by certain mycobacteria can be explained by inhibition of lysosomal movements observed after phagocytosis. J Exp Med 166:933.
35. Antony VB, Sahn SA, Harada RN, Repine JE (1983) Lung repair and granuloma formation. Chest 83:95s.
36. Bloch H (1948) The relationship between phagocytic cells and human tubercle bacilli. Am Rev Tuberc 58:662.
37. Bass SN, Spagnuolo PJ, Ellner JJ (1981) Augmented neutrophil adherence in active and remote tuberculosis. Am Rev Respir Dis 124:643.

38. Reiger M, Tranka L, Skvor J, Mison P (1979) Immunoprofile studies in patients with pulmonary tuberculosis: III Study of hemolytic complement in serum and phagocytic activity of blood neutrophils. Scand J Respir Dis 60:172.

39. Brown AE, Holzer TJ, Andersen BR (1987) Capacity of human neutrophils to kill *Mycobacterium tuberculosis.* J Infect Dis 156:985.

40. Jones GS, Amirault HJ, Andersen BR (1990) Killing of *Mycobacterium tuberculosis* by neutrophils: A nonoxidative process. J Infect Dis 162:700.

41. Knight-Shapiro CD, Harding GE, Smith DW (1974) Relationship of delayed type of hypersensitivity and acquired cellular resistance in experimental air borne tuberculosis. J Infect Dis 130:8.

42. Moore VL, Myrvik QN (1977) The role of normal alveolar macrophages in CMI. J Reticul Soc 21:131.

43. Johnson JD, Leehand W, King NL, Huges CG (1975) Activation of alveolar macrophages after lower respiratory tract infection. J Immunol 115:80.

44. Cantey JR, Hand WL (1974) Cell mediated immunity after bacterial infection of the lower respiratory tract. J Clin Invest 54:1125.

45. Mackaness GB (1974) Delayed hypersensitivity in lung diseases. Ann NY Acad Sci 221:312.

46. Dannenberg AM, Meyer OT, Esterly JR, Kambara T (1968) The local nature of immunity in tuberculosis, illustrated histochemically in dermal BCG lesions. J Immunol 100:931.

47. Kino T, Tsuji S (1972) Two depositories of immunocompetent cells for delayed hypersensitivity. The lung and the peritoneal cavity. Am Rev Respir Dis 105:832.

48. Lefford MJ (1975) Delayed hypersensitivity and immunity in tuberculosis. Am Rev Respir Dis 111:243.

49. Yamamura Y, Maeda H, Ogawa Y, Maeda J, Tomino T (1979) Experimental pulmonary cavity formation by mycobacterial components. Int Union Tuberc 54:171.

50. Freedman SO (1976) Circulating antibodies in pulmonary tuberculosis. Chest 70:1.

51. Youmans GP, Youmans AS (1965) Non-specific factors in resistance of mice to experimental tuberculosis. J Bacteriol 90:1675.

52. Ramseier H, Suter E (1964) An antimycobacterial principle of peritoneal mononuclear cells. I. The inhibition of tubercle bacilli by disrupted mononuclear cells obtained from normal and BCG immunized guinea pigs. J Immunol 93:511.

53. Rich AR (1951) The Pathogenicity of Tuberculosis. Springfield, IL: CC Thomas.

54. Mackaness GB (1968) The immunology of anti-tuberculosis immunity. Am Rev Respir Dis 97:337.

55. Youmans GP (1975) Relation between delayed hypersensitivity and immunity in tuberculosis. Am Rev Respir Dis 111:109.

56. Skamene E (1989) Genetic control of susceptibility of mycobacterial infections. Rev Infect Dis 11:s394.

57. Stanford JL, Shief MJ, Rook GAW (1981) How environmental mycobacteria predetermine the protective efficacy of BCG. Tubercle 62:55.

58. Ferraresi RW, Dedrick CT, Raffel S, Goihman-Yahr M (1969) Studies on macrophage inhibition test. I. Comparison of the skin and cell migration reactions during the course of development of delayed hypersensitivity. J Immunol 102:852.

59. Mills JA (1966) The immunologic significance of antigen induced lymphocyte transformation *in vitro*. J Immunol 97:239.

60. Maini RN (1974) A study of delayed hypersensitivity, lymphocyte transformation and leucocyte migration inhibition by *M. xenopi* and *M. tuberculosis* in patients harboring these organisms. Tubercle 55:269.

61. Rieger M, Tranka L, Skvor J (1979) Immunoprofile studies in patients with pulmonary tuberculosis. IV. Tuberculin skin reaction and test of inhibition of blood leucocyte migration. Scand J Respir Dis 60:355.

62. Spencer JC (1974) Local and systemic cell mediated immunity after immunization of guinea pigs with live or killed *M. tuberculosis* by various routes. J Immunol 112:1322.

63. Henney CS, Waldman RH (1970) Cell mediated immunity shown by lymphocytes from respiratory tract. Science 169:696.

64. Ivanyi J, Bothamley GH, Jackett PS (1988) Immunodiagnostic assays for tuberculosis and leprosy. Br Med Bull 44:635.

65. Daniel TM, Debanne SM (1987) The serodiagnosis of tuberculosis and other mybacterial diseases by enzyme-linked immunosorbent assay. Am Rev Respir Dis 135:1137.

66. Stanford JL, Bahr GM, Rook GAW, Shaaban MA, Chugh TD, Gabriel M, al-Shimali B, Siddiqui Z, Ghardani F, Shahin A, and Behbehani, K. (1990) Immunotherapy with *Mycobacterium vaccae* as an adjunct to chemotherapy in the treatment of pulmonary tuberculosis. Tubercle 71:87.

67. Nathan CF, Kaplan G, Levis WR, Nusrat A, Witmer MD, Sherwin SA, Job CK, Horowitz CR, Steinman RM, Cohn ZA (1986) Local and systemic effects of intradermal recombinant interferon-gamma in patients with lepromatous leprosy. New Engl J Med 315:6.

68. Kaplan G, Kiessling R, Teklemarniam S, Hancock G, Sheftel G, Job CK, Converse P, Ottenhoff THM, Becx-Bleumink M, Deitz M, Cohn ZA (1989) The reconstitution of cell-mediated immunity in the cutaneous lesions of lepromatous leprosy by recombinant interleukin-2. J Exp Med 169:893.

69. Jeevan K, Asherson GL (1988) Recombinant interleukin-2 limits the replication of *Mycobacterium lepraemurium* and *Mycobacterium bovis* BCG in mice. Infect Immun 56:660.

70. McDonough KA, Kress Y, Bloom BR (1993) Pathogenesis of tuberculosis: Interaction of *Mycobacterium tuberculosis* and macrophages. Infect Immun 61:2763.

71. Flynn JL, Goldstein MM, Triebold KJ, Koller B, Bloom BR (1992) Major histocompatibility complex class I-restricted T cells are required for resistance to *Mycobacterium tuberculosis* infection. Proc Natl Acad Sci USA 89:12013.

# 9

# The Immunological Cause, Prevention, and Treatment of Tuberculosis

*John Lawson Stanford*

## 1. Introduction

Mycobacteria, widely distributed in the environment though perhaps not ubiquitous because of their restriction by heat and ultraviolet light, vary in their frequency and species distribution from place to place. Although we know something of the physical conditions controlling their distribution, much more remains to be learned. Surface water pH plays a significant role and persons living in a region of acidic soil and drinking soft water can meet a quite different mix of mycobacterial species than those living in alkaline regions with hard water.

Different degrees of pathogenicity and virulence are exhibited by different species; some are primarily pathogens causing a well-described series of human disease conditions and others predominantly live in the environment and rarely or never give rise to cases of human disease. The fundamental mistake often made is of thinking that the major pathogens are the most important mycobacterial species—they are not. Those species living apparently unnoticed in the environment are really the most important. Although their presence does not impinge on the conscious mind, they are well recognized by the immune system. Under different sets of conditions and in individuals of different genotypes, silent contact with environmental mycobacteria primes the immune system, predisposes to success or failure of Bacille–Calmette–Guérin (BCG) vaccine, and determines natural protection or susceptibility to mycobacterial disease (1–3). Immune mechanisms resulting from contact with environmental mycobacteria also help determine the distribution of tuberculosis and leprosy, keep infection in a latent state for years (4), and ensure that the diseases afflict only a minority of the human population.

Besides these influences on mycobacterial disease, contact with environmental mycobacteria can have far-reaching effects on susceptibility to diseases caused by nonmycobacterial pathogens. This can be through the adjuvant activity of the mycobacterial cell wall, priming particular patterns of T lymphocyte maturation, and by the sharing of antigenic epitopes common to many genera.

Among the species sharing antigenic determinants with mycobacteria is *Homo sapiens,* certain proteins of which, notably the stress proteins, show 50% or more of homology with their mycobacterial counterparts (5). The nature of the immune response to such shared antigens is likely to lead to protection from, or induction of, autoimmune conditions such as rheumatoid arthritis (6), insulin-dependent diabetes melitus, and some forms of schizophrenia (7), as well as contributing to tissue damage in clinical tuberculosis. Antigens shared between man and mycobacteria can also be important in both causation and prevention of malignancies (8) and the bacteriomimetic nature of some tumor antigens is only just now receiving scientific attention. Preliminary evidence also suggests a relationship between regulatory activities of mycobacterial products and the chronic low-grade vasculitis that may underlie deposition of atheromatous plaques that lead to cardiovascular disease (9).

Of course, the mycobacteria are by no means the only outside influences that predetermine our lives, diseases, and deaths, but there can be little doubt that they are among the most potent. When measured against these influences, the actual diseases that mycobacterial pathogens cause by their invasion of the tissues seem almost insignificant, yet the great majority of immunological research currently undertaken is directed toward them.

The evolution of man extends back to the very earliest of times, at some point in which the mycobacteria also evolved. Since their appearance, other forms of life have had to develop mechanisms of defense against them, and thus contact with mycobacteria has been an evolutionary force molding the developing mechanisms of immunity, especially that of long-lived species such as man. In the same way, the immune system has been an evolutionary force on the bacterium in the selection of pathogenic forms. Without mycobacteria, and indeed many other bacterial genera, man as we know him would not be man. The considerable homology between some human stress proteins, of which the 60-kDa heat shock protein (HSP) has probably been most investigated, with those of mycobacteria appears to be evidence of a distant common ancestor, genetic exchange, or a remarkable example of parallel evolution. Against this very complex background we have to develop vaccines and immunotherapeutics capable of restricting, maximizing, or even replacing environmental influences.

## 2. The First Experience of Mycobacteria

During gestation, maternal IgG antibodies, including those with specificities for mycobacterial antigens, cross the placenta, and at the time of birth, antimycobacterial antibodies of the baby are similar to those of its mother. What part these play in protection of the infant is unknown, and it is possible that their real importance is regulatory, so that in some way not yet investigated, they modulate the subsequent development of the child's immunity. It is possible that cells too may cross the placental barrier, as T cells responsive to mycobacteria have been found in cord blood, although these might be the infant's cells responding to maternal regulatory influences. Babies born to strongly tuberculin-reactive mothers might be tuberculin positive during the first weeks of life. We have no notion of what the significance of these transplacental influences can have on later life.

Besides providing passive antibodies, and just possibly a few T cells, the mother's physiology prevents the baby from being exposed to a certain type of antibody, agalactosyl immunoglobulin (Gal [0]), the presence of which might be deleterious (10,11). This is achieved by the mother progressively losing Gal [0] antibodies during pregnancy, so that at the time of parturition, the percentage of this glycoform is extremely low, and very little can be detected in cord blood or in the newborn baby. The care with which maternal physiology protects the fetus from this antibody glycoform suggests that it has an immunoregulatory role that might prove deleterious to the developing immune system. The potential regulatory role of Gal[0] is the subject of much ongoing research. As well as these well-recognized participants in immunity, maternal hormones cross the placenta, and these too may be important in laying down a blueprint for the developing immune system.

One of the first acts of the mother after giving birth is to feed the baby, and the colostrum and mother's milk are also important to the developing immune system. Thus, it has been shown that the immune response to BCG vaccination at birth is different in breast-fed babies from those given substitutes (12). First thought to be due to maternal antibodies in milk modulating T-cell responses to BCG, it seems more likely that maternal adrenal steroid hormones or their metabolites absorbed by the infant can predispose the developing cellular immune system to produce Type 1 helper T cells (TH1) (13,14) laying down a lifetime of protective immunity to mycobacterial challenge only broken through in misfortune.

## 3. The Baby's Developing Immune System

At, or very soon after birth, the baby meets mycobacteria from the environment, or their heat-degraded antigens in food, and begins a relationship with them that lasts a lifetime. Within 2 months, maternal antibodies have mostly disappeared, and by 6 months, the baby is producing antibodies that will remain part of the repertoire of recallable responses until old age. The order in which antibodies of different classes appear does not seem to be known, but the selective specificity of the baby's antibodies is preferential toward protein antigens. Development of antibodies to carbohydrates is a later phenomenon in childhood. although antibodies to the lipoarabinomannan of mycobacteria develop very early.

Cellular immunity also develops early in life and can be detected by lymphocyte transformation within a few months of birth, although skin test positivity to tuberculin does not appear until much later in the developed world, unless tissue invasion by mycobacteria occurs. In countries where BCG is given at, or near, the time of birth, this constitutes the first mycobacterial invasion, and tuberculin positivity is induced (15).

Thus, the inheritance of immunity follows both Darwinian evolution through the immune response genes and Lamarkian evolution through the partial impression of maternal acquired characteristics.

## 4. Neonatal BCG Vaccination

Vaccination against tuberculosis by means of BCG is given in those countries, or in particular situations within countries, where childhood tuberculosis occurs. The purpose of such vaccination is to maximize the infant's developing immunity to mycobacterial challenge, by imprinting a TH1 antibacterial response.

The dose of BCG vaccine given near to birth is usually half of the recommended adult dose and is given by intradermal injection over one of the deltoid muscles. Thigh and buttock sites, previously used, have been discontinued because of their apparent association with chronic BCG osteitis (16). Following intradermal injection, the baby develops a macular response after 4–7 days which becomes indurated and can ulcerate, healing after a few weeks to a small scar, which is often hard to find in later life. Oral vaccination has been discontinued because it resulted in very variable tuberculin positivity and left no scar, although it seems to have induced a level of protective immunity equal to that following intradermal administration.

The half million live, and many more dead, bacilli that generally make up an infant's dose of BCG vaccine must form complexes with the maternal antibody very soon after injection. A few bacilli probably spread via the bloodstream to the spleen, liver, and lymph nodes, where their presence can be an essential part of the initiation of protective immunity. The majority of injected bacilli, following their coating with maternal antibody, are likely to be rapidly engulfed by phagocytic cells of the polymorph and macrophage series, which are immobilized, confining the bacilli to the local lesion. Cytokines like interleukin-1 (IL-1) attract T lymphocytes, some of which recognize bacterial antigen from engulfed bacilli, expressed on the surface of presenting cells together with class II HLA molecules. These antigen-presenting cells in the skin include dendritic and Langerhans cells as well as macrophages of the monocyte series. T cells responding to the presented bacterial epitopes undergo clonal expansion in the presence of IL-1 and begin to secrete a range of cytokines according to the particular maturation pattern they have followed (13). Usually, this is TH1 and the cytokines produced are IL-2, IL-12, gamma interferon (IFN-G), and tumor necrosis factor (TNF), probably dictated at this early time in life by transplacental maternal antibodies or adrenal hormones possibly absorbed from milk. BCG appears to work by maximizing whatever immune response was attained at the time of vaccination (2). Thus, it is important that BCG is given before the infant's response has been perturbed by challenge with a mycobacterial pathogen. Where this is likely to occur early in life, BCG must be given first, but where challenge is rare in early life, BCG can be given successfully at a much older age (17,18). Thus, in the United Kingdom, BCG vaccination is given to 13-year-old children unless they are likely to be in contact with tuberculosis at an earlier age.

The BCG vaccine given early in life varies in its long-term effects according to the time after birth that it is administered (19). If given within the first 4 weeks, it induces tuberculin positivity in about 90% of infants within a few weeks, but this can revert to negative relatively rapidly so that only 20% or so of 5-year-olds remain tuberculin positive (20). Subsequently, presumably as a result of contact with environmental mycobacteria, lost tuberculin reactivity is regained so that by the age of 11 or 12 years, 85% of such children have positive responses. Some interesting studies in Sri Lanka showed that wained tuberculin responsiveness could be restored to such children by revaccination at the age of 5 years, but that such children had the same tuberculin responsiveness at age 11 or 12 as those that were not revaccinated (21). This probably obviates the need for childhood revaccination which was commonly used in the past but has been given up in most countries.

The BCG vaccine given between 1 and 3 months of age induces tuberculin positivity retained by 80% of children at the age of 5 years, and by 85% or so tested at 12 years. A recent study from Finland (15) has shown that although the numbers of children skin test positive to tuberculin are similar whenever they were vaccinated during their first few months, those receiving BCG [Glaxo] during their first 4 weeks have significantly smaller tuberculin reactions 12 years later than have those who were vaccinated just a few weeks later (a finding not found after BCG [Copenhagen]). This is a quite remarkable observation and not at all easy to explain except on the basis of transferred maternal influences or as an early influence of environmental exposure. In an attempt to find out which vaccination time is associated with the development of the fewest cases of tuberculosis, case control studies in Finland are planned.

Although BCG vaccination enhances cell-mediated responses when given early in life, it has a regulatory effect on IgG antibody levels. In a comparative study between British and Finnish children, it has just been shown that the production of antibodies to mycobacterial lipoarabinomannan or whole sonicates of *Myco-*

*bacterium tuberculosis,* over the first 12 years of life is less in the Finnish children who all received BCG than in the nonvaccinated British children (22), although once again this could be an environmental influence. Despite contact with environmental mycobacteria, fewer than 5% of the British children are tuberculin positive aged 12 years in comparison with 85% of the Finnish children, reflecting the influence of neonatal BCG.

The lack of an increase in IgG antibodies accompanying tuberculin conversion after BCG vaccination could be evidence that the vaccine has successfully evoked a TH1 response rather than a TH2 response, as the latter results in marked IgG antibody production in experimental animals.

The BCG vaccine always affords some protection from tuberculosis when administered to the newborn (23), although the amount of protection induced varies from country to country and can depend on the exact time that it is administered, the particular BCG given, the immunological influence of the mother, and the environment into which the child is born. As well as an effect on susceptibility to tuberculosis, in some studies BCG vaccination has been found to reduce the frequency of childhood leukemia and solid tumors. This appears to occur only in those countries where BCG is effective in preventing tuberculosis (24).

## 5. BCG Vaccination Given to Older Children

It has been the practice of 40 years in the United Kingdom to offer BCG vaccination to tuberculin-negative 13-year-old school children. This policy was introduced so that the vaccine should be given shortly before the children reached the age at which tuberculosis commonly occurred 40 years ago (i.e., ages 15–25 years). It was assumed that the vaccines greatest effect would be over the 10 years after it was given, although there is little evidence that this is the case. A major trial of this procedure showed that about 80% of protection from clinical tuberculosis resulted, and this is equal to the best of other trials carried out in various parts of the world (3,25).

In the British situation, it seems that the great majority of children (more than 95%) are tuberculin negative nowadays at the age of 13 years, and therefore suitable for vaccination. Data also suggest that these eligible children still have TH1-associated cytokine release on challenge with BCG, and this can still be maximized by vaccination.

Some European countries have different policies; thus, Holland does not use BCG vaccination except in special at risk circumstances, and Sweden has recently given up vaccination at birth except for immigrant children, or children of families including tuberculosis patients (26). This was done because the incidence of tuberculosis in general had fallen to a low plateau in Sweden. Since stopping vaccination at birth, an increased incidence of mycobacterial cervical lymphadenitis has been observed due to infection with nontuberculosis mycobacteria (27). Presumably, such infections were prevented by BCG vaccination in the past. It is interesting that these infections of early childhood occur quite frequently in the United Kingdom, their maximum incidence being in children under 5 years of age (28). In Britain, most such cases are caused by *M. intracellulare, M. malmoense, M. avium,* and *M. scrofulaceum* in descending order of frequency.

The Western European experience is not the case in India, where there are great differences between different regions in the cellular immune responses following BCG vaccination, presumably reflecting the age at which infection or the environmental mycobacterial load drives children to switch from a predominantly Type 1 to Type 2 maturation pattern of their T cells responding to mycobacterial antigens. In some towns, such as Ahmednagar in the Maharashtra province of India, BCG vaccination appears to enhance a protective type of response, even if administration is delayed until the teen years (18). In other Indian towns, such as Agra in Uttar Pradesh, BCG has to be administered in the first 2 years of life to promote correlates of protective immunity (17).

## 6. Puberty and Immunity to Mycobacteria

With all the endocrine changes that accompany puberty, one would expect some to occur that influence the immune system. Studies in South Africa (29) and the author's own unpublished data clearly show these influences. If the development of tuberculin positivity is followed through increasing age, a "shoulder" and small dip occur at the relevant times of puberty of both boys and girls. Thereafter, the acquisition of tuberculin positivity in the sexes can be influenced by the different hormones and by the different occupations followed. In general, girls in their late teens and early twenties develop more cases of tuberculosis than do boys. At older ages, men exceed women in the prevalence of pulmonary and genitourinary forms of the disease, although the sexes are equal in development of other forms.

## 7. BCG Vaccination of Adults

Adults are not normally offered BCG unless they are in contact with patients with tuberculosis or their occupation is one that particularly exposes them to potential contact with such patients. An unpublished survey made some 10 years ago showed that about 75% of previously unvaccinated young adults (aged 18–35 years) entering the London Ambulance Service were tuberculin negative and suitable for vaccination. Such selected adults appear to respond to BCG just as do children, and the cellular response type predominating at the time of vaccination is maximized.

## 8. How BCG Vaccination Works

The BCG vaccine strain is of *M. tuberculosis,* probably originally of bovine type (30). Thus, it contains virtually the same antigens as virulent tubercle bacilli, packed in virtually the same adjuvant. Such an attenuated strain was selected on the basic premise that the nearer the vaccine strain was to the pathogen to which it was to induce protection, the better it would work. Jenner's views could still influence microbiologists some 200 years after his death! Even now, there are those seeking to enrich BCG with the secret specific ingredient of the tubercle bacillus, immune recognition of which will provide long-term protection from tuberculosis. Such an approach is flawed; with a mycobacterium, it is a matter of correctly stimulating the phagocytes rather than neutralizing a toxin. The phagocytes and their T-cell companions are not taxonomists; it is their role to provide the right kind of response, and the most economical way for them to do it is by recognizing mycobacteria and, for that matter, many more organisms by the antigens that they share (31). Thus, BCG protects against leprosy just as well as it does against tuberculosis.

When a person whose tissues have not been invaded with a mycobacterium and who has not been heavily bombarded by environmental mycobacteria is injected with BCG vaccine, a local response develops after about 7–10 days which normally pustulates and then heals. This results in the circulation of T-cells reactive to the antigens of tubercle bacilli so that 95% of vaccinees become skin test positive to tuberculin 8–12 weeks later. Shielded from other mycobacterial contact, this tuberculin positivity slowly fades over a year or two (32), only to return on exposure to environmental mycobacterial species, as exemplified by the findings in Sri Lanka (20). What the BCG has done is to lower the threshold of immunological recognition by inducing memory T cells for common mycobacterial antigens. Expression of tuberculin positivity is a constantly changing phenomenon; thus, a person skin test positive one year can be negative the next and positive the year after. A series of studies carried out on Gurka soldiers in the British army showed that their pattern of positivity to four new tuberculins (see below) measured in Aldershot, UK, had changed completely 2 years after they were transferred to Hong Kong, and this happened again 2 years after they had returned to Nepal (D. Jolliffe, personal communication). Thus, memory for species-specific antigens is less than 2 years, but that for common mycobacterial antigens is maintained, probably through environmental contact. This suggests that there may be a natural reversion to TH1 with time in the lack of contact with TH2-promoting pathogens. Thus, healthy persons skin tested 30 or 40 years after their successful treatment for tuberculosis quite often have small, well-regulated tuberculin reactions associated with protective immunity.

The sequence of events is quite different when a person whose tissues have been invaded by mycobacteria and who still retains persisters (4), or who has developed a predominantly TH2 response for other reasons, is vaccinated with BCG. There is induction of early tuberculin positivity which soon fades unless the person is infected with tubercle bacilli, and there is no enhanced recognition of environmental mycobacteria by their common mycobacterial antigens. Thus, in south India, where many people have been rendered, and are maintained, TH2 responsive to mycobacteria by their heavy environmental contact with *M. intracellulare* (33), post-BCG tuberculin positivity is soon lost (34). Indeed, the enhancement of TH2 predominance following BCG can increase susceptibility to tuberculosis in the short term, explaining the increase in cases observed in the early follow-ups of the BCG trial. Similarly, in Burma, where another major trial of BCG took place (35), TH2 hypersensitivity to *M. scrofulaceum* from the environment blocked the induction of protection from leprosy in those over 5 years in age (1).

Thus, BCG works by enhancing recognition of mycobacteria via their common antigens, and this effect is continuously boosted by casual environmental contacts. When a potentially pathogenic challenge with tubercle bacilli occurs, a successfully vaccinated person recognizes the challenge almost immediately, prevents the bacilli from multiplying, and eradicates them via a Type 1 cellular immune response, and can neutralize their pathogenetic secretions by complexing with antibodies to common epitopes on the secreted molecules.

## 9. When BCG Does Not Work

The controversy over the value of BCG has raged ever since the British trial achieved 78% protection from tuberculosis (25) and that held in Alabama and Georgia achieved only 14% (36). When the World Health Organisation and the Indian government published the results of their large trial of the vaccine in south India (37) in which no protection was found, many people despaired of the vaccine. Similarly, in leprosy, following behind the successful trial held in Uganda (38), in which about 80% protection from leprosy was recorded, the mere 18% protection found in Burma in the major trial run by WHO and the Burmese Government (35) was devastating. The results of most of the trials of BCG against tuberculosis have been reviewed (3) and it can be seen that they vary in efficacy from 80% to 0% and can even have a minus value.

A number of suggestions have been put forward to explain these differences, which are not recorded for other vaccines. They include explanations based on racial characteristics of the recipients, natural selection of persons at risk, dietary differences, variation in different manufacturers' vaccines, differences in study designs, and environmental differences. Although all of these can exert minor influences, environmental differences are the major cause.

It has already been pointed out that different qualitative and quantitative effects of contact with mycobacteria can imprint predominantly TH1 or TH2 responses to mycobacteria. BCG contains powerful adjuvants that maximize the pattern of T-helper-cell maturation predominating at the time of vaccination; thus, if an optimally protective response is already imprinted, this will be maximized (39). If a potentially tissue-damaging and less protective TH2 response is imprinted, this too will be maximized. Fortunately, BCG differs from its pathogenic progenitors in the imprint that it induces when given to those not committed to a response type. Thus, in the very young, it induces TH1 maturation, and in them, BCG always affords a degree of protection (23).

In simple terms, people share their environment with mycobacteria. These vary in their species and numbers. Contact with some species induce natural protective responses and others promote susceptibility factors (2). When these are maximized by BCG vaccination, they result in the major differences seen among the various trial results (3).

## 10. The Skin Test Response to Tuberculin and Its Significance

Tuberculin testing, introduced by von Pirquet, remains the only test in which the individual's response to a tissue challenge with antigens of mycobacteria can be directly measured in vivo. A negative test is commonly used to select persons suitable for BCG vaccination. Unfortunately, the test response is subject to many regulatory variables, from HLA-DR type (40) to the method of manufacture of the test reagent. Disregarding many of these variables, the test can be read as "positive" (skin around injection site indurated by infiltration of immunocompetent cells, and the results of their released cytokines) or "negative" (no palpable response, despite infiltration being demonstrable on biopsy in some cases) (41). Positive reactions can then be qualitatively classified on the basis of evidence of local incipient necrosis associated with microvascular changes. Tests producing incipiently necrotic reactions are said to be of "Koch-type," on the assumption that they mimic in microcosm the tissue-destructive changes occurring around tuberculous lesions, which their occurrence often heralds (42). Positive tests producing local infiltration without appreciable evidence of tissue death (non-Koch-type responses) can represent protective mechanisms, and such positive responses often follow BCG vaccination in places where it is protective (1,2,24).

Besides overt tuberculosis, tubercle bacilli can rest in the tissues as live persisters, a state of latent tuberculosis (4). The Koch-type response to tuberculin can also indicate such a condition. When these parameters are applied to the ambulance recruits referred to earlier, some observations can be made from the results. Whereas the great majority of recruits with old BCG scars produced positive reactions of non-Koch type, those without such scars were either tuberculin negative or produced positive reactions with a higher incidence of Koch-type responses (Fig. 9.1).

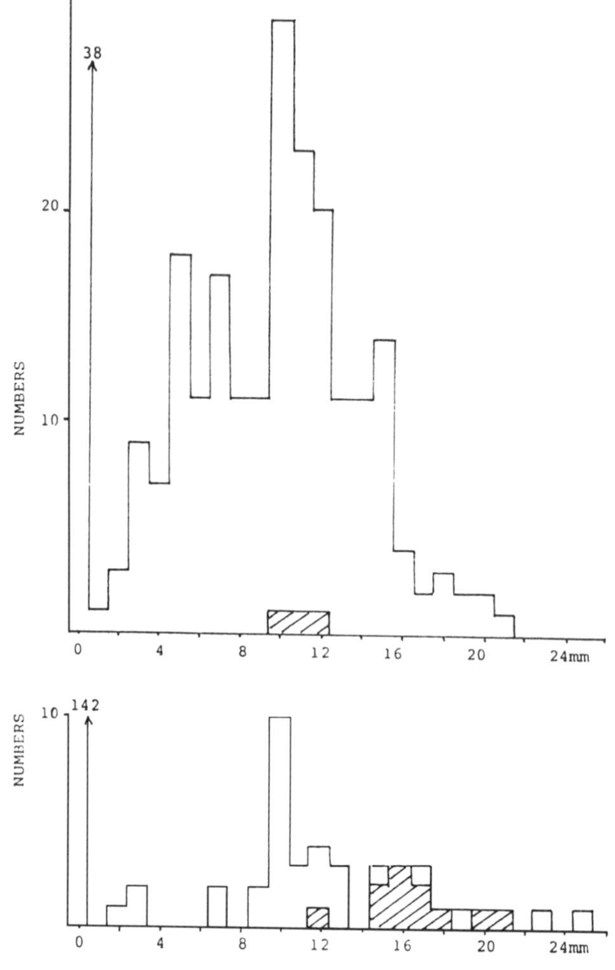


Figure 9.1. Tuberculin skin test results on 431 recruits to the London Ambulance Service, with a mean age of 24 years. Results for the 247 persons with scars of BCG vaccination given 10–12 years earlier are shown in the upper figure. 87% produced responses of 2 mm or more diameter of induration, with a mean diameter of 9.9 ± 4.0 mm. Only three persons produced reactions of Koch-type (hatched region) with insipient necrosis. Results for 184 persons without BCG scars are shown in the lower figure. Only 23% produced positive responses, of whom 11 persons produced Koch-type responses (hatched region) with insipient necrosis. The mean size of the Koch responses was 16.6 ± 2.3 mm, and of non-Koch responses, it was 11.2 ± 4.8 mm.

*Statistics:*

Nos. T + ve, BCG − ve c.f. T + ve BCG + ve: $p < .0000$* Nos. Koch reactors among T + ve BCG − ve c.f. T + ve BCG − ve: $p < .0000$.* Sizes of non-Koch responses BCG − ve c.f. BCG + ve: not significant. Sizes of Koch reactors c.f. non-Koch reactors, BCG + ve: $p < .001$.**

*Note:* *Fisher's exact test; **Student's *t*-test.

Unfortunately, BCG is not effective in preventing activation of tuberculosis in those already infected with persister bacilli, merely maximizing their preexistent Koch-type response. Thus, the vaccine would be ineffective in the third of the world's population estimated by WHO to be already infected with tubercle bacilli, although it might help in holding bacilli in the persister state.

A major factor associated with activation of latent tuberculosis is HIV seropositivity. It is interesting to see that the proportion of British HIV-positive individuals developing tuberculosis is similar to the proportion of ambulance recruits with Koch-type responses to tuberculin. HIV seropositive patients lack response to common mycobacterial antigens (43), as do patients with mycobacterial diseases.

## 11. Antigens of Mycobacteria and Categories of Responders

Most mycobacterial species contain antigens of three taxonomic significances (44). Common, or group i, antigens are shared by all mycobacterial species and by nocardiae, and it is among these that sequence homology with human proteins is known to exist and might be important (5). Then, each species has group iv antigens specific to itself, although showing variation between subspecies/serotypes. Slowly growing species share group ii, slow-grower-associated antigens, and most rapidly growing species share group iii, fast-grower-associated antigens (Fig. 9.2).

Man is capable of making skin test responses to groups i, ii and iv* antigens, at least, but there appear to be differences in the types of response that can be made. Whereas the species-specific, group iv antigens can evoke Koch-type or non-Koch-type reactions, responses to group i, common antigens always appear to be of non-necrotizing type. Thus, species-specific responses can denote TH1 or TH2 maturation, and responses to common antigens seem to be mediated only via TH1 cells and can be used to assess the status of protective immunity and the apparent efficacy of BCG vaccination programs (45).

Using a panel of four tuberculins made from different carefully selected mycobacterial species, people can be separated into three responder categories according to whether they react to group i or group iv antigens, or whether their responses are suppressed by group i antigens (46). The reagents used are usually prepared from *M. tuberculosis, M. scrofulaceum,* both slow growers, *M. vaccae* and another fast grower, or *M. leprae.* The categories are as follows:

Category 1: positive to all four reagents, recognizing group i antigens
Category 2: negative to all four reagents due to regulatory effects (except in the young, when it may denote lack of environmental experience)
Category 3: positive to some reagents and negative to others, recognizing group iv antigens.

---

*It is normal practice to use small case roman numbers for these antigenic groups, reserving upper case roman numbers for Runyon's groups.

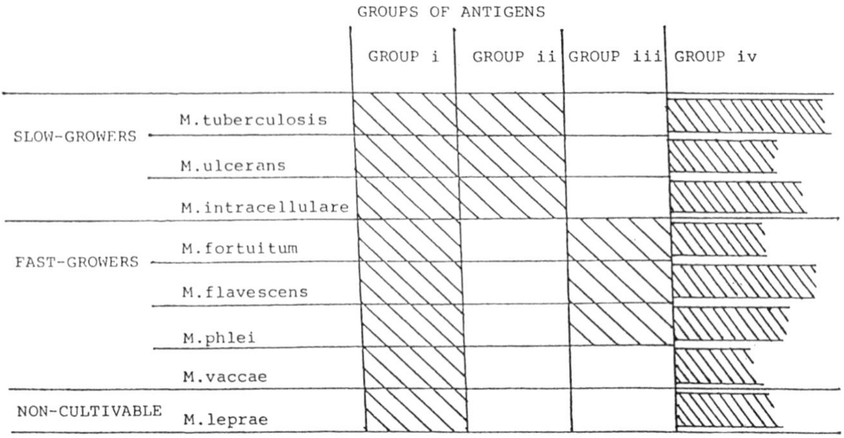

GROUPS OF ANTIGENS

GROUP i | GROUP ii | GROUP iii | GROUP iv

SLOW-GROWERS
M.tuberculosis
M.ulcerans
M.intracellulare

FAST-GROWERS
M.fortuitum
M.flavescens
M.phlei
M.vaccae

NON-CULTIVABLE
M.leprae

Figure 9.2. The antigens of mycobacteria demonstrable with high titre rabbit antisera and sonicates of bacteria by double diffusion in gel. Representative species of the two major subgenera, and of the only named noncultivable species are shown.
Antigens of group i are common to all mycobacterial species, and some are common to all other bacterial genera and mammalian species including man. Important among this group are the stress proteins, and immune responses to this group are strongly associated with protective immunity.
Antigens of group ii are shared by the members of the slow-growing subgenus. Skin test responses can be made to them, but the contribution that they might make to protective immunity or immunopathology remains unknown.
Antigens of group iii are shared by most fast-growing mycobacterial species, with the apparently unique exception of *M. vaccae.* Their relevance to immune mechanisms is unknown.
Antigens of group iv are species specific, being almost completely restricted to individual species. It is thought that responses to these confer specificity to skin test reactions, and Koch-type immune responses to them are a major part of mycobacterial immunopathology.

A fourth category of persons reacting to group ii antigens can be distinguished in some circumstances (47).

## 12. BCG and the Establishment of Tuberculin Positivity

Most manufacturers expect to produce a BCG that will induce more than 90% tuberculin positivity if recipients are tested 8 weeks after vaccination. Some manufacturers have tried to produce a vaccine producing the largest tuberculin responses under the mistaken idea that the larger the response, the better the immunity. In fact, the best evidence of future protection is a small well-controlled response without evidence of tissue damage (25), which is likely to indicate priming of antibacterial immunity; large responses tend to associate with priming of the tissue damaging and less protective TH2 response.

Tuberculin responsiveness after vaccination with BCG soon wanes and disappears in individuals living in regions where contact with mycobacteria is low. However, it rapidly returns when such organisms are met, showing that the individual has memory T cells for group i, common mycobacterial antigens, having been primed to react to encounters that might be below the immune recognitive threshold for nonvaccinated individuals. Our studies in the Lebanon illustrate this (48), providing a parallel with events following BCG vaccination at birth which is often followed by an extended "silent phase" during which tuberculin positivity disappears, only to return with later contact with environmental species.

In areas of moderate contact with environmental mycobacteria such as the United Kingdom, post-BCG tuberculin positivity persists, partly due to prevaccination priming and partly to the lowered threshold of recognition being frequently broken through eliminating the "silent phase." In the above situations, the size of tuberculin response remains relatively small and is of the non-necrotizing, immunoprotective type.

In areas of high contact with environmental mycobacteria, the situation is more complicated. To be effective, BCG has to be given early in life, before a potentially tissue damaging, Koch-type response has been primed by the environment. Thus, in Agra (17), we found that skin test indicators of protective immunity only followed vaccination in children of 2 years or less.

One of the "surprise findings" of the Chingleput BCG trial was that within 2 years of vaccination, a high proportion of individuals had reverted to tuberculin negativity, whereas others retained large-sized responses (34,37). The logical explanation for this is that excessive environmental load first enhances the BCG-exposed Koch-type response and then regulates it to negativity as part of an homeostatic mechanism designed to protect the individual from unnecessary tissue damage in response to injections of mycobacterial antigen lower than those assimilated continuously from environmental sources (33). Much the same is seen in patients with extensive tuberculous disease when they become anergic to small doses of tuberculin. This disease-associated anergy can usually be broken through by challenge with larger tuberculin doses or by reducing the antigen load by treating the patient! It is the practice of many physicians to demonstrate this as part of their diagnostic procedure for tuberculosis. First, they test with a low dose of tuberculin, let us say 1 in 10,000 in the old parlance, then they test with 1 in 1000, and on to 1 in 100 to try to break through this important homeostasis.

Even where it is most successful, BCG does not result in 100% protective immunity for a number of reasons. Leaving aside the obvious explanations of poor vaccine and poor procedure, there remain individuals who, although tuberculin negative, are either already infected with persister tubercle bacilli or are environmentally primed to make a Koch-type response to BCG with its low protective efficacy. To some extent this is illustrated by the studies on persons who subsequently developed tuberculosis in the placebo group of the United Kingdom BCG trial. These were found to be chiefly among those who had no response to tuberculin at the time of the pretrial skin test and those who had large responses at this time (25). In the BCG recipients, it was found that those with absolutely negative reactions to postvaccination tuberculin tests were just as protected as those who were tuberculin positive. A rational explanation for this came many years later when Swanson Beck and his colleagues demonstrated that in a proportion of completely negative tuberculin test sites, needle biopsies showed that precisely the same proportions and numbers of cells had migrated into the test site (41) but had not produced palpable induration or visible erythema. Although it was not proven at that time, regulation of cytokine release offers the best explanation, and it is the effect of cytokines, rather than the cell infiltrate, that is responsible for the appreciable changes of a positive tuberculin reaction.

## 13. The Establishment of Overt and Latent Tuberculosis

Persons potentially having Koch-type, predominantly T-cell Type 2 responses to species-specific antigens might be readily susceptible to tubercle bacilli, or already infected with them. Those with predominantly Type 1 T-cell responses to common mycobacterial antigens appear to be better protected, and a large challenge dose, or intercurrent immunosuppressive episode, might be needed before tubercle bacilli can successfully invade. Once infection is established and the disease process has begun, a mixed Type 1 plus Type 2 response with predominance of the latter develops (49). Under this circumstance, tissues infected with tubercle bacilli or contaminated with their antigens necrose to eventually form caseum. The mechanism by which this occurs is partly destruction of microvascular endothelium, probably by TH2-enhanced toxicity of tumor necrosis factor, loss of blood supply to the tissue, and cytokine-mediated death of cells of the macrophage series presenting antigens on their surface. These pathological processes are accompanied by an increase in the proportion of the Gal[0] antibody glycoform (50).

Although the mechanism by which tubercle bacilli perturb immunity and allow the development of clinical disease is not precisely known, an attractive hypothesis is that TNF-enhancing activity (TEA) (51) of *M. tuberculosis* allows TNF to interfere with the activity of adrenal cortical trophic hormone (ACTH) released from the pituitary gland. This influences the release of adrenal cortical hormones. The balance between circulating cortisol and dehydroepiandrosterone (DHEA) appears to determine the maturation pattern of helper T cells into TH1 and TH2 (52). The reduced levels of DHEA associated with active tuberculosis and some other infections including HIV result in the TH2 predominance allowing cell destruction, immunopathology, and progressive disease.

## 14. Immunoprophylaxis

From the foregoing, it can be seen that protective immunity from tuberculosis requires T-cell maturation of predominantly Type 1 cytokine pattern. Although BCG can maximize this if already established, it can also maximize a Type 2 pattern of T-cell maturation if this is present, increasing susceptibility. What is needed for a new vaccine against tuberculosis is one that will not only maximize preexisting Type 1 maturation but which will switch off Type 2 and replace it with Type 1. Such a vaccine should be efficaceous in both the uninfected person and in eliminating bacilli in those with latent tuberculosis (preventive immunotherapy).

Robert Koch had originally thought that his tuberculin might provide the basis of a vaccine for tuberculosis. This was not the case, although it was successful as an immunotherapeutic in some circumstances (53). The success of other vaccines based on induction of high levels of toxin-neutralizing or other antibodies reasonably led to the expectation that this would also be true for tuberculosis. The experience of Jenner's remarkable success with *Vaccinia* also suggested that a live attenuated vaccine might be necessary, and it was this that led Calmette and Guerin to the development of their bacillus—BCG. Even with an undoubted stimulus to cell-mediated immunity such as their attenuated strain of bovine tubercle bacillus, the induction of protective immunity to tuberculosis often fails. In all probability, BCG will have been relegated to history before we fully understand why it can be so successful in some places and utterly fail in others (3).

One of the major problems in our comprehension of immunity to mycobacterial disease has been slowness to realize that it is the nature of the response that is essential to antibacterial cellular mechanisms rather than the specificity of the antigens toward which the response is directed. Such efficacy which BCG shows does not depend on the closeness of its antigens to those of virulent tubercle bacilli but on its adjuvant capacity to enhance preexisting pathways of response. Thus, it is the result of genetics, acquired maternal immune mechanisms, and experience of the environment that decides whether BCG is an effective vaccine in a given situation. Of course, there can also be variation in the adjuvant efficacy of different manufacturers' products further obscuring the issue. No amount of engineering antigens unique to *M. tuberculosis* into BCG can logically produce an improved vaccine; what is needed is a modification of its adjuvant power so that only beneficial aspects of cellular immunity are enhanced and disadvantageous ones are switched off. This raises the question: Why use BCG at all?

The dogma of early 20th-century immunology, based on Jenner, that a live vaccine was necessary to stimulate cellular immunity must also be questioned. Probably true for virus vaccines, where the antigens and adjuvants are not present in the virus but encoded in the genome and needing expression before they can have an effect, it is more questionable when applied to bacterial vaccines. Those experiments showing that the amount of live BCG is smaller than the amounts of killed BCG or killed tubercle bacilli required to induce immune responses in animals prove nothing. After all, BCG multiplies up to larger numbers in the tissues before inducing sufficient immunity to eradicate itself, and the comparison with dead bacilli should be with this enhanced number of live ones not with the size of the original inoculum. Similarly, in the original mouse model for leprosy, a small challenge of $10^4$ live bacilli will multiply by 2 logs and then stop, and injection of $10^6$ killed leprosy bacilli will prevent replication of a subsequent small live challenge dose.

Bretscher's hypothesis (39) and his experiments suggest that protection following BCG can be increased by reducing the dose 10-fold or 100-fold. This is probably true for naive animals and neonatal man, but it has yet to be shown that reducing dose size can revert potential susceptibility to a potential protective immunity. In mice, the experiments of Hernandez-Pando (54) have shown that inoculation with $10^7$ killed *M. vaccae* induces Type 1 cytokines and inoculation with $10^9$ induces Type 2 cytokines.

What then is the future for vaccination against tuberculosis? Are there antigens that need to be recognized? The answer to the last question is probably yes, but these are likely to be the bacterial versions of self-antigens such as the heat shock or stress proteins. For a long-lived mammalian species such as man, the T-cell repertoire is much more complete for such substances than it is for antigens utterly specific to pathogens, to which the natural response is via B cells and antibodies. If the T-cell response to antigens shared with self were necrotizing, autoimmunity would destroy us; yet this rarely happens. The immune mechanism directed toward such shared antigens, notably the stress proteins (6), is apoptosis for many cell types and destruction of engulfed organisms in macrophages. Enhancement of such a mechanism should eradicate infection with intracellular bacterial pathogens, control harmful inflammations such as those of autoimmune disease, and stop the growth of malignant tumors. Thus, under some circumstances, we have seen BCG to be highly effective in preventing tuberculosis and leprosy, and in the same regions, effective in preventing leukemias and solid tumors of childhood (24). Indeed, the prevention of tuberculosis comes very close to the regulation of self, and when we have a really effective vaccine against tuberculosis, it will have unexpected additional beneficial effects.

A killed preparation of a selected strain of *M. vaccae* (NCTC 11659) was originally developed as an additive to BCG and possible replacement for BCG (55). Preliminary studies of combined vaccination with $10^6$ live BCG plus $10^7$ killed *M. vaccae* have been very successful (48,56–58), although the high cost of a vaccine trial have so far prevented a thorough investigation. Even smaller studies of vaccination with $10^8$ killed *M. vaccae* alone have been carried out, and one of vaccination with $10^9$ killed *M. vaccae* is in progress. Follow-up of some studies has continued for 10 years, and the beneficial effects over those of BCG alone appear to persist. With the current problems of treatment of tuberculosis, the emphasis of interest in *M. vaccae* has swung away from its potential use as a vaccine and turned to its potential as an immunotherapeutic.

## 15. Immunotherapy

For immunotherapy of tuberculosis, an agent is needed that will strongly promote Type 1 T cells, and switch off Type 2 maturation. Such a treatment should reduce deaths during treatment, stop immune-mediated tissue damage in lesions, and correct the disadvantageous effects of Type 2-associated cytokine release. Thus, patients should show more rapid recovery of body weight, and symptoms such as cough, chest pain, fever, night sweats and hemoptyses should resolve more rapidly. The treatment should lead to immune destruction of intracellular tubercle bacilli, whatever their pattern of susceptibility to drugs, with a consequent faster disappearance of live bacilli from the sputum and reduced likelihood of infecting others. These changes should be accompanied by rapid correction of biochemical, hematological, immunological, and radiological parameters. Effective immune destruction of bacilli and correction of the disease process should allow a marked reduction in the period of chemotherapy currently required to treat the disease, no need for directly observed treatment, and more money and facilities freed for effective case finding. If immunotherapy can be developed that will provide all this at a reasonable price, the whole face of tuberculosis worldwide will be changed.

What of tuberculous disease pathology; does this exemplify the alternative mechanism of tissue destruction to apoptosis?

Shwartzman's phenomenon and Koch's phenomenon appear to have many similarities. Where a foreign antigen contaminates vascular endothelium, this becomes sensitive to death mediated by tumor necrosis factor (TNF) and TH2 cytokines. Such sensitization can be directly mediated by foreign substances or is perhaps orchestrated by the balance of hormones and their metabolites (59). Endocrine control of immunity is by no means limited to this effect; the first step in determination of the nature of a cellular immune response will depend on the pathway of development of T lymphocytes. Understanding of the control of maturation of T cells has greatly progressed following the description of T helper cells 1 and 2 (13) and the recognition of Type 1 and Type 2 immunity with their characteristic patterns of cytokine release. Type 1 immunity seems associated with non-tissue-damaging responses to group i, common mycobacterial antigens [including the mycobacterial variants of stress proteins (5)], and is the likely mechanism of apoptosis of damaged cells and of overcoming infection with tubercle bacilli. The mechanisms of Shwartzman's and Koch's phenomena in tuberculosis seem to be a combination of Types 1 and 2 immunity to group iv antigens specific to tubercle bacilli. What is needed of an immunotherapeutic for tuberculosis is a stimulus to switch off maturation of Type 2 T cells and to reinforce that of Type 1 cells including those responding to group i, common mycobacterial antigens.

## 16. History of Immunotherapy for Tuberculosis

Once the infectious nature of tuberculosis was firmly established by Villemin in 1868 (60), scientists of the day started to think of rational approaches to the treatment of the disease. Following the demonstration of tubercle bacilli by Koch in 1882 and their successful culture, bacterial products became available as potential immunotherapeutics. Early experiments had shown that antisera alone did not provoke a cure in experimental animals, giving rise to the mistaken idea that antibodies have no part to play in immunity to tuberculosis. Thus, the use of bacterial products directly injected into man became the popular approach to treatment in the preantibiotic days and was particularly so after Koch had described his phenomenon (42,61). This is the ability of an already infected animal to cast off and heal small challenge doses of live tubercle bacilli injected intradermally. The same phenomenon occurs when proteinaceous extracts of tubercle bacilli are injected, and it was this observation that led to the tuberculin test. The same observation was developed by Koch (1891) for the first immunotherapeutic for tuberculosis effective in some cases of clinical disease, notably lupus vulgaris (53). As an interesting aside, in the same year, Coley (1891), in New York, developed his toxins as immunotherapy for cancer based on the same immunological mechanism (62).

Providing the tuberculous lesions were small or superficial, repeatedly injecting tuberculin (at first called Koch's brown fluid) resulted in the necrosis of the lesions. Dead tissue plus live bacilli were sloughed from superficial lesions, and internal lesions became encapsulated by fibrous tissue, cutting off oxygen from the bacilli and stopping their replication, effectively returning them to a kind of persister state. The reaction producing this effect was considerable, with the patient developing fever and sometimes going into shock, leading to death within a few hours of injection (63). This reaction, the "tuberculin shock syndrome," was due to the release and circulation of tumor necrosis factor (TNF) and other cytokines in a person whose tissue was primed to excessive sensitivity to them. Local reactions around deep lesions in vital organs could also lead to a slower death through organ failure. Thus, Koch's tuberculin therapy was a two-edged sword, healing some and killing others. Because Koch produced his tuberculin from heat-sterilized culture medium in which tubercle bacilli had been grown, it was rich in heat-stable secreted substances, focusing attention for the first time on the importance of secreted antigens of tubercle bacilli, a concept to which mycobacteriologists have recently returned (64). Successful in Koch's own hands, his tuberculin therapy was too uncertain a tool for most clinicians, and by the beginning of this century, it had few remaining exponents.

Over the few years following Koch's efforts, immunotherapy was extensively investigated for leprosy, and many generally unsubstantiated claims for its efficacy in treating various forms of this disease can be found in the literature. One of the difficulties in their interpretation in those days was the lack of a sound conceptual framework for the immunopathology of mycobacterial disease. Seminal as it should have been, Koch's description of his phenomenon held back progress because it was assumed that the mechanism of the phenomenon, of immunopathology of tuberculosis, and of antibacterial protective immunity were one and the same. It was not until the 1970s that protective immunity to tubercle bacilli became distinguished as a separate pathway from that of Koch's phenomenon and immunopathology (65).

After Koch, the first major step toward successful immunotherapy was made by Friedmann, also in Berlin. Following his discovery of the schildkrotentuberkelbazillus from diseased captive sea turtles in the Berlin Zoo, Friedmann worked on the concept that antigens shared with other mycobacterial species should be the basis of treatment for tuberculosis. Thus, he was the first to recognize the importance of group i, common mycobacterial antigens, in this context. Friedmann's bacillus is now known as *M. chelonae,* and although he considered it nonpathogenic for man, it is now known to cause important and very difficult-to-treat disease, especially in iatrogenically immunosuppressed patients. Friedmann was still fettered by Koch's concepts and believed that in order to be effective, a vaccine or an immunotherapeutic had to be alive and had to give rise to a "limited tuberculous focus" in order to promote immunity. Rather remarkably, Friedmann's preparation is still available for use in the treatment of a number of diseases under the registered name of Anningzochin (Laves-Arzeimittel GmbH) (66).

The next major steps forward were taken by Spahlinger in Geneva with the development of his "vaccine" and "serum" therapies. Spahlinger (67) believed that immunotherapeutic agents should be prepared from tubercle bacilli grown on media made as similar as possible to those pertaining in the host tissues. Thus, if the preparations were for treating man, he enriched the culture medium with human serum, and if for treating cattle, he enriched them with bovine serum. In order not to damage the "toxins" produced by any mechanism of killing the bacilli, Spahlinger incubated his cultures for long periods and then left them hermetically sealed in the dark at room temperature for a year or more. If subculture showed no growth, the material was directly injected into the patient, or into black Irish hunters (horses) that he favored for antiserum production. Spahlinger appreciated the importance of the presence of stress proteins of tubercle bacilli, and he heat shocked his bacilli during culture (68) and used them to raise equine antisera for the treatment of bone and joint tuberculosis. Spahlinger's special forte seems to have been the treatment of very advanced patients whose doctors despaired of their recovery. Such patients were treated, apparently with success, with his sera up until his death in London in the early 1970s. It seems a quirk of fate that Spahlinger's life should just have encompassed, and in the same city, the early work on the strain of *M. vaccae* described below.

A sad episode in the history of therapy of tuberculosis involved the ostracism and financial ruin of a Mr. Stevens by the British Medical Association in 1912 and 1914 (69). Undoubtedly, described as a quack, Stevens discovered the use of Umckaloabo, the dried root of a plant, while in South Africa at the beginning of this century. Clinical description of its use strongly suggest that it inhibited TNF in some way, years before the discovery of cytokines and the current interest in their regulation. Recent inquiry in the region around east London (RSA), from where Stevens claimed it originated, finds the herb still being used for the treatment of tuberculosis and other conditions by modern native healers (P. Onyebujoh, personal communication).

## 17. The Development of *Mycobacterium vaccae*

Thought to prime children for a successful outcome of BCG vaccination against leprosy, *M. vaccae* was developed as an additive to BCG which did two things. First, it lowered the threshold for immune recognition of casually met mycobacterial species, resulting in increased skin test positivity to them. Second, it appeared to ensure protection rather than susceptibility after vaccination, preferentially enhancing the maturation of TH1 cells.

As a model immunotherapeutic for tuberculosis, killed *M. vaccae* exemplifies many of the possibilities and it has become the first major immunotherapeutic for tuberculosis to be developed since the introduction of antituberculosis drugs. A species described by Bonicke and Juhasz (70) from soil samples taken from the vicinity of cattle, no special note of it was made until it was found to be present in the environment of the region of Uganda where BCG had been shown to be highly effective in the prevention of leprosy (38). A soluble skin test reagent—Vaccin—prepared from the species produced reactions in leprosy patients and their contacts that might have been expected of a similar reagent prepared from *M. leprae* (71). Although not unique in this, responses to Vaccin showed better than 90% correlation with those to Leprosin A, when this became available (72). At first, it was assumed that *M. vaccae* shared some group iv, species-specific antigens with *M. leprae* not shared by other organisms, but this possibility remains unproven (73). What has been shown, however, is that these two species are more immunogenic when killed than they are when alive, a very useful characteristic that does not seem to apply to other mycobacterial species (74).

## 18. New Tuberculins

The development of the range of "new tuberculins" (75) made from filtered sonicates of nonheated mycobacterial species with their much greater content of group iv, species-specific antigens resulted in better recognition and understanding of tissue damage around mycobacterial antigens. It had been known for 60 years that there could be qualitative differences between the responses made to old tuberculin or purified protein derivative by different individuals. Some believed that this differentiated immunity from allergy; others believed that it was simply a quantitative difference in the amount of reaction shown by an individual. These qualitative differences were much more easily differentiated when new tuberculins were used and were soon found to show antigenic restriction. The soft, non-necrotizing induration (non-Koch-type response) correlating with protective immunity could be directed toward antigens of groups i, ii, or iv, whereas the hard, craggy and tender induration (Koch-type response) associated with incipient necrosis was directed to group iv antigens of certain species only (2,33,42,75). There is also a tendency for the non-necrotizing responses to be smaller in diameter than those that are necrotizing. The potentially necrotizing response is found almost exclusively to species known to have pathogenic capacity and is particularly seen in the response to tuberculin of tuberculosis patients. As such, it appears to be an example of the mechanism leading to tissue death in tuberculosis lesions exhibited at the skin test site around tiny quantities of injected antigen. Not only does this qualitative differentiation of response to tuberculin acquire diagnostic value, it has a prognostic value, identifying persons with increased susceptibility to tuberculosis or with latent infection with persisters and it provides a tool for the investigation of potential immunotherapeutics.

A series of studies of mixed skin tests were carried out in which new tuberculins prepared from different species were mixed together and injected intradermally into volunteers. Such mixtures were shown to evoke two types of regulation of skin test responsiveness (76–79). In general, mixtures containing reagents prepared from fast-growing mycobacterial species locally suppressed the response to reagents prepared from slow-growing species included in the mixture. Thus, many leprous tissues fail to respond to concurrent infection with *M. intracellulare* or *M. scrofulaceum.* Mixtures containing extracts of certain fast-growing species (notably including *M. vaccae*) could also affect responses to extracts of potentially pathogenic slow-growing species injected on the other arm. If an injection on one arm could regulate reaction on the other, why should it not regulate response at disease sites in the chest?

The sonicates used in these mixed skin test experiments induced immune regulation which had disappeared when tests with reagents of the slow growers were repeated 3 months later. Experiments in animals and later observations in man showed that if whole, killed *M. vaccae* was injected instead of soluble sonicate, the immunoregulatory effect was longer lasting. This prolongation might be due to constant boosting from the environment in individuals with a lowered threshold for immune recognition of such organisms, or to a direct effect of the injected bacilli.

## 19. The Results of Immunotherapy with *M. vaccae*

A series of pilot studies have been carried out to determine optimum manufacture, dose, duration, and effects of immunotherapy for tuberculosis with *M. vaccae.* The very first studies after animal evaluation were carried out on long-term-treated leprosy patients in Spain (80), and in leprosy and tuberculosis patients in London (81). These were followed by in depth studies on tuberculosis in Kuwait (82,83), a trial in tuberculosis in The Gambia, and studies on multidrug-resistant tuberculosis in Iran (84). Subsequent studies on tuberculosis have been carried out in Argentina (85), Nigeria (86), Romania (87,88), and Vietnam. A major Phase III trial is almost finished in Durban, South Africa (89) on pulmonary tuberculosis for drug regulatory purposes, and a further regulatory trial is progressing in London. Rather than reviewing the individual studies, an account follows of the findings to date.

Studies in a guinea pig model showed heat-killed *M. vaccae* to be more effective than radiation-killed *M. vaccae,* which were more effective than live bacilli (90). Dose-ranging studies in mice showed that $10^5$–$10^7$ bacilli was an optimal dose for promoting better immune recognition of tiny doses ($10^4$) of leprosy bacilli.

Starting with an intradermal injection of $10^7$ radiation-killed *M. vaccae,* studies in fully treated lepromatous leprosy patients showed that the dose had to be $10^9$ to break through the threshold for enhanced responsiveness to skin tests with soluble antigens of *M. leprae* (Leprosin A) and that this could be further improved by adding a 1/10th dose of tuberculin to the injection (80). Subsequently, it has been shown that this dose enhances the rate of removal of leprosy bacilli during the course of multiple-drug therapy (MDT) for multibacillary (MB) leprosy, and removes PCR-detectable genomes of leprosy bacilli from long-treated patients with paucibacillary (PB) leprosy (91). Clinically, erythema nodosum leprosum (ENL) is treated and prevented, anterior uveitis aleviated, and blood flow to the finger pulps and skin temperature increased. Ten years after injection of *M. vaccae,* responsiveness to Lepromin has been found to be enhanced.

Dose-ranging studies in tuberculosis found the threshold for induced recognition of group i, common mycobacterial antigens (subsequently recognized as a possible marker for a TH1 maturation pattern) to be $10^9$ *M. vaccae.* Autoclaved bacilli were more effective than irradiated bacilli, and the addition of tuberculin imparted little or no improvement in the treatment of tuberculosis (83). In the earliest studies of *M. vaccae* in tuberculosis, the intradermal injection was given 4–6 weeks after starting full courses of effective chemotherapy. After the study in The Gambia had shown that during the course of chemotherapy deaths were reduced after injection of *M. vaccae,* this has been given at progressively earlier times until it is now advocated at the time of diagnosis on the first day of chemotherapy (92). The most recent pilot studies to be set up are for exploring the possibility of reducing the period of chemotherapy required for treatment success. Eventually, it is hoped to reduce this to the period of early bactericidal activity (EBA) (93) of perhaps just a few weeks or days.

Markers for success found in the various studies, as well as reduced mortality, have resulted in increased therapeutic success, faster disappearance of acid fast bacilli (AFB) from the sputum, faster regain of body weight, fall in ESR, fall in Gal[0], fall in C-reactive protein (CRP), and faster and more complete resolution of radiological changes including cavities. Examples of some of these are shown in Table 9.1.

Table 9.1 Abbreviated results of some preliminary studies of *M. vaccae* in the immunotherapy of tuberculosis.

(a) Some results obtained in randomized, placebo, controlled studies of immunotherapy for pulmonary tuberculosis with a single injection of killed *M. vaccae* carried out in Africa, Asia, and Europe. The time of intervention in the different studies varied from 1 to 6 weeks after starting chemotherapy. The quality of chemotherapy varied between studies, and one study was of patients with chronic tuberculosis.

| ***M. vaccae* group** | | **Placebo group** |
|---|---|---|
| Deaths during treatment | | |
| 10/389 (2.6%) | $p < .0001$ | 38/420 (9.0%) |
| Cure[a] at the end of treatment | | |
| 329/382 (86%) | $p < .00001$ | 302/415 (72%) |

(b) Results a year after starting a 6-month course of chemotherapy for newly diagnosed or chronic pulmonary tuberculosis in Romania. Patients were randomized to receive an injection of *M. vaccae* or placebo 1 month after starting chemotherapy.

| | | |
|---|---|---|
| Closure of cavities on x-ray by the end of treatment | | |
| 41/120 (34%) | | 34/120 (28%) |
| Reduction in cavity surface areas of those still with cavities | | |
| 54% | $p < .006$ | 26% |
| Regain of body weight | | |
| 6.79 kg | $p < .005$ | 4.45 kg |
| Fall in ESR | | |
| 77.5% | | 64% |

[a]Cure refers to both bacteriological and clinical cure.

An interesting observation has been the improvement in the patients sense of well-being just a few days after *M. vaccae* has been injected. This has been so striking to some clinicians that it is difficult to keep studies blinded to them. The explanation for it is likely to be a change in adrenal secretion of steroids (59).

There is a small amount of evidence that in some cases, an injection of *M. vaccae* will cure tuberculosis without the use of chemotherapeutic agents. This comes from a study in Nigeria (86) in which at one point no drugs were available, and 10 patients received an injection of *M. vaccae* and did not return for chemotherapy when this became available. A year later, five of these patients were traced and found to be well and with their tuberculosis cured despite claiming that they had received no chemotherapy at all. Other evidence comes from patients with multidrug-resistant tuberculosis for whom the only drugs available were those to which they were resistant. A proportion of such patients have been cured seemingly by injections of *M. vaccae* alone (73).

In general, multidrug-resistant tuberculosis poses special problems to immunotherapy, not because of the drug resistance of the bacilli but because of the long period that patients have often had disease and the proportion of the bacilli that are multiplying in an extracellular situation (94). Such patients require injections of *M. vaccae* to be repeated at 2-month intervals, and even then, only a proportion are cured even after six injections.

## 20. How Does Immunotherapy with *Mycobacterium vaccae* Work?

This appears to be multifactorial with a change in the secretion of adrenal cortical hormones apparently being the first step. Effects are seen within 2–4 days, and probably result from increased production of dehydroepiandrosterone and its metabolites. Second, the Erythrocyte Sedimentation Rate (ESR) begins to fall, and Acid-Fast Bacilli (AFB) might fall in the sputum. This reduction in AFB is probably due to a reduction in intracellular replication and intracellular destruction of tubercle bacilli. The mechanism of intracellular killing of tubercle bacilli in man remains a contentious issue, but it is probable that heavily infected macrophages are killed by cytotoxic CD8 + T cells and their debris taken up by fresh macrophages activated to kill live bacilli, possibly by the burst of oxygen metabolism shortly following phagocytosis. Extracellular bacilli might have to await reengulfment by activated cells before they can be killed, as there is no reason to think that immunotherapy can directly influence extracellular tubercle bacilli, although it might hasten their phagocytosis by antibody changes, resolution of fibrous tissues around cavities, and prevention of further tissue destruction by the Koch phenomenon.

## 21. Immunotherapy and Future Treatment of Tuberculosis

Immunotherapy with *M. vaccae* is not an alternative strategy to chemotherapy but a powerful complement to it that should lead to complete domination over the disease. The best use of such a powerful new tool has yet to be determined, but it seems likely that it will enable a massive reduction in the period most patients need for successful treatment. The pilot studies in their development stage are designed to investigate a halving of chemotherapy from 6 months to 3 months when an injection of *M. vaccae* is given at the start of chemotherapy. However, it is appreciated that this is unlikely to be sufficient in relapsed patients, or those with a history of 2 years or more of disease in whom many bacilli are extracellular. In such patients, it might be necessary to repeat the immunotherapeutic injection at intervals of 2 months or so until lesions have resolved and all extracellular bacilli have been phagocytosed and killed. Although experience so far suggests that at least six injections can be given without side effects of the later injections being any worse than following the first, experience of more injections than this is not yet available and unexpected problems can arise.

If the first studies of reducing chemotherapy are successful, the length of chemotherapy will be successively reduced until only that necessary for EBA is given, perhaps for no more than a week or two. The principle by which this can be achieved is that among the intracellular bacilli destroyed by immunotherapeutic action should be the slowly metabolizing persisters for which the majority of modern chemotherapy is administered. A major reduction in period of chemotherapy will avoid the need for directly observed therapy (DOT) currently advocated by WHO. It will overcome the problem of treatment noncompliance, releasing health care workers for more vigorous case finding. Financial savings should contribute to the health budgets for other diseases.

Reducing the period during which viable bacilli are shed will secondarily reduce numbers of new infections among contacts. The briefer period required for treatment will erode the prejudice around tuberculosis so that patients do not hide their condition, further reducing numbers of new cases of infection.

Although yet to be proven, it is likely that vaccination with *M. vaccae,* unlike that with BCG, will be effective in destroying persisters in those silently infected with tubercle bacilli and with latent disease. Studies of this are planned in HIV-seropositive, tuberculin-positive persons in whom activation of tuberculosis due to immunosuppression is at its highest. If this can be achieved, then the captain of all the men of death may meet his master.

## 22. *Mycobacterium vaccae* in the Treatment of Nonmycobacterial Disease

As an adjuvant particularly enhancing maturation of TH1 cells, *M. vaccae* can have very widespread activities, of which only a number might be suspected so far. The first to be discovered by chance was psoriasis in India (95). A leprosy patient suffering from ENL received an injection of *M. vaccae* and his coincident psoriasis showed marked improvement. Subsequently, 10 Indian patients with psoriasis, but without leprosy, were given an injection of *M. vaccae* and 9 of them showed similar improvement, and so did the last patient after a second injection. Two years later, all 11 patients were almost completely free of psoriatic lesions. Following the Indian experience, 48 Iranian patients with psoriasis were treated with 2 injections of *M. vaccae* and 44 showed marked sustained improvement. A study in London showed moderate effects in some patients, but this was less impressive than that found in India and Iran. A small study on Indonesians with psoriasis in Holland showed marked improvement, and a recently completed randomized and blinded study in Argentina showed statistically significant improvements in psoriasis patients, although, again, this was less impressive than in India and Iran. Whether these geographic differences reflect the different genetic makeup of the patients or differences in the aetiology of psoriasis remain unknown.

Other skin conditions that are improved by injections of *M. vaccae* include Pityriasis rosea and athletes foot, but its investigation has not been extended in these conditions.

Theoretically, at least, it should be possible to modulate a whole series of difficult-to-treat conditions such as rheumatoid arthritis (96) and inflammatory bowel diseases with immunotherapy based on *M. vaccae.* The next few years should prove the truth of this.

Circulatory conditions alleviated by *M. vaccae* include Raynaud's disease and, in leprosy patients, anterior uveitis and peripheral circulation, the latter detectable 18 months after the *M. vaccae* was injected for its effect on leprosy. Currently, studies are starting to investigate the potential use of *M. vaccae* in the prevention of arteriosclerosis, now thought to be precipitated by autoimmune reactions to stress proteins damaging the vascular endothelium in certain human genotypes. High levels of antibodies to the human 65-kDa stress protein have recently been associated with arteriosclerosis (9), and this has been found to be lowered by immunotherapy with *M. vaccae* in leprosy patients. A preliminary experiment in a rat model of myocarditis in *Trypanosoma cruzi* infection has shown reduced inflammatory responses in the heart muscle after injection of *M. vaccae.* In support of the potential value of *M. vaccae* in human Chagas' disease has been the recent observation in Argentina of loss of skin test reactivity to group I, common mycobacterial antigens in persons in the silent, seropositive phase of the disease (97).

The theories of Clerici and Shearer (98) about maturation of T helper cells in relation to susceptibility to HIV seropositivity when taken with the evidence for lack of skin responsiveness to group I, common mycobacterial antigens in HIV positive persons suggests a potential role for *M. vaccae* in the prevention/treatment of this disease (99). Preliminary studies have started to investigate this.

## 23. Immunotherapy for Cancer

As mentioned early in this chapter, in the very same year, 1891, that Koch described his experiments in the treatment of tuberculosis with tuberculin therapy, Coley, in the United States, described the development of his "toxin" (62) produced from *Streptococcus pyogenes* and *Serratia marcescens.* Coley's immunotherapy was initially very successful in the treatment of patients with tumors of mesodermal origin. Its efficacy became less in patients toward the end of Coley's life in the 1930s, and a suggested reason for this is that strong tuberculin positivity (Koch-type responses) had become much less common in the United States by that time (62). The descriptions of cancer patients responding to Coley's treatment are remarkably similar to those of tuberculosis patients in their response to Koch's treatment. The mechanism by which both treatments worked are likely to have been the same, maximization of the phenomenon by which microvascular endothelium contaminated with "foreign" antigen (tumor specific or tubercle bacillus specific), became sensitive to TNF and necrosed. This resulted in anoxic death of the tumor or tuberculous lesion.

Studies of the effect of BCG vaccine against tumors have produced conflicting but interesting results. Neonatal vaccination with BCG in countries where this is effective in preventing tuberculosis is also associated with a reduction in childhood cases of leukemia and solid tumors. This does not occur where BCG is ineffective against tuberculosis, suggesting that the mechanism effective against tuberculosis is also that effective against tumors (8,24). Use of BCG in immunotherapy of tumors has occasionally been successful but not frequently. This might be because the scarification technique often employed ensured that a Koch-type response ensued, which would have the same limited effects as those of Coley's toxin.

If the inference is correct that the immune mechanism needed to treat tuberculosis is the same as that needed to treat tumors, then immunotherapy for tuberculosis and for cancer should be one and the same. Studies have already begun to investigate this possibility.

### References

1. Stanford JL, Shield MJ, Rook GAW (1981) How environmental mycobacteria may predetermine the protective efficacy of BCG. Tubercle 62:55–62.
2. Stanford JL, Rook GAW (1983) Environmental mycobacteria and immunization with BCG. In Easmon CSF and Jeljaszewicz J, eds. Medical Microbiology, vol. 2. London: Academic Press.
3. Fine P (1995) Variation in protection by BCG: Implications of and for heterologous immunity. Lancet 346:1339–1345.

4. Grange JM (1992) The mystery of the mycobacterial persistor. Tubercle Lung Dis 73:249–251.
5. Young DB (1992) Heat-shock proteins: Immunity and autoimmunity. Curr Opin Immunol 4:396–400.
6. Bahr GM, Rook GA, al-Saffar M, Van-Embden J, Stanford JL, Behbehani K (1988) Antibody levels to mycobacteria in relation to HLA type: Evidence for non-HLA-linked high levels of antibody to the 65 kD heat shock protein of *M. bovis* in rheumatoid arthritis. Clin Exp Immunol 74:211–215.
7. Kilidirias K, Latov N, Strauss DH, Goring AD, Hoshim GA, Gorwan JM, Sodiq SA (1992) Antibodies to the 60 kDa heat shock protein in patients with schizophrenia. Lancet 340:569–572.
8. Grange JM, Stanford JL, Rook GAW (1995) Tuberculosis and cancer: Parallels in host responses and therapeutic approaches? Lancet 345:1350–1352.
9. Wick G, Schett G, Amberger A, Kleindienst R, Qingbo Xu (1995) Is atherosclerosis an immunologically mediated disease? Immunol Today 16:27–33.
10. Rook GAW, Steele J, Brealey R, Whyte A, Isenberg D, Sumar N, Nelson N, Bodman JL, Young KB, Roitt A, Williams IM, Scrogg P, Edge I, Ackwright CJ, Ashford P, Womold D, Rudd M, Redman P, Divek C, Rodemocker TW (1991) Changes in IgG glycoform levels are associated with remission of arthritis during pregnancy. J Autoimmun 4:779–794.
11. Pilkington C, Lefvert A-K, Rook GAW (1995) Neonatal myasthenia gravis and the role of agalactosyl IgG. Autoimmunity 21:131–135.
12. Pabst HF, Godel J, Grace M, Cho H, Spady DW (1989) Effect of breast-feeding on immune response to BCG vaccination. Lancet 339:295–297.
13. Mossman TR, Cherwinski H, Bond MW, Giedlin MA, Coffman RL (1986) Two types of murine helper T-cell clone. 1. Definition according to profiles of lymphokine activities and secreted proteins. J Immunol 136:2348–2357.
14. Romagnani S (1991) Human TH1 and TH2 subsets: doubt no more. Immunol Today 12:256–257.
15. Tala-Heikkila M, Stanford JL, Misljenovic O, Bleiker MA, Tala E (1994) The effects of two different BCG vaccines given in early infancy on tuberculin sensitivity of school-age children. Tubercle Lung Dis 75(Suppl 3):58.
16. Lotte A, Wasz-Hockert O, Poisson N, Dumitrescu N, Verron M, Couvet E (1984) BCG complications. Adv Tuberculosis Res 21:107–193.
17. Stanford JL, Cunningham F, Pilkington A, Sargeant I, Series H, Bhatti N, Bennett E, Mehrotra ML (1987) A prospective study of BCG given to young children in Agra, India—A region of high contact with environmental mycobacteria. Tubercle 68:39–49.
18. Stanford JL, Sheikh N, Bogle G, Series H, Mayo P (1987) Protective effect of BCG in Ahmednagar, India. Tubercle 68:169–176.
19. Tala-Heikkila M, Stanford JL, Misljenovic O, Bleiker MA, Tala E (1994) The effects of two different BCG vaccines given in early infancy on tuberculin sensitivity of school-age children. Tubercle Lung Dis 75(Suppl 1):58–59.
20. Seth V, Kukreja N, Sundaram KR, Seth SD (1982) Waning of cell mediated immune response in preschool children given BCG at birth. Indian J Med Res 76:710–715.

21. Karalliedde S, Katugaha LP, Uragoda CG (1987) Tuberculin response of Sri Lankan children after BCG vaccination at birth. Tubercle 68:33–38.

22. Pilkington C, Tala-Heikkila M, Rook GAW, Stanford JL, Costello AMdeL (1996) The effect of BCG vaccination on IgG responses to mycobacterial antigens. Tubercle Lung Dis Pilkington, (personal communication) Submitted.

23. Ten Dam HG, Hitze KL (1980) Does BCG protect the newborn and young infants? Bull WHO 58:37–41.

24. Grange JM, Stanford JL (1990) BCG vaccination and cancer. Tubercle 71:61–64.

25. Hart PD'A, Sutherland I (1977) BCG and vole bacillus vaccines in the prevention of tuberculosis in adolescence and early adult life: Final report to the Medical Research Council. Br Med J ii:293–295.

26. Romanus V (1983) Childhood tuberculosis in Sweden. Tubercle 64:101–110.

27. Katila ML, Brander E, Backman A (1987) Neonatal BCG vaccination and mycobacterial cervical adenitis in childhood. Tubercle 68:291–296.

28. Grange JM, Yates MD, Pozniak A (1995) Bacteriologically confirmed non-tuberculous mycobacterial lymphadenitis in South-East England: A recent increase in the number of cases. Arch Dis Childhood 72:516–517.

29. Gatner EMS, Rubinstein E (1981) The pattern of age-specific tuberculin hypersensitivity in two groups of South African schoolchildren. Tubercle 62:181–185.

30. Grange JM, Gibson JA, Osborn TW, Collins CH, Yates MD (1983) What is BCG? Tubercle 64:129–139.

31. Rook GAW (1991) Mobilising the appropriate T-cell subset: The immune response as taxonomist? Tubercle 72:253–254.

32. Bahr GM, Stanford JL, Rook GAW, Rees RJW, Frayha GJ, Abdelnoor AH (1986) Skin sensitisation to mycobacteria amongst school children, prior to a study of BCG vaccination in North Lebanon. Tubercle 67:197–203.

33. Shield MJ (1983) The importance of immunologically effective contact with environmental mycobacteria. In: Ratledge C and Stanford JL, eds. The Biology of the Mycobacteria, vol 2. London: Academic Press.

34. Tripathy SP (1987) Fifteen-year follow-up of the Indian BCG prevention trial. Bull Int Union Tuberculosis Lung Dis 62:69–72.

35. Bechelli LM, Lwin K, Garbajosa PG, Gyi MM, Uemura K, Sundaresan T, Tamondong C, Matejka M, Sansarricq H, Walter J (1974) BCG vaccination of children against leprosy: Nine-year findings of the controlled WHO trial in Burma. Bull WHO 51:93–99.

36. Comstock GW, Palmer CE (1966) Long-term results of BCG vaccination in the Southern United States. Am Rev Respir Dis 93:973–978.

37. Tuberculosis Prevention Trial Madras (1980) Trial of BCG vaccines in South India for tuberculosis prevention. Indian J Med Res 72(Suppl):1–74.

38. Brown JAK, Stone MM, Sutherland I (1966) BCG vaccination of children against leprosy in Uganda. Br Med J 1:7–14.

39. Bretscher PA (1991) A strategy to improve the efficacy of vaccination against tuberculosis and leprosy. Immunol Today 13:342–345.
40. van Eden W, de Vries RRP, Stanford JL, Rook GAW (1983) HLA-DR3 associated genetic control of response to multiple skin tests with new tuberculins. Clin Exp Immunol 52:287–292.
41. Gibbs JH, Ferguson J, Brown RA, Kenicer KJA, Potts RC, Coghill G, Beck JS (1984) Histometric study of the localisation of lymphocyte subsets and accessory cells in human Mantoux reactions. J Clin Pathol 37:1227–1234.
42. Stanford JL (1991) Koch's phenomenon: Can it be corrected? Tubercle. 72:241–249.
43. Khoo SH, Wilkins EGL, Fraser IS, Hamour AA, Stanford JL (1996) Lack of skin test reactivity to common mycobacterial antigens in human immunodeficiency virus-infected individuals. Thorax; 51:932–935.
44. Stanford JL, Grange JM (1974) The meaning and structure of species as applied to mycobacteria. Tubercle 55:143–151.
45. Stanford JL, Eshetu Lemma (1983) The use of a sonicate preparation of *Mycobacterium tuberculosis* (new tuberculin) in the assessment of BCG vaccination. Tubercle 64:275–282.
46. Lockwood DNJ, McManus IC, Stanford JL, Thomas A, Abeyagunawardana DVP (1987) Three types of response to mycobacterial antigens. Eur J Respir Dis 71:348–355.
47. McManus IC, Lockwood DN, Stanford JL, Shaaban MA, Abdul-Ati M, Bahr GM (1988) Recognition of a category of responders to group II, slow-grower associated antigens amongst Kuwaiti senior school children, using a statistical model. Tubercle 69:275–281.
48. Bahr GM, Stanford JL, Rook GAW, Rees RJW, Abdelnoor AM, Frayha JH (1986) Two potential improvements to BCG and their effect on skin test reactivity in the Lebanon. Tubercle 67:205–218.
49. Rook GAW, al-Attiyah R (1991) Cytokines and the Koch phenomenon. Tubercle 72:13–20.
50. Rook GA, Onyebujoh P, Wilkins E, Ly HM, al-Attiyah R, Bahr G, Corrah T, Hernandez H, Stanford JL (1994) A longitudinal study of percent agalactosyl IgG in tuberculosis patients receiving chemotherapy, with or without immunotherapy. Immunology 81:149–154.
51. Filley EA, Rook GAW (1991) Effect of mycobacteria on sensitivity to the cytotoxic effects of tumour necrosis factor. Infect Immun 59:2567–2572.
52. Daynes RA, Meikle AW, Araneo BA (1991) Locally active steroid hormones may facilitate compartmentalisation of immunity by regulating the types of lymphokines produced by helper T-cells. Res Immunol 142:40–45.
53. Koch R (1890) An address on bacteriological research delivered before the International Medical Congress held in Berlin, August 1890. Br Med J 2:380–383.
54. Rook GAW, Hernandez-Pando R (1996) The pathogenesis of tuberculosis. Annu Rev Microbiol 50: 259–284.

55. Stanford JL (1991) Improving on BCG. Acta Pathol Microbiol Immunol Scand 99:103–113.
56. Ganapati R, Revankar CR, Lockwood DN, Wilson RC, Price JE, Ashton P, Ashton LA, Holmes RM, Bennett C, Stanford JL (1989) A pilot study of three potential vaccines for leprosy in Bombay. Int J Leprosy Other Mycobact Dis 57:33–37.
57. Ghazi-Saidi K, Stanford JL, Stanford CA, Dowlati Y, Farshchi Y, Rook GA, Rees RJ (1989) Vaccination and skin test studies on children living in villages with differing endemicity for leprosy and tuberculosis. Int J Leprosy Other Mycobact Dis 57:45–53.
58. Stanford JL, Stanford CA, Ghazi-Saidi K, Dowlati Y, Weiss SF, Farshchi Y, Madlener F, Rees RJ (1989) Vaccination and skin test studies on the children of leprosy patients. Int J Leprosy Other Mycobact Dis 57:38–44.
59. Baker R, Zumla A, Rook GAW (1996) Tuberculosis, steroids and immunity. Quart J Med 89:387–394.
60. Villemin JA (1868) Études expérimentales et cliniques sur tuberculose. Paris: Ballière et fils.
61. Koch R (1891) Fortsetzung über ein Heilmittel gegen Tuberculose. Dtsch Med Wochenschr 17:101–102.
62. Starnes CO (1992) Coley's toxins in perspective. Nature 357:11–12.
63. Anonymous (1890) Professor Koch's remedy for tuberculosis; Austria. Br Med J 2:1490.
64. Abou-Zeid C, Smith I, Grange JM, Ratliff TL, Steele J, Rook GAW (1988) The secreted antigens of *Mycobacterium tuberculosis* and their relationship to those recognised by the available antibodies. J Gen Microbiol 134:531–538.
65. Rook GAW (1978) Three forms of delayed skin test response evoked by mycobacteria. Nature 271:64.
66. Friedmann FF (1958) The Friedmann Preparation: Anningzochin. Hannover: Laves Arzneimittel GmbH. Ronnenberg.
67. MacAssey L, Saleeby GW, eds. (1934) Spahlinger contra tuberculosis 1908–1931. London: John Bale, Sons and Danielsson Ltd.
68. Spahlinger H (1922) Note on the treatment of tuberculosis. Lancet 1:5–8.
69. Sechehaye A (1916) The Treatment of Pulmonary and Surgical Tuberculosis with Umckaloabo (Stevens' Cure). London: B. Fraser and Company.
70. Bonicke R, Juhasz SE (1964) Beschreibung der neuen Species, *Mycobacterium vaccae* n. sp. Zentralblatt Bakeriol Parasitenkunde Infektionkrankheiten Hygiene 192:133–135.
71. Paul RC, Stanford JL, Carswell JW (1975) Multiple skin testing in leprosy. J Hyg 75:57–68.
72. Stanford JL, Rook GAW, Samuel NM, Madlener F, Khamenei AA, Nemati T, Modabber F, Rees RJW (1980) Preliminary studies in search of correlates of protective immunity carried out on some Iranian leprosy patients and their families. Leprosy Rev 51:303–314.

73. Stanford JL, Convit J, Godal T, Kronvall G, Rees RJW, Walsh GP (1975) Preliminary taxonomic studies on the leprosy bacillus. Br J Exp Pathol 56:579–585.

74. Rook GAW (1980) The immunogenicity of killed mycobacteria. Editorial. Leprosy Rev 51:295–301.

75. Shield MJ, Stanford JL, Paul RC, Carswell JW (1977) Multiple skin testing of tuberculosis patients with a range of new tuberculins, and a comparison with leprosy and *Mycobacterium ulcerans* infection. J Hyg 78:331–348.

76. Stanford JL, Nye PM, Rook GAW, Samuel N, Fairbank A (1981) A preliminary investigation of the responsiveness or otherwise of patients and staff of a leprosy hospital to groups of shared or specific antigens of mycobacteria. Leprosy Rev 52:321–327.

77. Nye PM, Price JE, Revankar CR, Rook GA, Stanford JL (1983) The demonstration of two types of suppressor mechanism in leprosy patients and their contacts by quadruple skin-testing with mycobacterial reagent mixtures. Leprosy Rev 54:9–18.

78. Morton A, Nye P, Rook GA, Samuel N, Stanford JL (1984) A further investigation of skin-test responsiveness and suppression in leprosy patients and healthy school children in Nepal. Leprosy Rev 55:273–281.

79. Nye PM, Stanford JL, Rook GA, Lawton P, MacGregor M, Reily C, Humber D, Orege P, Revankar CR, Terencio-de-las-Aguas J (1986) Suppressor determinants of mycobacteria and their potential relevance to leprosy. Leprosy Rev 57:147–157.

80. Stanford JL, Terencio de Las Aguas J, Torres P, Gervasioni B, Ravioli R (1987) Studies on the effects of a potential immunotherapeutic agent in leprosy patients. Quad Cooper Sanitaria 7:201–206.

81. Pozniak A, Stanford JL, Johnson NM, Rook GAW (1987) Preliminary studies of immunotherapy of tuberculosis in man. Proceedings of the International Tuberculosis Congress, Singapore, 1986. Bull Int Union Tuberculosis 62:39–40.

82. Stanford JL, Bahr GM, Rook GA, Shaaban MA, Chugh TD, Gabriel M, al-Shimali B, Siddiqui Z, Ghardani F, Shahin A, Behbehani K. (1990) Immunotherapy with *Mycobacterium vaccae* as an adjunct to chemotherapy in the treatment of pulmonary tuberculosis. Tubercle 71:87–93.

83. Bahr GM, Shaaban MA, Gabriel M, Al-Shimali B, Siddiqui Z, Chugh TD, Denath FM, Shahin A, Behbehani K, Chedid L, Rook GAW, Stanford JL (1990) Improved immunotherapy for pulmonary tuberculosis with *Mycobacterium vaccae.* Tubercle 71:259–266.

84. Farid R, Etemadi A, Mehvar M, Stanford JL, Dowlati Y, Velayati AA (1994) *Mycobacterium vaccae* immunotherapy in the treatment of multi-drug-resistant tuberculosis: a preliminary report. Iranian J Med Sci 19:37–39.

85. Vacirca A, Dominino J, Valentini E, Hartopp R, Bottasso OA (1994) A pilot study of immunotherapy with *M. vaccae* against tuberculosis. Tubercle Lung Dis 75(Suppl 3):47–48.

86. Onyebujoh P, Abdulmumini T, Robinson S, Rook GAW, Stanford JL (1995) Immunotherapy for tuberculosis in African conditions. Respir Med 89:199–207.

87. Corlan E, Marica C, Macavei C, Stanford JL, Stanford CA (1997) Immunotherapy with *Mycobacterium vaccae* in the treatment of newly diagnosed pulmonary tuberculosis in Romania. Respir Med 91:13–19.
88. Corlan E, Marica C, Macavei C, Stanford JL, Stanford CA (1997) Immunotherapy with *Mycobacterium vaccae* in the treatment of chronic or relapsed pulmonary tuberculosis in Romania. Respir Med 91:21–29.
89. Stanford JL, Rook GAW, Fourie B, Onyebujoh P, Carswell JW (1994) The use of killed *Mycobacterium vaccae* preparation in clinical tuberculosis: A major trial in South Africa. Tubercle Lung Dis 75(Suppl 3):79.
90. Stanford JL, Cordess G, Rook GAW, Barnass S, Lucas S (1988) Immunotherapy of tuberculosis in mice and guinea pigs. Bull Int Union Tuberculosis Lung Dis 62:10–11.
91. Rafi A, Donoghue HD, Stanford JL (1995) Application of the polymerase chain reaction for the detection of *Mycobacterium leprae* DNA in specimens from treated leprosy patients. Int J Leprosy 63:42–47.
92. Stanford JL, Stanford CA, Rook GAW, Grange JM (1994) Immunotherapy for tuberculosis—Investigative and practical aspects. Clin Immunother 1:430–440.
93. Michison DA (1985) The action of antituberculosis drugs in short course chemotherapy. Tubercle 66:219–225.
94. Stanford JL, Stanford CA, Etemadi A, Farid R, Marica C, Corlan E (1994) Immunotherapy for multi-drug resistant tuberculosis. J Am Med Assoc. Southeast Asia 10(Suppl 3):42–46.
95. Ramu G, Prema GD, Balakrishnan S, Shanker Narayan NP, Stanford JL (1990) A preliminary report on the immunotherapy of Psoriasis. Indian Med Gazette 124:381–382.
96. Thompson SJ, Butcher PD, Patel VKR, Rook GAW, Stanford JL, van der Zee R, Elson CJ (1991) Modulation of pristane-induced arthritis by mycobacterial antigens. Autoimmunity 11:35–43.
97. Bottasso OA, Ingledew N, Keni M, Morini J, Pividori JF, Rook GAW, Stanford JL (1994) Cellular immune response to common mycobacterial antigens in subjects seropositive for *Trypanosoma cruzi*. Lancet 344:1540–1541.
98. Clerici M, Shearer GM (1993) A TH1 to TH2 switch is a critical step in the etiology of HIV infection. Immunol Today 14:107–111.
99. Stanford JL, Onyebujoh PC, Rook GA, Grange JM, Pozniak A (1993) Old plague, new plague, and a treatment for both? (letter). AIDS 7:1275–1277.

# 10

# Animal Models for Tuberculosis Research

*Frank M. Collins*

## 1. Experimental Models of Human Tuberculosis

Robert Koch identified *Mycobacterium tuberculosis* as the causative agent of phthisis in 1882 by serially culturing this organism on blood serum and then infecting guinea pigs and rabbits with the resulting growth to reproduce the clinical disease and satisfy his famous set of postulates (1). Shortly after this, Trudeau used a group of tuberculous rabbits to demonstrate the role played by environmental factors in promoting this highly contagious disease within an isolated animal colony (2). This small study was important because it demonstrated the protective effects of a healthy environment in a relevant animal model of the human disease. Since then, many different experimental protocols have been developed, many of them bearing little relationship to the naturally acquired human disease (3). Although such studies have provided useful research data, the results obtained in different laboratories have often been difficult to compare objectively, as they were obtained using disparate experimental methodologies (4). Few comparative studies have been carried out in any sort of systematic manner, but those which have been published indicate an extreme variability in results coming from different laboratories, each of which was using its own test system. One of the most important of these variables was the experimental host (5).

## 2. Experimental Animals

### *2.1 Primates*

Although *M. tuberculosis* can induce tuberculous lesions in a wide variety of experimentally infected animal species, it normally only infects primates (6).

There is no evidence that virulent tubercle bacilli are able to survive outside the body for any length of time, the infection normally spreading person to person by direct droplet infection as a result of coughing or sneezing by an individual suffering from open cavitary lung disease (7). The tuberculosis reservoir in the community will be those untreated individuals with active, cavitary disease who will shed enormous numbers of viable organisms into their surroundings. Most normal adults are remarkably resistant to tuberculous challenge, with less than 5% of the tuberculin converters going on to develop clinically significant disease (8). On the other hand, *M. tuberculosis* is highly pathogenic to nonhuman primates, almost invariably resulting in a life-threatening systemic disease (6). Thus, protection studies carried out in Rhesus monkeys might be skewed simply due to the extreme susceptibility of these animals to tuberculosis, and any results they produce might be of limited validity when extrapolated to human populations. The other limiting factor in such studies is their high cost. Despite this, nonhuman primates might have to be used in protection and treatment studies which cannot be ethically carried out in HIV-infected human volunteers.

### *2.2. Rabbits*

This host species was once widely used in studies of tuberculosis pathogenesis because, unlike rodents, rabbits develop cavitary lung disease following exposure to virulent *M. tuberculosis* or *M. bovis* (9). Interestingly, these animals resist infection by human tubercle bacilli, developing infections which more closely resemble that seen in people. Tuberculous rabbits can pose a substantial infectious threat to other animals housed in close proximity and special care must be taken to protect laboratory staff from infection. Because of the higher costs associated with rabbit studies, these animals have not been widely used in tuberculosis research in recent years, despite the availability of strains with high and low innate resistance to tuberculous challenge, a characteristic used to advantage by Lurie in his pathogenesis studies (9). However, fully inbred rabbits of the type required for adoptive immunity studies are not widely available. Despite these disadvantages, a number of investigators have used the rabbit as a source of alveolar macrophages for phagocytosis studies, comparing lung lavage cells harvested from normal and Bacille–Calmette–Guérin (BCG) vaccinated animals (10). In this type of study, it must be remembered that rabbits are subject to intercurrent infections caused by *Pasteurella multocida* and *Bordetella bronchiseptica* (snuffles) as well as *Yersinia pseudotuberculosis,* all of which will result in severe lung involvement and death. Care in the selection of specific pathogen-free breeding stock is essential to success in both in vivo and in vitro studies involving cells harvested from this host species.

### 2.3. Guinea Pigs

Hartley-strain guinea pigs have been the experimental animal of choice for virulence testing almost from the beginning. These outbred animals are exquisitely susceptible to tuberculous challenge, so that provided one is patient enough, death will occur following a challenge dose of only one or two viable *M. tuberculosis* (11). Virulence is usually assessed in the laboratory from the mortality rate (or mean time to death) in groups of animals infected with relatively large doses of tubercle bacilli given by the intramuscular route (12). Such studies will become increasingly important as the number of multidrug-resistant strains of *M. tuberculosis* (MDR-TB) continues to rise and there is a need to monitor them for infectiousness and virulence. This can best be measured as the rate of growth following the introduction of a small bacterial inoculum into the lung (or even in a designated lobe) determined after some arbitrary time interval (4). Although most of these studies can be carried out using outbred Hartley-strain guinea pigs, some investigators are using inbred strains 2 or 13 in their increasingly sophisticated immunological studies of the acquired antituberculous response seen in these animals (13). Inbred guinea pigs are more resistant to tuberculous challenge than their outbred counterparts, a characteristic which might be of advantage if the test is designed to mimic the response developed against the naturally acquired human disease (V. Montalbine and F.M. Collins, unpublished data). However, the reason for this difference in susceptibility is unclear.

### 2.4. Rats

Compared to mice and guinea pigs, the rat seems remarkably resistant to tuberculous challenge, although it is possible to induce progressive systemic disease with a relatively large intravenous inoculum of virulent tubercle bacilli (14). In recent years, relatively little work has been carried out in this host species despite some technical advantages they provide (15).

### 2.5. Mice

This species has become the prime experimental host for immunological and chemotherapeutic studies involving the tubercle bacillus (14). This is partly due to economic factors but also because of the availability of a large number of inbred strains exhibiting a spectrum of genetic and cellular characteristics needed in the increasingly sophisticated immunological approaches used in current tuberculosis research (16). Although mice can be readily infected with tubercle bacilli in the laboratory, they are not naturally subject to tuberculosis when living in the wild. The only exception to this rule seems to be *M. microti,* which was originally isolated from field voles and which is capable of causing progressive, caseating lung lesions in most rodent species (17). This organism bears a close resemblance to *M. tuberculosis* but does not cause progressive lung disease in humans and has been used as an alternative to BCG vaccine in Britain (18).

Mouse strains vary extensively in their innate resistance to a tuberculous challenge (19). This characteristic correlates with the presence of the *Bcg*$^r$ gene which affects the level of nonspecific macrophage activation in BCG-infected mice some time prior to the development of acquired cell-mediated resistance (20). The *Bcg*$^r$ gene has also been reported to affect susceptibility to a number of nontuberculous mycobacteria (21), although not to *M. tuberculosis,* which is said to depend on the presence of a *Tbc*-1 gene (22). The relevance of these genes to the expression of innate antituberculous resistance in humans is unclear.

Inbred immunodeficient mice have been used in tuberculosis research, to study the cellular mediators of primary resistance, as well as adoptive immunity in animals infused with T cells harvested from syngeneic, immune donors (23,24). In addition to congenitally athymic ($nu^+/nu^+$) mice, studies have been carried out in beige ($bg^+/bg^+$) mice which lack natural killer (NK) cell activity and are highly susceptible to *M. avium*-complex challenge (25,26). Recently, BALB/c mice with severe combined immunodeficiency ($SCID^+/SCID^+$), which lack the ability to express humoral and cellular immunity and which can be fatally infected by a modest dose of live BCG have been investigated (27). These severely immunosuppressed animals are highly susceptible to a spectrum of bacterial and parasitic diseases and, even when maintained under specific pathogen-free (SPF) conditions, often die as a result of a *Pneumocystis* pneumonia (28). However, they might provide a useful animal model with which to study the development of systemic *M. avium*-complex disease of the type seen in many AIDS patients.

Recently, a number of gene knockout (GKO) mice have been developed by disrupting specific genes by means of homologous recombination. The first of these mice lack the $\beta 2$ microglobulin gene and are unable to sensitize the CD8 + T cells normally involved in the resolution of the tuberculous infection (29). Since then, other groups have shown that disruption of the gamma-interferon gene renders the host exquisitely susceptible to a *M. tuberculosis* challenge and the host succumbs to a fulminating disease resembling military tuberculosis (30).

## 3. Microorganisms

### *3.1. Mycobacterium tuberculosis*

The standard strain for experimental tuberculosis studies has been *M. tuberculosis* H37Rv [Trudeau Mycobacterial Culture (TMC) #102; ATCC strain #27294], originally selected on the basis of subtle colonial morphological differences and its virulence for rabbits, guinea pigs, and mice (31). The avirulent variant (H37Ra: TMC #201, ATCC #25177) fails to produce progressive disease in experimental animals, although it might induce some acquired antituberculous resistance in the immunodepleted host (32). H37Rv was maintained on Proskauer and Beck medium for over 35 years with only a minimal loss of virulence. *M. tuberculosis* Erdman (TMC #107, ATCC #35801) is highly virulent for both mice and guinea pigs, even after the organism has been repeatedly cultured in Tween-containing media. It has been widely used in protection studies for this reason (14). All of these strains have been preserved as frozen or lyophilized seed lots since 1970 in order to prevent further genetic drift or loss of virulence (33). When cultured on Lowenstein–Jensen egg medium, they produce rough, irregular, heaped-up colonies which are difficult to emulsify evenly in saline. This is due to the presence of cord factor which binds the organisms into large sheets of cells which are notoriously difficult to break up mechanically. As a result, H37Rv suspensions can be very difficult to standardize in terms of viable units. Smoother suspensions can be obtained if the organism is grown in liquid medium containing small amounts (0.025–0.05%) of the detergents Tween-80 or Triton WR 1339. Even under these circumstances, the suspension should be exposed to a brief ultrasonic dispersion in order to break up any clumps. Washing the suspension prior to inoculation can remove the detergent, resulting in rapid and extensive clumping. On the other hand, continuous cultivation in detergent-containing media can result in a progressive loss of virulence which can only be regained when the organism is repeatedly passaged through mice (34). For this reason, stock cultures should not be cultured in Tween-containing media more than five times before returning to the original seed culture.

Stock mycobacterial suspensions are prepared by inoculating a frozen seed into 10 ml of Proskauer–Beck, modified Sauton's, or Middlebrook 7H9 liquid medium which has been enriched with 0.025% Tween-80 and 10% (v/v) of dextrose–albumin–catalase (ADC) additive (35,36). This primary culture is used to inoculate 100 ml of fresh medium in a 500-ml roller bottle which is incubated at 37°C until it reaches the mid-logarithmic growth phase (usually 8–10 days). The resulting suspension is diluted with an equal volume of fresh medium and dispensed into 1-ml ampoules and frozen at $-70$°C (37). Viable (34) and total (38) cell counts are carried out on randomly selected ampoules 24 h later and the mean viability calculated. Stock frozen suspensions should have a minimum of 50% viability and contain at least $5 \times 10^8$ colony-forming units (CFU) per milliliter.

Challenge inocula are prepared by rapidly thawing a frozen ampoule and diluting it in Tween-saline to the required number of viable bacilli, after which the suspension is briefly sonicated immediately prior to inoculation in order to break up any remaining clumps. Known numbers of viable bacilli can be introduced by the intravenous route for reproducible in vivo growth curves in normal and vaccinated animals (34,39).

### 3.2. *Mycobacterium bovis*

The other major causative agent of human tuberculosis is *M. bovis* which normally infects ungulates (cattle, deer, goats, and sheep) but which can spread to children drinking unpasteurized, contaminated milk and who often develop a severe tuberculous lymphadenitis (40). Experimental studies with bovine tubercle bacilli have not been widely reported in recent years, which is somewhat surprising because *M. bovis* strains Branch and Ravanel are among the most virulent laboratory strains of tubercle bacillus presently available, producing fulminant systemic disease in rabbits, guinea pigs, and mice (14). Relatively little is known about the growth characteristics of this organism in cattle, although infection studies have been carried out in rabbits and guinea pigs (9,40).

Much more is known about the growth characteristics of the attenuated strain of *M. bovis* (Bacille–Calmette–Guerin or BCG) which was developed following prolonged, serial culture of a virulent isolate of *M. bovis* in a potato–glycerol–ox bile medium (41). The attenuated strain was unable to induce progressive disease in calves or guinea pigs but protected them from a subsequent virulent challenge (14). Protection was monitored in terms of increased survival (mean time to death) in the vaccinated versus control animals following a normally lethal challenge infection (42). A better assessment of BCG potency can be achieved by enumerating the viable bacterial population within the lungs and spleens before and after receiving a sublethal aerogenic challenge infection (39,43). Even under carefully controlled conditions, consistant and reproducible protection can prove to be an elusive goal, due to the multitude of experimental variations used by different testing laboratories (44). One of the most important of these variables turns out to be the host species, a component which has all too often been overlooked in the past (45,46).

### 3.3. *Mycobacterium avium complex*

This is a group of opportunistic pathogens which are widely distributed throughout the natural environment (soil, dust, and water) and which were initially considered to be of little pathologic significance for humans (47,48). Prior to the emergence of the AIDS epidemic, patients with systemic *M. avium*-complex (MAC) disease usually had some other underlying lung disease or had been receiving prolonged immunosuppressive chemotherapy (49). The resulting infections were relatively indolent and seldom caused life-threatening disease (48). However, these organisms exhibit high levels of resistance to most conventional antituberculous drugs, and once established within the lung, they can be very difficult to dislodge (50).

Relatively little is known about the pathogenesis or the immunology of these infections (49). Most MAC serovars are nonpathogenic for guinea pigs and mice, although some are virulent for chickens (51). Several strains of *M. avium-intracellulare* can cause progressive disease in laboratory mice, although the size of the lethal intravenous inoculum and the prolonged time to death raise fundamental questions regarding the meaning of "virulence" with respect to these organisms (52,53). Many AIDS patients develop life-threatening systemic *M. avium*-complex (usually serovar 4 or 8) disease, although we still know very little about the epidemiology of this disease, largely because there is no realistic animal model of human MAC disease (49). This has severely restricted research into the colonization/invasion of mucosal surfaces by these organisms, especially in the immunosuppressed host. Such information is badly needed in order to develop effective preventive strategies against the disseminated MAC disease which often develops in AIDS patients and which is known to substantially increase the mortality rate for this complex disease (54).

## 4. Animal Models for Tuberculosis Research

Despite a great deal of work, we still know surprisingly little about the virulence factors responsible for the induction of active lung disease in the naturally infected individual. Despite its infectiousness, *M. tuberculosis* will induce active, clinically significant disease in only a few percent of tuberculin converters (8). This is because most adults are innately resistant to tuberculosis, developing a latent form of this disease in which small numbers of virulent tubercle bacilli can persist virtually indefinitely in residual caseated nodules within the tissues (7). The cellular interactions responsible for this latent disease are largely unknown, although its maintenance seems to be T-cell mediated (49).

Recent studies of the molecular mediators of acquired antituberculous resistance have built on recent advances made in the molecular biology of the tubercle bacillus (55). As a result, we can now identify a number of the mycobacterial antigens associated with the induction of protective T-cell-mediated immunity in the infected host and have cloned the genes responsible for their production (56). Despite the elegance of these molecular approaches, they still depend on animal test systems to demonstrate their relevance to disease prevention, treatment, and control. Some of these time-honored test systems have not changed substantially in more than half a century and might bear little relationship to the naturally acquired disease they purport to represent. As a result, they may be measuring irrelevant or misleading parameters of the immune response (14). More relevant models of the human disease might depend on a better understanding of the pathogenesis and immunology of this chronic pathogen, especially in the immunosuppressed patient (49). The problem might not be so much the development of novel test systems but rather to decide, on the basis of limited information, which of the many available systems provides the most suitable surrogate of the immune response to this human pathogen. In vaccine development, there have been almost as many assay procedures as there are investigators in the field (45). Some of the permutations are listed in Table 10.1 where as many as 20,000 combinations of dependent and independent variables can exist for this one laboratory assay, making it virtually impossible to develop a universal standardized assay procedure (5). As a result, new recombinant vaccines might have to be tested in human volunteers if we are to demonstrate that they are more effective than currently available commercial BCG vaccine preparations (57). In addition, a better animal model of nontuberculous lung disease will be needed to quantitate the immunoprophylactic and immunotherapeutic effectiveness of the new reagents currently being developed in the laboratory for possible use against *M. avium*-complex disease in AIDS patients.

Table 10.1. Variables associated with vaccine testing

| |
|---|
| **Animals** |
| a. Mouse: Outbred Swiss white. Inbred $Bcg^s$, $Bcg^r$, $nu^+/nu^+$ |
| b. Guinea pig: Outbred Hartley, inbred strain 2 or 13 |
| c. Rabbit: Outbred susceptible, resistant (*P. multocida*-free) |
| d. Primate: Rhesus, Cynomolgus |
| **Organisms** |
| a. *M. tuberculosis:* Erdman, H37Rv, H37Ra |
| b. *M. bovis:* BCG Pasteur, Tice, Glaxo |
| c. *M. avium* complex: Translucent, opaque colony |
| d. Recombinant: *M. smegmatis,* BCG |
| e. Killed adjuvanted vaccine: Whole cells, subunit |
| **Assay variables** |
| a. Size and route of vaccination |
| b. Growth/survival of vaccinating inoculum in vivo |
| c. Active versus adoptive versus passive immunization |
| d. Time interval between vaccination and challenge |
| e. Size and route of challenge |
| f. Time to sacrifice and the number of assays |
| g. Differentiation of residual vaccine and challenge organisms |
| h. Choice of test organs. Lungs, spleen, lymph node (LN), Bone marrow (BM), blood |
| i. Tuberculin skin or footpad testing. Lymphocyte transformation test (LTT) |
| j. Antibody responses |
| k. Lymphokine profiles. IL-1, IL-2, IL-4, IL-12, IFN-$\gamma$, TNF-$\alpha$. |
| **Assessment criteria** |
| a. Survival/mortality after an arbitrary time period |
| b. Reduced growth of challenge in test organs |
| c. Latency and reactivation |
| d. Growth/inactivation in macrophage monolayers |
| e. Lymphokine profiles compatible with Th1 response. |

Table 10.2. Variables in testing antituberculous drugs

**Animals**
- a. Mice: Inbred *Bcg*$^s$, *Bcg*$^r$, *bg*$^+$/*bg*$^+$, *nu*$^+$/*nu*$^+$, SCID
- b. Guinea pig: Hartley
- c. Primates: Rhesus

**Cells**
- a. Human PBMC cultured in vitro
- b. HeLa cells (nonprofessional phagocytes)
- c. Macrophage cell line: J774
- d. Mouse Peritoneal exudate cells (PEC), alveolar macrophages (AM), bone marrow-derived macrophages (BM), splenic macrophages

**Organisms**
- a. *M. tuberculosis:* H37Rv, H37Ra, or clinical isolates (MDR-TB)
- b. *M. avium* complex: Translucent, opaque colony
- c. *M. smegmatis:* Avirulent rapid grower
- d. *M. leprae:* Mouse footpad. Nonculturable in vitro

**Drug regimen**
- a. Optimum dosage: Single versus mixed drug regimens. Synergy versus antagonism
- b. Route of drug treatment: IV, SC, IM, IP, oral (gavage, drinking water, chow)
- c. Aqueous solubility of hydrophilic versus hydrophobic drugs
- d. Depot and slow release vehicles

**Assessment criteria**
- a. Drug concentration in blood and tissues
- b. Survival/mortality
- c. Growth versus bacteriostatic versus bactericidal activity at different drug concentrations
- d. Reactivation infection and/or drug resistance

Many of these technical problems are again encountered when we consider testing antimycobacterial drugs for their efficacy in the treatment of mycobacterial disease. A simple mortality/survival comparison in treated versus control groups has given way to an increasingly sophisticated methods to assess bactericidal versus bacteriostatic activity in a variety of animal models (58). This has been the inevitable outcome of increasingly complex drug mixtures in the treatment of tuberculosis which has led to unexpected synergistic and antagonistic effects which could influence the in vivo activity of any new drug combinations. These complex drug interactions require the use of large numbers of animals and a careful statistical analysis of the resulting data. In an attempt to avoid this, some investigators are turning to infected macrophage cultures exposed to increasing concentrations of drug as a simpler, more reproducible screening procedure (59,60). The use of macrophage monolayers might be quicker and cheaper, but these drugs will ultimately have to be tested in experimental animals and then in human volunteers in order to demonstrate their clinical efficacy against the naturally acquired disease.

Table 10.3. The Everyman animal model for tuberculosis research

**Organisms**
- a. *M. tuberculosis:* H37Rv
- b. *M. bovis:* BCG Pasteur, Tice, Glaxo
- c. *M. avium* complex (serovar 4)
- d. Clinical isolates: MDR-TB

**Animals**
- a. SPF C57BL/6 ($Bcg^s$) or DBA/2 ($Bcg^r$) mice
- b. SPF strain 2 guinea pigs.

**Vaccination route**
- a. Intravenous: $10^5$–$10^6$ CFU
- b. Aerogenic: $10^2$–$10^3$ CFU
- c. Oral: $10^8$ CFU in 10% bicarbonate

**Vaccine growth *in vivo***
- a. Sacrifice at 0 and 4 weeks; lungs, spleen, lymph nodes

**Tuberculin test (4 weeks)**
- a. 5 $\mu$g PPD SC in mouse footpad (24 h)
- b. 1 $\mu$g PPD ID in guinea pigs (24–48 h)
- c. Splenic T-cell LTT response to PPD
- d. Lymphokine production (IL-1, IL-2, IL-4, IFN-$\gamma$, TNF-$\alpha$)

**Challenge**
- a. *M. tuberculosis:* Erdman; $10^2$ aerosol at 3 months

Or

**Drug treatment**
- a. Single drug given IM daily or orally in drinking water to half of test group beginning 1 or 3 months after infection
- b. Multidrug regimen given orally for 1 or 3 months
- c. Check organs and blood for drug concentration

**Assessment criteria**
- a. Survival/mortality at 6 months
- b. Organ homogenates cultured at 0, 2, 4, 8, and 12 weeks after challenge or treatment
- c. Treat survivors with cortisone (300 mg/kg) and culture organs for reactivation and/or drug resistance

Some of the variables associated with *in vivo* drug testing are listed in Table 10.2. New, more effective protocols are needed for the treatment of *M. avium* infections in immunosuppressed patients and these tests will involve nude and beige mice exposed to an oral challenge with a variety of nontuberculous mycobacteria (26). Thus, it seems unlikely that any one experimental model will prove to be suitable for drug testing against these innately resistant organisms. Nevertheless, a list of desirable features in such a test system are included in Table 10.3, more in the hope of stimulating discussion than in providing the final word on the best laboratory test for assaying vaccines and antibacterial agents. Some sort of consensus regarding the most desirable features to be incorporated into any future test protocol could help to simplify the comparison of data from different laboratories as well as ensuring that promising new drugs are not overlooked simply because of technical inadequacies in the laboratory procedures used to assess them.

## 5. Conclusions

Despite concern over the efficacy of BCG vaccine under field-test conditions (61) this is still the only vaccine which is currently licensed for use in the prevention of pulmonary tuberculosis in human populations (7). In fact, live BCG vaccine is the most widely used immunogen in the world today, with nearly 3 billion doses administered to infants and school children, with a surprisingly low incidence of serious side effects (62). The safety and effectiveness of this attenuated vaccine was demonstrated innumerable times before it was used in 1921 to vaccinate several infants at high risk of developing miliary tuberculosis (41). Despite this early success, formal proof of its protective value did not appear until some 30 years later (61). Over the years, the level of protection achieved in a number of carefully controlled field trials has fluctuated wildly, leading some epidemiologists to question the value of communitywide BCG vaccination programs (63). Some of this confusion could be due to further attenuation of some BCG substrains as a result of prolonged cultivation on laboratory media by different vaccine producers (64). Although it is clear that the immunizing efficiency of different BCG substrains varies considerably in both experimental animals and human populations (65,66), a recent meta-analysis of published protection data indicates an average of 50% protection achieved in both infants and adults receiving several different live BCG vaccines (67). Although this level of protection might seem modest in comparison to that achieved by a number of other vaccines, it is nevertheless sufficient to justify the continued use of BCG as the cornerstone of tuberculosis control programs in many parts of the world (68). However, with the emergence of increasing numbers of MDR-TB strains in the United States, vaccination of tuberculin-negative health care workers, social workers, and prison guards who work in close contact with AIDS patients potentially infected with these organisms is being actively considered (69).

The BCG vaccine has the serious disadvantage that it lacks immunotherapeutic effectiveness in individuals who are incubating the disease or are suffering from active drug-resistant lung disease (70). Such individuals constitute an important infectious reservoir within the community and can be a highly infectious source of drug-resistant disease in a prison, nursing home, or AIDS ward setting (71). Development of therapeutic vaccines are urgently needed if we are to treat such outbreaks effectively (72). Such multifactorial preparations will require careful evaluation using a variety of animal test systems if we are to predict their likely behavior in human patients with any confidence. Again, a standardized animal test protocol would be very helpful in this type of study.

Effective drugs used in the treatment of pulmonary tuberculosis have been available for half a century (73), but the animal models used in the assessment of their in vivo activity have changed very little over this same time period. With the increasingly complex treatment regimens in current clinical use, this is proving to be a severe technical limitation (74). In addition, the increasing number of MDR-*M. tuberculosis* and *M. avium*-complex isolates coming from hospitals, prisons, and homeless shelters throughout this country (75–77) emphasizes the need for new, mycobactericidal drugs which must first be screened for in vitro and in vivo activity in the laboratory. Suitable surrogates are needed to rapidly discriminate bactericidal from bacteriostatic activity, synergism, from antagonism, sensitivity from resistance. It is especially important to quantitate the emergence of drug resistance in organisms subject to these complex drug combinations in animals exposed to increasingly shorter treatment protocols. It seems unlikely that any one assay procedure of the type outlined in Table 10.3 can achieve all of these goals, given the complexity of the host–bacterium–drug interactions involved. However, it is the absence of a realistic animal model of disseminated *M. avium*-complex disease which makes it likely that many of the combined chemotherapy and immunotherapy studies will have to be carried out in human volunteers (78).

**References**

1. Koch R (1882) Aetiology of tuberculosis. Berlin Klin Wochenshr 19:221.
2. Trudeau EL (1887) Environment in its relation to the progress of bacterial invasion in tuberculosis. Am J Med Sci 94:118.
3. Wiegeshaus EH, Smith DW (1968) Experimental models for the study of immunity in tuberculosis. Ann NY Acad Sci 154:194.
4. Smith DW, Harding GE (1977) Approaches to the validation of animal test systems for assay of protective potency of BCG vaccines. J Biol Stand 5:131.
5. Smith DW (1984) Disease in guinea pigs. In: Kubica GP, Wayne LG, eds. The Mycobacteria: A Sourcebook, Vol. 2, p. 925. New York: Marcel Dekker.
6. Good RC (1984) Diseases in nonhuman primates. In: Kubica GP, Wayne LG, eds. The Mycobacteria: A Sourcebook, Vol. 2, p. 903. New York: Marcel Dekker.
7. Collins FM (1991) Pulmonary tuberculosis: The immunology of a chronic infection. In: Cryz SJ, ed. Vaccines and Immunotherapy, p. 140. New York: Permagon Press.
8. Collins FM (1988) Immunology of mycobacterial infection. In: Escobar MR, Utz JP,

eds. The Reticuloendothelial System: A Comprehensive Treatise, Vol. 10, p. 125. New York: Plenum Press.

9. Lurie MB (1964) Resistance to Tuberculosis: Experimental Studies in Native and Acquired Defense Mechanisms. Cambridge, MA: Harvard University Press.
10. Heisse ER, Myrvik QV, Leake ES (1965) Effect of BCG on levels of acid phosphatase, lysozyme and cathepsin in rabbit alveolar macrophages. J Immunol 95:128.
11. Webb GB, Williams WW, Barber MA (1909) Immunity production by increasing numbers of bacteria beginning with one viable organism. J Med Res 20:1.
12. Collins FM, Smith MM (1969) A comparative study of the virulence of *Mycobacterium tuberculosis* measured in mice and guinea pigs. Am Rev Respir Dis 100:631.
13. Phalen SW, McMurray DN (1993) T-lymphocyte response in a guinea pig model of tuberculous pleuritis. Infect Immun 61:142.
14. Collins FM (1984) Protection against mycobacterial disease by means of live vaccines tested in experimental animals. In: Kubica GP, Wayne LG, eds. The Mycobacteria: A Sourcebook, Vol. 2, p. 787. New York: Marcel Dekker.
15. Lefford MJ, McGregor DD, Mackaness GB (1973) Immune response to *Mycobacterium tuberculosis* in rats. Infect Immun 8:182.
16. Smith DW, Wiegeshaus EH (1989) What animal models can teach us about the pathogenesis of tuberculosis in humans. Rev Infect Dis 11(Suppl 2):S385.
17. Wells AQ (1937) Tuberculosis in wild voles. Lancet 232:1221.
18. Anonymous (1972) BCG and Vole bacillus vaccines in the prevention of tuberculosis in adolescence and early adult life. Bull WHO 46:371.
19. Lynch CL, Pierce-Chase CH, Dubos RJ (1965) A genetic study of susceptibility to experimental tuberculosis in mice infected with mammalian tubercle bacilli. J Exp Med 121:1051.
20. Buschman E, Apt AS, Nickonenko BV, Moroz AM, Averbakh MH, Skamene E (1988) Genetic aspects of innate resistance and acquired immunity in inbred mice. Springer Semin Immunopathol 10:319.
21. Orme IM, Stokes RW, Collins FM (1986) Genetic control of natural resistance to nontuberculosus mycobacterial infections in mice. Infect Immun 54:56.
22. Nickonenko BV, Apt AS, Moroz AM, Averbakh MM, Skamene E (1985) Genetic analysis of susceptibility of mice to H37Rv tuberculosis infection: Sensitivity versus relative resistance. Prog Leuk Biol 3:291.
23. Collins FM, Stokes RW (1987) *Mycobacterium avium*-complex infections in normal and immunodeficient mice. Tubercle 68:127.
24. Orme IM, Collins FM (1986) Cross-protection against nontuberculous infections by *Mycobacterium tuberculosis* memory immune T lymphocytes. J Exp Med 163:203.
25. Ueda K, Yamazaki S, Someya S (1976) Experimental mycobacterial infection in congenitally athymic 'nude' mice. J Reticuloendothel Soc 19:77.
26. Gangadharam PR (1986) Murine models for mycobacterioses. Semin Respir Infect 1:250.

27. North RJ, Izzo AA (1993) Granuloma formation in severe combined immunodeficient (SCID) mice in response to progressive BCG infection. Am J Pathol 142:1959.
28. Schultz LD, Sidman CL (1987) Genetically determined murine models of immunodeficiency. Annu Rev Immunol 5:367.
29. Flynn JL, Goldstein MM, Triebold KJ, Koller B, Bloom BR (1992) Major histocompatibility complex class-1-restricted T-cells are required for resistance to *Mycobacterium tuberculosis* infection. Proc Natl Acad Sci USA 89:12013.
30. Cooper AM, Falton DK, Stewart TA, Griffin JP, Russell DG, Orme IM (1993) Disseminated tuberculosis in interferon-$\gamma$ gene disrupted mice. J Exp Med 178:2243.
31. Steenken W, Gardner LV (1946) History of H37 strain of tubercle bacillus. Am Rev Tuberc 54:62.
32. Steenken W, Gardiner LV (1943) Vaccinating properties of avirulent dissociates of five different strains of tubercle bacilli. Yale J Biol Med 15:393.
33. Kim HK, Kubica GP (1972) Long term preservation and storage of mycobacteria. Appl Microbiol 24:311.
34. Collins FM, Miller TE (1969) Growth of a drug-resistant strain of *Mycobacterium bovis* (BCG) in normal and immunized mice. J Infect Dis 120:517.
35. Stokes RW, Collins FM (1990) Passive transfer of immunity to *Mycobacterium avium* in susceptible and resistant strains of mice. Clin Exp Immunol 81:109.
36. Grover AA, Kim HK, Wiegeshaus EH, Smith DW (1967) Host-parasite relationships in experimental airborne tuberculosis. II. Reproducible infection by means of an inoculum preserved at $-70$°C. J Bacteriol 94:832.
37. Collins FM, Wayne LG, Montalbine V (1974) The effect of cultural conditions on the distribution of *Mycobacterium tuberculosis* in the spleens and lungs of specific-pathogen free mice. Am Rev Respir Dis 110:147.
38. Collins FM, Morrison NE, Dhople AM, Watson SR (1980) Microscopic counts carried out on *Mycobacterium leprae* and *M. tuberculosis* suspensions. Int J Leprosy 48:402.
39. Wiegeshaus EH, McMurray DN, Grover AA, Harding GE, Smith DW (1970) Host–parasite relationships in experimental airborne tuberculosis. III. Relevance of microbial enumeration to acquired resistance in guinea pigs. Am Rev Respir Dis 102:422.
40. Thoen CO, Karlson AG, Hines EM (1984) *Mycobacterium tuberculosis* complex. In: Kubica GP, Wayne LG, eds. The Mycobacteria: A Sourcebook, Vol. 2, p. 1209. New York: Marcel Dekker.
41. Collins FM (1982) The immunology of tuberculosis. Am Rev Respir Dis 125:42.
42. Siebenmann CO, Barbara C (1974) Quantitative evaluation of the effectiveness of Connaught freeze-dried BCG vaccine in mice and guinea pigs. Bull WHO 51:283.
43. Pierce CH, Dubos RJ, Schaefer WB (1953) Multiplication and survival of tubercle bacilli in the organs of mice. J Exp Med 97:189.
44. Smith DW, Harding GE, Chan J, Edwards M, Hank J, Muller D, Sobhi F (1979) Potency of 10 BCG vaccines as evaluated by their influence on the bacillemic phase of experimental airborne tuberculosis in guinea pigs. J Biol Stand 7:179.

45. Wiegeshaus EH, Harding GE, McMurray DN, Grover AA, Smith DW (1971) A co-operative evaluation of test systems used to assay tuberculosis vaccines. Bull WHO 45:543.
46. Wiegeshaus EH, Smith DW (1989) Evaluation of the protective potency of new tuberculosis vaccines. Rev Infect Dis 11(Suppl 2):S484.
47. von Reyn CF, Waddell RD, Eaton T, Arbeit RD, Maslow JN, Barber TW, Brindle RJ, Gilks CF, Lumio J, Lahdevirta J, Ranki A, Dawson D, Falkinham JA (1993) Isolation of *Mycobacterium avium* complex from water in the United States, Finland, Zaire and Kenya. J Clin Microbiol 31:3227.
48. Wolinsky E (1979) Nontuberculous mycobacteria and associated diseases. Am Rev Respir Dis 119:107.
49. Collins FM (1989) Mycobacterial disease, immunosuppression and acquired immunodeficiency syndrome. Clin Microbiol Rev 2:360.
50. Davidson PT (1979) The management of disease with atypical mycobacteria. Clin Notes Respir Dis 18:3.
51. Schaefer WB, Davis CL, Cohn ML (1970) Pathogenicity of transparent, opaque and rough variants of *Mycobacterium avium* in chickens and mice. Am Rev Respir Dis 102:499.
52. Collins FM, Morrison NE, Montalbine V (1978) Immune response to persistent mycobacterial infection in mice. Infect Immun 20:430.
53. Wolinsky E (1981) When is an infection disease? Rev Infect Dis 3:813.
54. Horsburgh CR, Selik RM (1989) The epidemiology of disseminated tuberculous mycobacterial infection in acquired immunodeficiency syndrome (AIDS). Am Rev Respir Dis 139:4.
55. Young DB (1991) What next in basic research? Int J Leprosy 59:95.
56. Jacobs WR, Snapper SB, Tuckman M, Bloom BR (1989) Mycobacteriophage vector systems. Rev Infect Dis 11(Suppl 2):S404.
57. Stanford JL (1991) Improving on BCG. Acta Pathol Microbiol Immunol Scand 99:103.
58. Mitchison DA (1992) Understanding the chemotherapy of tuberculosis—Current problems. J Antimicrob Chemother 29:477.
59. Dhillon J, Mitchison DA (1992) Activity *in vitro* of rifabutin, FCE 22807, rifapentine and rifampin against *Mycobacterium microti* and *M. tuberculosis* and their penetration into mouse peritoneal macrophages. Am Rev Respir Dis 145:212.
60. Sbarbaro JA, Iseman MD, Crowle AJ (1992) The combined effect of rifampin and pyrazinamide within the human macrophage. Am Rev Respir Dis 146:1448.
61. Fine PEM (1989) The BCG story: Lessons from the past and implications for the future. Rev Infect Dis 11(Suppl 2):S353.
62. Lotte A, Wacz-Hochert O, Poisson N (1984) BCG complications. Estimations of the risks among vaccinated subjects and statistical analysis of their main characteristics. Adv Tuberc Res 21:107.
63. Comstock GW (1988) Identification of an effective vaccine against tuberculosis. Am Rev Respir Dis 138:479.

64. Osborne TW (1983) Changes in BCG strains. Tubercle 64:1.
65. Rouillon A, Waaler H (1976) BCG vaccination and epidemiological situation: A decision making approach to the use of BCG. Adv Tuberc Res 19:64.
66. Horwitz JB, Bunch-Christensen K (1972) Correlation between tuberculin sensitivity after 2 months and 5 years among BCG vaccinated subjects. Bull WHO 47:49.
67. Colditz GA, Brewer TF, Berkey CS, Wilson ME, Burdicke E, Fineberg HV, Mosteller F (1994) Efficacy of BCG vaccine in the prevention of tuberculosis. J Am Med Assoc 271:698.
68. Styblo K (1989) Overview and epidemiological assessment of the current global tuberculosis situation with an emphasis on control in developing countries. Rev Infect Dis 11(Suppl 2):S339.
69. Greenberg PD, Lax KG, Schechter CB (1991) Tuberculosis in house staff. A decision analysis comparing the tuberculin screening strategy with the BCG vaccination. Am Rev Respir Dis 143:490.
70. Stanford JL, Bahr GM, Byass P, Corrah T, Dowlati Y, Lucas S, Shaaban M, Torres P (1990) A modern approach to the immunotherapy of tuberculosis. Bull Int Union Tuberc 65:27.
71. Iseman MD, Madsen LA (1989) Drug-resistant tuberculosis. Clin Chest Med 10:341.
72. Stanford JL, Grange JM (1993) New concepts for the control of tuberculosis in the twenty-first century. J R Coll Physicians (London) 27:218.
73. Ryan F (1992) The Forgotten Plague: How the Battle against Tuberculosis was Won—and Lost. Boston: Little Brown and Co.
74. D'Esopo ND (1982) Clinical trials in pulmonary tuberculosis. Am Rev Respir Dis 125:85.
75. Beck-Sague C, Dooley SW, Hutton MD, Otten J, Breeden A, Crawford JT, Pitchenik AE, Woodley C, Canthen G, Jarvis WR 1992. Hospital outbreak of multidrug-resistant *Mycobacterium tuberculosis* infections. J Am Med Assoc 268:1280.
76. Centers for Disease Control (1993) Transmission of multidrug-resistant tuberculosis among immunocompromised persons in a correctional system—New York, 1991. Morbid Mortal Wkly Rep 41:507.
77. Centers for Disease Control (1992) Prevention and control of tuberculosis among homeless persons. Morbid Mortal Wkly Rep 41(RR-5):13.
78. Etemadi A, Farid R, Stanford JL (1992) Immunotherapy for drug resistant tuberculosis. Lancet 340:1360.

# 11

# Speciation of Mycobacteria in Clinical Laboratories

*L.B. Heifets and P.A. Jenkins*

## 1. Introduction

*Every herring is a fish,*
*but not every fish is a herring.*
An old Yiddish proverb

Although *Mycobacterium tuberculosis* has been the only mycobacteria to be found in clinical specimens in the past, today clinical laboratories have to deal with more than 25 other mycobacterial species that can be detected in human specimens. Some of these nontuberculous mycobacteria have been known for 50 years, and some have been discovered only recently. Despite this substantial period of dealing with these species, the mind of the clinical microbiologists often has been captured and influenced by the experience from work with tubercle bacilli. It took some time and effort to recognize that the first-discovered mycobacterium other than *M. tuberculosis* represented new independent species and not just an "atypical mycobacteria." But, even today, some scientists and clinicians have difficulty recognizing that the approaches to these infections must be different from those established for tuberculosis. This is particularly relevant to such issues as drug-susceptibility testing and treatment regimens. One example is the tendency to include isoniazid and pyrazinamide in the treatment regimen for a patient with any mycobacterial infection. These agents are most effective in the therapy of tuberculosis but useless against *M. avium, M. intracellulare, M. fortuitum, M. chelonae,* and many other nontuberculous mycobacteria. It is clear now that the rapidly growing mycobacteria (*M. fortuitum, M. chelonae*), in regard to the selection of drug for therapy or methods of the drug-susceptibility tests, have less in common with *M. tuberculosis* than they have with nonmycobacterial aerobes. At the same time, a similar situation with *M. avium* complex has not yet been fully recognized. In fact, various organisms that belong to the genus *Mycobacterium,* have different media requirements, temperature preferences, and growth rates, as well as antimicrobial agents to which they are susceptible.

From the point of view of a clinical microbiologist, the most common feature uniting all mycobacteria is their acid-fastness. Due to the low proportion of the nontuberculous mycobacteria found in the past among the sputum specimens submitted to the laboratory, detection of acid-fast bacilli (AFB) in sputum was almost pathomnemonic to the diagnosis of tuberculosis. This approach is still more or less valid for many developing countries that have very high rates of tuberculosis, for example in Africa. The situation is changing now, and the diagnostic value of the AFB-positive sputum smear is diminishing dramatically. Therefore, despite the clarity of the situation, there is a need to repeat again and again the simple truth that tubercle bacilli are indeed mycobacteria, but not all mycobacteria and not all acid-fast-rods are tubercle bacilli.

Although bacteriological diagnosis of tuberculosis and differentiation between *M. tuberculosis* and nontuberculous mycobacteria can be a part of the protocol in any properly equipped clinical laboratory, speciation of mycobacteria other than tubercle bacilli is a subject for only specialized mycobacteriology laboratories. In the United States, a three-level concept of mycobacteriology laboratory services has been developed. Under this concept, Level 1 laboratories are limited to performing AFB smear examination only (from either concentrated or nonconcentrated specimen). Assignment of Level 2 allows culture isolation, differentiation of *M. tuberculosis* from other mycobacteria, and drug-susceptibility testing with the first-line drugs. Level 3 allows the same testing as Level 2 plus identification of nontuberculous mycobacteria and drug-susceptibility tests with all drugs. This three-level concept, established about 20 years ago by U.S. government agencies, was quite reasonable as long as the proportion of nontuberculous mycobacteria among the clinical isolates was minimal and the initial drug resistance in *M. tuberculosis* was a rare occasion. With the recent emergence of the new epidemic of *M. avium*-complex infections, especially among patients with AIDS, differentiation between *M. tuberculosis* and nontuberculous mycobacteria became an issue in the laboratory diagnosis of tuberculosis. The problem was aggravated during the last years by increasing rates of drug resistance among the U.S. tuberculosis patients, and especially because of the recent outbreaks of multidrug-resistant tuberculosis. Extremely high mortality rates during these outbreaks were associated with delays of laboratory reports on identification of the isolates and detection of resistance to the antimicrobial agents included in the treatment regimens. These delays, with a turnaround time reaching 3 months or longer, were attributed to the multistep transfer of the specimens/cultures through the three-level laboratory service systems. This situation stimulated extensive research in developing rapid methods for identification and detection of drug resistance. Unfortunately, some of these studies are detached from the actual needs of the clinical laboratory. At the same time, proper application of the currently available technology should allow the precise bacteriological diagnosis of tuberculosis within 2–3 weeks and detection of drug resistance within 3–4 weeks for 90–95% of new tuberculosis patients. This is quite feasible if the raw specimen is submitted directly to a laboratory specializing in mycobacteriology which is properly equipped and has qualified personnel to accomplish all necessary tests within the shortest turnaround time. Therefore, it is very clear that the three-level systems has to be replaced with the concept of direct specialized services.

The methods described in this chapter are oriented toward such specialized mycobacteriology laboratories, with the assumption that the modern standard equipment approved (or under approval) for use in clinical laboratories is available. We do not provide detailed descriptions of the technologies that are not standardized yet for application in the clinical laboratory, as these methods are being addressed in other chapters of this publication.*

The following are issues addressed in this chapter:

1. Bacteriological diagnosis of tuberculosis
2. Protocols for speciation of nontuberculous mycobacteria
3. Conventional identification test techniques
4. AccuProbe technology
5. Polymerase chain reaction (PCR) and other amplification techniques
6. Cell-wall lipids analyses

## 2. Bacteriological Diagnosis of Tuberculosis

### 2.1. General Considerations

The *M. tuberculosis* complex includes the following species: *M. tuberculosis, M. africanum, M. bovis, M. bovis* BCG (Bacille–Calmette–Guérin), and *M. microti.* Tuberculosis in humans caused by the first three is the only mycobacterial infection transmissible from person to person. Therefore, identification of these species in clinical specimens and/or differentiation from nontuberculous mycobacteria is of paramount importance for timely public health measures and for initiation of appropriate antimicrobial therapy. Some of the rapid methods described below can identify the *M. tuberculosis* complex without speciation among the members of this complex. Such a report usually gives sufficient grounds to start all measures necessary for a case of tuberculosis, because in most countries, the frequency of isolation of members of the complex other than *M. tuberculosis*

*Some recent developments might have taken place after the manuscript had been submitted.

is very rare. Nevertheless, identification of the *M. tuberculosis* complex requires further speciation. It is necessary for legal reasons, for epidemiological assessment, and for proper adjustments in the drug regimen (particularly, replacement of pyrazinamide with another drug if it is *M. bovis*). Identification of *M. bovis* BCG in a clinical specimen (usually the result of complications following BCG vaccination or treatment) confirms that a patient is not contagious, which excludes the necessity of public health actions. The type of specimens submitted to the laboratory depends on the clinical manifestation of tuberculosis. The most common are sputum, bronchial and gastric washings from patients with suspected pulmonary involvement. Blood or bone marrow specimens should be submitted when a disseminated mycobacterial infection is suspected, especially in an HIV-positive individual. Other specimens that can be submitted, depending on the suspected site of infection, are cerebrospinal fluid, pleural exudate, joint exudate, urine, stool, and tissue samples. Detection of AFB in a raw specimen can be used in conjunction with typical clinical symptoms and radiological findings for a provisional diagnosis of tuberculosis. Final bacteriological diagnosis of tuberculosis requires isolation of a pure culture of *M. tuberculosis,* but recent development of the polymerase chain reaction (PCR) or other DNA amplification methods will make identification of *M. tuberculosis* in a raw specimen possible in the near future.

### 2.2. *Mycobacterium tuberculosis* Complex

In smears from raw sputum specimens, stained by Ziehl–Neelsen, tubercle bacilli are usually seen as slightly curved deeply red rods, 1–4 $\mu$m long and 0.2–0.6 $\mu$m thick, often appearing beaded. The acid-fastness and, consequently, the intensity of red color can be diminished if the organisms are resistant to isoniazid or isoniazid and rifampin. If the morphology of acid-fast bacilli found in sputum smears stained with auramine O and examined with a fluorescence microscope is questionable, the slide should be restained by the Ziehl–Neelsen method and examined under oil-immersion microscopy. Smear examination of a raw sputum specimen is essential for timely presumptive clinical diagnosis, providing results to the physician within 24 h, but it should be taken into account that the limits of sensitivity of microscopy is about 10,000 AFB/ml of the specimen. Also, no attempt should be made to speculate in the laboratory report whether the AFB found in a raw specimen represent *M. tuberculosis* or nontuberculous mycobacteria.

Tubercle bacilli grown in culture media have the same morphology as those seen in raw specimens, practically indistinguishable from other mycobacteria. At the same time, most of the *M. tuberculosis* and *M. bovis* strains have a tendency to form microcolonies appearing as tight serpentine cord formations (1,2) and these cords are especially distinctive in smears made from broth cultures. It should be taken into account that cording can be produced by some *M. kansasii* strains as well, although the formations are not as tight as those formed by *M. tuberculosis*. On the other hand, lack of cords in the culture does not completely exclude *M. tuberculosis*. The attraction of smear examination from broth cultures is that these results can be available within a week, speeding the presumptive diagnosis of tuberculosis. At the same time, the presence of cords should not be used alone as an official identification of the *M. tuberculosis* complex.

Members of the *M. tuberculosis* complex are nonsporulating, gram-positive, obligate aerobes, with a long doubling time of 12–20 h, depending on the cultivation conditions. An important distinctive feature of these mycobacteria is that they can grow on culture media at 34–38°C only. They grow poorly in acidic environment at pH below 6.0 and they cannot be cultivated at all below pH 5.4. Clumpy growth in various types of liquid media can be detected within 7–10 days, and diffuse growth of some strains in media containing Tween-80 reveals microcolonies (clumps) in smears made from these cultures. Visible colonies of *M. tuberculosis* appear on Lowenstein–Jensen (L-J) and other egg-based media after 3–8 weeks of cultivation and on 7H10/7H11 Middlebrook agar medium in 2–6 weeks. It is typical to observe nonpigmented (buff color on L-J) colonies with rough surface, irregular edges, and wrinkles when they merge. *M. bovis* more often than *M. tuberculosis* can produce smooth colonies with a wet surface. They can appear on L-J slants as very small, transparent smooth colonies resembling water droplets. This poor growth has been attributed to the inhibitory effect of glycerol in standard media, previously defined as dysgonic colonies of *M. bovis* versus eugonic colonies of *M. tuberculosis* on L-J medium. The outdated term "dysgonic" was introduced at a time when *M. bovis* was thought to be a variant of *M. tuberculosis*. What was described as "dysgonic" colonial morphology is now known as "smooth colonies" for many nontuberculous mycobacteria. *M. bovis* grows better on L-J than on agar medium, but it takes 6–8 weeks of cultivation to recover visible growth. In areas where the probability of finding *M. bovis* is significant, an additional unit of L-J medium containing 0.4% pyruvate but without glycerol should be used to improve the recovery rate of these organisms (3).

Along with the cultural characteristics, some biochemical tests are used for identification of these organisms. The drug-susceptible ("wild") *M. tuberculosis* strains usually show a positive niacin test, positive nitrate reduction test, positive pyrazinamidase test, and growth on agar medium containing 1 $\mu$g/ml or 5 $\mu$g/ml of thiophen-2-carboxylic acid hydrazide (TCH). Most of the *M. bovis* strains show opposite results with these four tests: negative niacin and nitrate reduction, and pyrazinamidase tests, and no growth in the presence of TCH. *M. africanum* is similar to *M. bovis* in three of four of these tests but is positive in the pyrazinamidase test. Other biochemical tests are similar for all members of the *M. tuberculosis* complex. One of them, the negative heat-stable (68°C) catalase test is important in differentiation of the *M. tuberculosis* complex from most of the nontuberculous nonchromogenic mycobacteria (although it is also negative for *M. gastri* and *M. malmoense*). Other tests for these organisms are arylsulfatase (2 weeks)—neg., semiquantitative catalase—neg., no growth in the presence of 5% NaCl, urease—pos., Tween-hydrolysis—pos. for about 70% of the isolates.

### 2.3. *Conventional Methods to Identify Mycobacterium tuberculosis*

Identification of *M. tuberculosis* at the time when neither of the rapid methods described below had been developed was based on the assessment of all typical features described above. This "conventional" protocol is still in effect in areas where the rapid methods are not available. Diagnosis of tuberculosis in many developing countries is still based only on detection of AFB in sputum smears, and this approach is motivated by a lack of funds and by the need to detect as many as possible of the most infectious patients. Smear examination remains an important test in more advanced laboratories as well, but only for presumptive diagnosis and only as a step in the bacteriological testing of the patient's specimen. Finding AFB in sputum smears signals the setting up of the direct drug-susceptibility test, which gives the most effective way for an expedited detection of drug resistance, with a turnaround time of 3 or 4 weeks (4). It has to be stressed that the efficiency of detecting AFB can be increased if the smear is made from a concentrated (processed) specimen and if it is examined with a fluorescence microscope. Following presumptive diagnosis of tuberculosis by detection of serpentine cording in a smear of a broth culture, the next steps in identification are based on growth at different temperatures and colonial morphology. Because these features also represent the first steps in speciation of nontuberculous mycobacteria as well, any unknown isolate should be inoculated on a set of solid media to be cultivated at different temperatures. At the National Jewish Center in Denver, we inoculated a set of five 7H11 agar plates to be incubated at 25°, 37°C, and 42°C, including foil-wrapped plates at 25°C and 37°C for evaluation of photochromogenicity. Detection of typical for *M. tuberculosis* nonpigmented colonies on plates cultivated at 37°C, with no growth at 25°C and 42°C is a sign for performing the tests listed in the previous section.

### 2.4. *Rapid Identification of Mycobacterium tuberculosis Using Nucleic Acid Probes with Primary Broth Cultures*

One of the options for rapid identification of the *M. tuberculosis* complex employs a nucleic acid probe performed with a culture in a liquid medium that has been inoculated with a raw specimen. The attraction to this approach is that the accumulation of a sufficient bacterial harvest can be achieved within a much shorter time period than that on solid media. Liquid media have not been used for primary isolation in the past due to the interference of nonmycobacterial contaminants surviving digestion–decontamination processing and overgrowing the slowly multiplying tubercle bacilli. Development of selective liquid media, first BACTEC 7H12 broth (Becton Dickinson Diagnostic Instrument Systems, Sparks, MD) and recently of the selective 7H9 broth in other systems, provided the solution to this problem. Cultures grown in liquid media can be used with the commercially available AccuProbe (Gen-Probe, San Diego, CA) within 1–2 weeks after inoculation, although there are certain limitations in the use of this technique, including the need to repeat the AccuProbe test with a culture from a solid medium if false-negative results are suspected in a test with a broth culture. A culture on solid medium is usually available, because a liquid medium should never be used as the only means of culture isolation. For example, at the National Jewish Center in Denver we inoculate four units of media with every processed sputum specimen: BACTEC 12B, 7H11 plain and selective agar, and an L-J slant (4). If the results of a nucleic acid probe identify the isolate as the *M. tuberculosis* complex, further speciation should follow, which, in fact, is just differentiation between *M. tuberculosis* and *M. bovis,* based on the results of the four tests mentioned in the previous section: niacin production, nitrate reduction, pyrazinamidase, and susceptibility to TCH. Final identification of *M. tuberculosis* can require 2 weeks or more, but the first report that the *M. tuberculosis* complex has been identified is sufficient for immediate public health measures and administration of an appropriate treatment regimen. *M. bovis* is relatively rare in the United States and many European countries, and its identification by the four tests gives sufficient time for proper adjustment of the treatment regimen. Several issues should be taken into account when evaluating the results of the four differentiation tests. The niacin and nitrate reduction tests can be negative not only for *M. bovis* but for some isoniazid-resistant *M. tuberculosis* strains as well. The pyrazinamidase test is positive only for pyrazinamide-susceptible *M. tuberculosis* strains and cannot differentiate between *M. bovis* and pyrazinamide-resistant *M. tuberculosis.* Therefore, results of all four tests should be taken into consideration, even if they have been performed with particular observance of all requirements.

### 2.5. BACTEC NAP Differentiation Test

Para-nitro-$\alpha$-acetylamino-$\beta$-hydroxypropiophenone (NAP) inhibits the growth of the *M. tuberculosis* complex and usually does not affect the growth of nontuberculous mycobacteria. The NAP test is a part of the BACTEC technology, and vials containing a paper disk impregnated with 5 mg of NAP are manufactured by Becton-Dickinson Diagnostic Instrument Systems. The test can be performed with a culture grown in 7H12 broth (in 12B vials), when the daily radiometric Growth Index (GI) reading reaches 50 or greater. Depending on the actual GI reading, the culture is appropriately diluted and distributed between two vials—

with and without the NAP disk. Inhibition of growth in the NAP-containing vial while the daily GI reading increases in the NAP-free vial indicates the presence of the *M. tuberculosis* complex in the culture. Substantial growth in both vials indicates nontuberculous mycobacteria. The test requires about 5 days.

### 2.6. *Polymerase Chain Reaction for Mycobacterium tuberculosis Identification*

This approach is designed for identification of the *M. tuberculosis* complex directly in a raw specimen, because PCR can detect just a few bacterial cells. PCR and other amplification methods have been developed with several technological modifications, but none of them is fully standardized yet for the clinical laboratory. Results can be available within 24 or 48 h upon arrival of the specimen, and positive results indicate that the specimen contains the *M. tuberculosis* complex. This can be especially helpful for testing specimens that usually contain a small number of bacteria, like cerebrospinal fluid. Unfortunately, PCR can also be positive when only dead tubercle bacilli are present, and it cannot distinguish active tuberculosis from the signals in a specimen from an individual who had tuberculosis in the past. A positive PCR test identifies the *M. tuberculosis* complex but does not distinguish *M. tuberculosis* from *M. bovis*. Therefore, final identification of *M. tuberculosis* will still require performance of the four conventional differentiation tests mentioned above. Negative PCR results are not final because of the probability, although perhaps very low, of a false-negative result nor does it rule out nontuberculous mycobacteria. This means that a final negative report will rely on cultivation and the final reading after 8 weeks of cultivation. Because of these shortcomings, other options besides testing a raw specimen will have to be considered. Current data indicate that in contrast to raw specimens, PCR has 100% specificity and 100% sensitivity when applied to a pure culture of *M. tuberculosis,* even at the very initial phases of growth. Therefore, PCR with a young culture in a selective liquid medium can represent a rational approach toward rapid identification. For example, it can be done with a BACTEC culture as soon as the growth index just becomes positive (GI $> 15$), which requires less than a week of cultivation for most of the isolates. Finally, PCR cannot replace cultivation, which is still necessary for the final speciation of *M. tuberculosis* versus *M. bovis,* as well as for drug-susceptibility testing, examination of the isolates for epidemiological analysis [restriction fragment-length polymorphism (RFLP)], and for monitoring the patient's response to treatment during the course of therapy.

### 2.7. *Cell-Wall Lipid Chromatography for Mycobacterium tuberculosis Identification*

Any of the methods to analyze the mycobacterial cell-wall lipids, [high-performance liquid chromatography (HPLC), gas–liquid chromatography (GLC), or thin-layer chromatography (TLC)] require substantial bacterial harvest from solid media, requiring at least 3 weeks of cultivation. Therefore, although some of these methods have been suggested for rapid identification of mycobacteria, they are, in fact, only rapid techniques, not really rapid methods. They do not represent any advantage over other methods described above for the identification of *M. tuberculosis.* At the same time, they are quite applicable as a tool for speciation among the nontuberculous mycobacteria, which will be discussed in the following sections of this chapter.

## 3. Protocols for Speciation of Nontuberculous Mycobacteria

### 3.1. General Considerations

Although the priority in mycobacterial speciation should be given to differentiation between *M. tuberculosis* and nontuberculous mycobacteria, in reality, when a raw specimen is directed to a specialized mycobacteriology laboratory, one cannot separate this priority from the initiation of speciation of nontuberculous mycobacteria. This approach is the most efficient for dealing not only with *M. tuberculosis* but also for shortening the turnaround time in speciation of the nontuberculous mycobacteria. In the three-level laboratory service system in the United States, Level I and II laboratories are not encouraged to speciate the nontuberculous mycobacteria. Under these conditions, after identification of *M. tuberculosis* isolates, pure cultures of nontuberculous mycobacteria are forwarded to Level III laboratories for their speciation. Therefore, there are two basic protocols: One deals with the unknown raw specimens, and the second is for speciation among nontuberculous isolates.

### 3.2. Unknown Raw Specimens

With the progress in molecular biology of mycobacteria, some attempts are being made to identify by PCR not only *M. tuberculosis* but the nontuberculous mycobacteria as well. One such direction of research is an attempt to develop a PCR test for the genus *Mycobacterium,* in which case all raw specimens will be screened to determine whether they contain any mycobacteria at all. Because 90–95% of sputum specimens normally do not contain mycobacteria, such screening would allow limiting the subsequent processing and cultivation to only those that are positive (e.g., 5–10% of all specimens). This approach would decrease the operational cost and volume of work and would make the laboratory much more efficient. That can happen only when a genus probe is 100% sensitive, meaning that all negative specimens can be discarded. Although this approach has definite merits, rapid identification of selected nontuberculous mycobacterial species in raw specimens is much less important than identification of *M. tuberculosis,* inevitably leading to a compromise in accuracy and other shortcomings compared with speciation of the pure cultures. Therefore, speciation of nontuberculous mycobacteria in raw specimens should be followed by isolation of pure cultures. We have referred earlier to the protocol used at National Jewish Center for Immunology and Respiratory Medicine in Denver for the primary isolation of mycobacteria from sputum specimens, which includes inoculation of four units of media: BACTEC 12B vial, biplates containing 7H11 plain and selective agar, and an L-J slant. The combination can be different, but it is important, taking into account speciation of nontuberculous mycobacteria, that along with a selective liquid medium, which is necessary for rapid detection of *M. tuberculosis,* an agar medium is used as well, preferably in plates. Examination of the colonies on agar plates gives the best assessment of their morphology, which is essential for selection of speciation procedures. Use of L-J medium as the only solid medium is not only less efficient for evaluation of the colonial morphology and detection of mixed cultures, but it is also much less efficient than the agar medium for isolation of some nontuberculous mycobacteria. Isolation of mycobacteria from blood specimens can be limited to one or two units of medium only: a plain 7H11 agar plate and, if rapid detection is important, a liquid medium (e.g., BACTEC 12B). L-J medium is not needed because the blood specimens, usually taken from AIDS patients, most likely contain *M. avium* rather than *M. tuberculosis.* Selective 7H11 medium is not needed when the isolation is done from a sterile specimen. Isolation from blood is most efficient when the specimen is concentrated after lysing, which also gives an opportunity to quantitate the bacterial load by the colony count on agar plates (5).

Primary cultures from skin lesions should be cultivated at 30–32°C, and the media should be supplemented with hemin when *M. haemophilum* is suspected.

### *3.3. Speciation of the Isolates*

Nucleic acid probes can be used for rapid identification of the pure cultures. The Gen-Probe Company (San Diego, CA) manufactures the following probes for the AccuProbe technology: *M. tuberculosis* complex, *M. avium, M. intracellulare, M. avium* complex (that covers identification of *M. avium, M. intracellulare,* and the so-called "x"-strains), *M. kansasii,* and *M. gordonae.*

Screening unknown cultures with the *M. avium* complex and the *M. tuberculosis* complex probes efficiently minimizes the number of cultures subjected to the battery of conventional identification tests (Fig. 11.1). For the same purpose, a third probe, *M. gordonae,* can be added in laboratories observing a substantial proportion of this species found in certain geographic areas. Although it is important for rapid detection of *M. tuberculosis* to probe a BACTEC 12B culture, timing is not an issue for the *M. avium* complex or *M. gordonae.* It should be also taken into account that the AccuProbe technique is quite sensitive when the

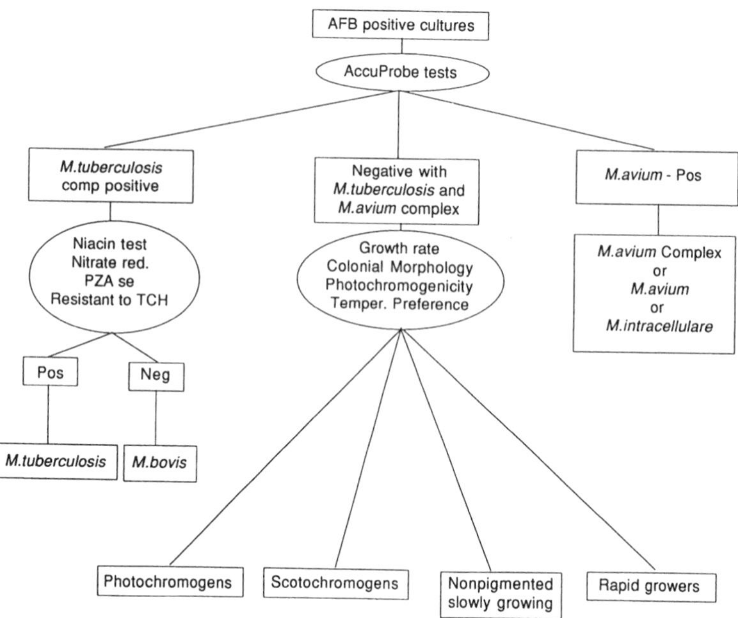


Figure 11.1. An initial phase of speciation of mycobacteria in laboratories equipped with the AccuProbe Technology.

BACTEC culture is used with the *M. tuberculosis* probe but not with the *M. avium* probe (more details below). Therefore, it is rational to test with the BACTEC culture against the *M. tuberculosis* probe only. If the results are negative, a harvest from solid medium can be used with two or three probes, including a repeated *M. tuberculosis* complex and an *M. avium* complex. A test with separate *M. avium* and *M. intracellulare* probes can be done later, usually as a part of research protocols pursuing the analysis of differences between these two infections. Photochromogens warrant testing with the *M. kansasii* probe if the results of the conventional tests are questionable. Generally, all cultures that are negative with the *M. tuberculosis* complex and *M. avium* complex probes (and in some areas with the *M. gordonae* probe) are subjects for speciation by a battery of conventional biochemical tests or cell wall lipids analysis, or both. The lipid analysis by HPLC or GLC can be done instead of or after the AccuProbe tests with the cultures grown on solid media, depending on the expertise and established protocols. Application of HPLC or GLC may not provide clear answers and final speciation of all mycobacterial isolates. More details about these technologies as applied in a clinical laboratory are given below. Availability of the HPLC or GLC system does not exclude the application of conventional speciation techniques to isolates that cannot be clearly identified by lipid analysis.

The first step in conventional speciation is evaluation of cultural characteristics, such as growth rate, temperature preference, colonial morphology, and chromogenicity. These features can be determined by inoculating several units of medium to be incubated wrapped and unwrapped at different temperatures. At the National Jewish Center in Denver we inoculate five 7H11 agar plates, two of which, wrapped in aluminum foil and unwrapped, are cultivated at 25°C, two (also wrapped and unwrapped) at 37°C, and one at 42°C. Growth is checked at 1, 3, and 6 weeks. When nonpigmented colonies appear, the unwrapped plates are exposed to light for 2 h followed by additional 24 h reincubation. If the colonies become pigmented, they are compared with those grown on the still-wrapped plates. Further speciation flow depends on which of the four Runyon groups the isolate belongs (Fig. 11.1), helping to focus on the most essential tests. Runyon's grouping is not a taxonomic or clinical classification of mycobacteria but is used just as a convenient step in speciation. If grouping is not clear, equal attention should be given to all speciation procedures. A description of the procedures is given in a another section of this chapter, and we provide here only the most convenient protocols for speciation within each of the four groups.

### *3.4. Conventional Speciation of the Photochromogens*

Development of a yellow pigment for most of the photochromogenic mycobacteria occurs after the grown colonies have been exposed to light for 2–3 h followed by additional overnight incubation. When *M. simiae* or *M. asiaticum* are suspected based on other test results, and the colonies do not turn yellow, the photochromogenicity test should be repeated with 6–8 h exposure to light followed by 48 h of reincubation. Cultures grown at 25°C and 37°C should be subjected to the photochromogenicity test at both temperatures. The test can be done with cultures grown on L-J slant as it is described in the CDC manual (6), but at the National Jewish Center in Denver, we perform this test with cultures grown on 7H11 agar plates. A plate sealed in a plastic bag is exposed to light from a 60-W bulb placed 20–25 cm from the culture, while a parallel wrapped plate remains on the bench for the same period of time. After reincubation, both plates are examined at magnification of 30× to 60× under a dissecting stereomicroscope. The examination of the colonies include assessment of pigmentation, shape (domed, flat), transparency or opacity, surface (rough or smooth, dry or wet, aeral hyfae), and edges (regular, irregular, with an apron or without). Some scotochromogens can intensify their color after light exposure, and therefore photochromogenic colonies are defined as those pigmented only after light exposure while those kept in the dark remain white, cream, or buff. The most reliable results are obtained at the temperature most favorable for the isolate. For example, the preferred temperature for *M. kansasii* is usually 37°C, whereas for *M. marinum,* it is often 25°C. At least one species, *M. szulgai,* is known as being photochromogenic at 25°C but scotochromogenic at 37°C. It should be taken into account that some colonies of the photochromogenic species might not produce any pigment. For example, there are some strains of *M. kansasii* that produce only nonchromogenic colonies, and some strains that are scotochromogenic. Therefore, although photochromogenicity is very important for placing the isolate into this group, other features must be considered as well. With these cautions in mind, the photochromogenic species differentiation depends on the following test results: growth temperature preference, niacin production, nitrate reduction, semiquantitative and thermostable (68°C) catalase tests, arylsulfatase (14 days), pyrazinamidase, urease, and Tween-80 hydrolysis (10 days). The optimal flow of

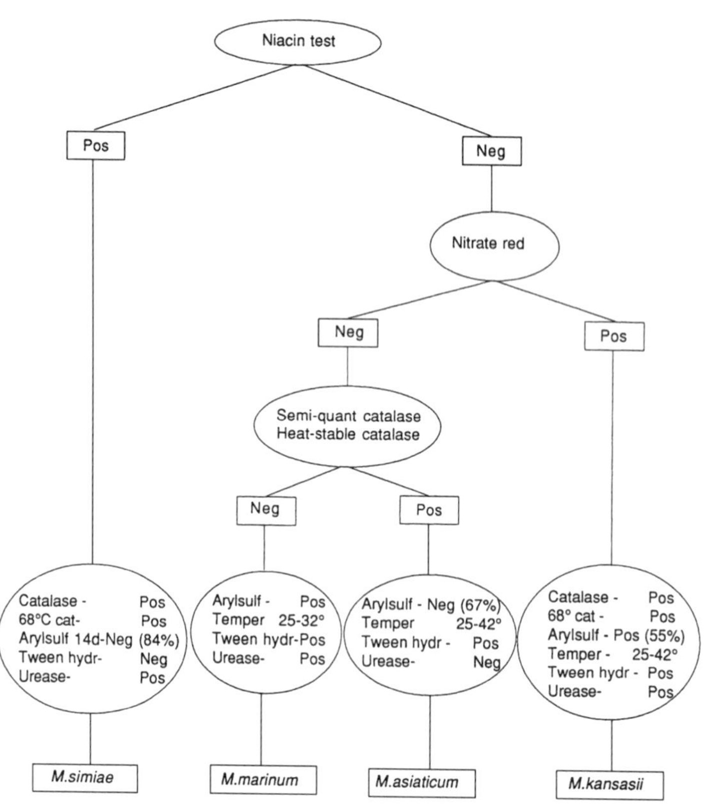


Figure 11.2. An optional workflow in speciation of the photochromogenic mycobacterial isolates.

differentiation among the isolates with clear photochromogenicity is shown in Fig. 11.2.

*Mycobacterium kansasii* is photochromogenic at 25°C and 37°C, but better growth occurs at 37°C. Crystals of $\beta$-carotene can be seen in old cultures. The colonies are rough or smooth and wet with irregular edges. The niacin test is usually negative, whereas the nitrate reduction, Tween-80 hydrolysis test at 10 days, urease, semiquantitative and heat-stable catalase tests are all positive. The pyrazinamidase test is positive for most isolates, and arylsulfatase test at 14 days can be either positive or negative. Most of the isolates are resistant to kanamycin and susceptible to thiacetazone, features which are important for differentiation from *M. marinum.*

*Mycobacterium marinum* most often produces smooth flat colonies with irregular edges. The preferred temperature is 25–32°C, but photochromogenicity with the subcultures can be detected at both 25°C and 37°C. Whereas the most common specimen for *M. kansasii* isolation is sputum, *M. marinum* is usually isolated from skin lesions, lymph nodes, and other nonrespiratory specimens. Otherwise, the isolates have very much in common with the features of *M. kansasii* described above. The following test results are most essential for differentiation from *M. kansasii.* Most of the *M. marinum* isolates show negative nitrate reduction (95% probability)* and negative semiquantitative catalase (80% probability). They are usually susceptible to kanamycin and resistant to thiacetazone.

*Mycobacterium simiae* usually produces smooth opaque domed colonies with regular edges when grown on agar medium. The optimal temperature is 37°C, but the colonies appear on agar plates cultivated at 25°C as well. Lightly yellow colonies appear at both temperatures, but a prolonged, up to 8 h, exposure to light followed by at least 2 days of incubation is required to detect pigmentation. A unique feature of this photochromogen is the positive (with 85% probability) niacin test, which can be detected better on mature, 6-week-old cultures on L-J slants. When *M. simiae* is suspected, the L-J slant culture should be wrapped in foil during the whole period of cultivation, as pigmentation of the colonies can interfere with the niacin test results. With the same probability (85%), the nitrate reduction, 10-day Tween-80 hydrolysis, and 14-day arylsulfatase tests are negative. Semiquantitative and heat-stable catalase and pyrazinamidase tests are positive with most isolates (95–99%).

*Mycobacterium asiaticum* has cultural characteristics similar to *M. simiae.* Negative niacin and urease tests (99%) and a positive 10-day Tween-80 hydrolysis test (91%) help differentiate it from *M. simiae.*

*Mycobacterium genavense* is a fastidious organism, originally described as a novel, unidentified mycobacterium causing a disseminated infection and death in

---

*The probability of this and other tests throughout this chapter are given according to the summary by Kent and Kubica (6).

a person with AIDS (7). A distinguishing feature is that the organism fails to grow on the conventional solid media and grows poorly in liquid media. Therefore, the first identification was based on PCR amplification of the DNA sequences from tissue, leucocyte extracts, and blood cultures in BACTEC 13A broth from 11 AIDS patients using primers complementary to 16S rRNA (8). Further efforts by other authors resulted in subcultivation on 7H11 agar supplemented with mycobactin J (9) or on 7H10 agar containing charcoal and yeast extracts (10). The organism is listed here because of its genetic closeness to *M. simiae* (8,10). Photochromogenicity has never been reported, probably because of poor growth on solid media (9). Strongly positive results were shown in urease, semiquantitative, and heat-stable catalase tests, and negative results in niacin, nitrate reduction, Tween hydrolysis, and arylsulfatase tests. Similar results were obtained in another report (10), showing, in addition, negative tellurite reduction and positive pyrazinamidase tests. It is still unclear whether *M. genavense* should be classified as a new species or just a subspecies (10).

*Mycobacterium szulgai* can produce either rough or smooth colonies, photochromogenic at 25°C but scotochromogenic at 37°C. Other distinctive features include positive nitrate reduction and pyrazinamidase tests, positive (94%) urease test, and positive semiquantitative (99%) and heat-stable catalase (81%) tests. Tween-80 hydrolysis and arylsulfatase tests (10 days) can be either positive or negative.

### 3.5. Speciation Among the Scotochromogens

Mycobacteria that produce pigmented colonies on solid media shielded from light during the whole period of cultivation are considered scotochromogens. Colonies of so-called nonchromogenic mycobacteria, for example, some *M. avium* complex isolates, can turn yellow, especially the opaque colonies (this issue is discussed in more detail in Section 2.6). *M. szulgai* is sometimes classified as a scotochromogen because it is photochromogenic at 25°C only. Some of the *M. kansasii* isolates can be scotochromogenic. All these and other exceptions should be taken into account when grouping by chromogenicity as first step in speciation. Most of the scotochromogens are slowly growing organisms, and their colonies usually become visible on 7H10/7H11 agar medium after 2 weeks of cultivation at the optimal temperature. One of the species of this group, *M. xenopi,* grows better at 42°C, showing very light yellow pigmentation after 4–6 weeks of cultivation. On the other hand, colonies of *M. flavescens* can become fully mature within 1 week. Differentiation among the scotochromogens can be based on the test results shown in Fig. 11.3.

*Mycobacterium gordonae* is the most common contaminant in sputum specimens, usually coming from water-supply systems. The colonies usually grow within 2 weeks, usually have smooth surface and regular edges, flat or slightly

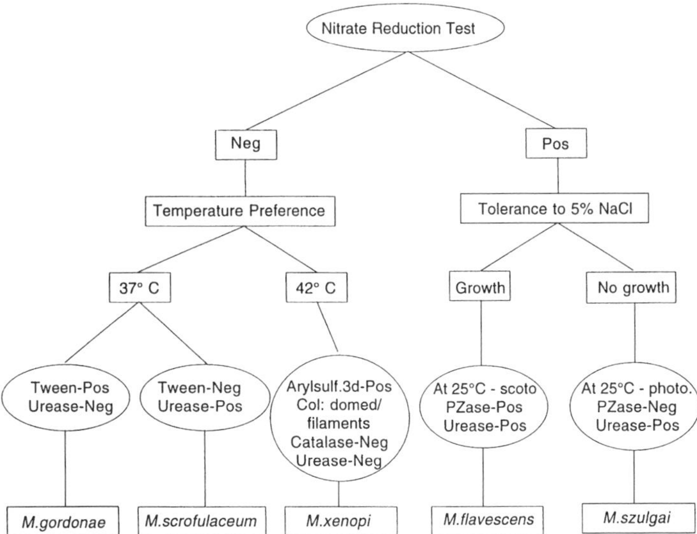


Figure 11.3. Workflow in speciation of scotochromogens.

domed, yellow to orange, and the color becoming more intense with time. The best growth appears at 25–32°C. Important identification tests are positive 10-day Tween-80 hydrolysis (99%) and negative urease (85%) and nitrate reduction tests. *M. gordonae* does not grow on media containing 5% NaCl.

*Mycobacterium scrofulaceum* can be isolated from cervical lymph nodes, sputum, or blood. The colonies are yellow to orange, usually smooth, slightly domed, and rarely rough. The preference temperature growth is 37°C, but some isolates grow better at 42°C. The distinctive features are positive semiquantitative catalase (93%) and urease (99%) tests and negative Tween-80 hydrolysis (98%) and nitrate reduction tests; there is no growth on media containing 5% NaCl.

*Mycobacterium avium-intracellulare-scrofulaceum* Intermediate (MAIS) is a scotochromogenic intermediate between the *M. avium* complex and *M. scrofulaceum.* Because some subcultures of the *M. avium* complex can have pigmented opaque colonies, differentiation is based on the fact that one of two tests, urease or catalase, is positive, along with a more intense color of the MAIS colonies than those of *M. avium.* The *M. avium* complex probe in the AccuProbe test does not react with MAIS.

*Mycobacterium flavescens* is a typical scotochromogen but can also be considered a rapid grower because some strains produce fully grown colonies within 1 week of cultivation. Nevertheless, its identification is more successful by differentiation within the group of scotochromogens. The colonies are smooth or rough, yellow-orange or yellow-gray, and usually flat with irregular edges. Different isolates can grow at temperatures ranging from 25°C to 42°C. An important distinctive feature is that most of the isolates (71%) can grow on Lowenstein–Jensen medium containing 5% NaCl. Other important features are positive nitrate reduction (96%), Tween hydrolysis (97%), and urease (71%) tests.

*Mycobacterium xenopi* is characterized by very slow growth on any medium, with best growth appearing at 42°C. After 4–6 weeks of cultivation, the colonies are clearly domed, almost hemispheric, smooth, some with small rough "bumps" on the surface, and some having a thin apron (filamentous extensions) around them. Pigmentation is cream or light yellow and can appear only after a prolonged incubation of up to 6 weeks. A unique feature is a positive 3-day arylsulfatase test (preferably at 42°C), whereas other tests used for identification among the scotochromogens, including nitrate reduction, Tween hydrolysis, urease, and semiquantitative catalase are negative. It does not grow in the presence of 5% NaCl.

*Mycobacterium selatum.* Some *M. xenopi*-like isolates have been suggested as a new species of mycobacteria based on the restriction fragment-length polymorphism (RFLP) analysis of the amplified sequence of the *Hsp65* gene multilocus enzyme electrophoresis and 16S rRNA sequence analysis (11–13). These isolates initially attracted the attention of the scientists because they gave false-positive results with the *M. tuberculosis* probe in the AccuProbe test. It has been reported that 8 of 20 strains were positive with the *M. tuberculosis* probe (13). The authors of this report have described *M. selatum* as a nonchromogen, which did not grow at 45°C and had a positive 3-day arylsulfatase test. At least one of these isolates was referred to the National Jewish Center in Denver. The AccuProbe test with the *M. tuberculosis* probe was positive, but all biochemical tests were identical to those described above for *M. xenopi,* including very light yellow or creamy color of the colonies, good growth at 42°C, and positive 3-day arylsulfatase test. Therefore, we identified the culture as *M. xenopi.* This judgment corresponds with the original report (11) about the similarity of the mycolic acid patterns of these isolates and *M. xenopi.* It is very likely that *M. selatum* is just a subspecies of *M. xenopi.*

### 3.6. Speciation Among the Slowly Growing Nonchromogenic Mycobacteria

Despite the name of this group, some organisms, for example, the *M. avium* complex, can produce pigmented colonies (see below). On the other hand, some photochromogens (*M. simiae* and some strains of *M. kansasii*) as well as some scotochromogens (*M. xenopi*) can appear nonpigmented, especially when immature growth is being examined. One of the most important tests for identification within this group is the 10-day Tween hydrolysis test, which is negative for the *M. avium* complex only. This test, repeated at 14 days, is important to differentiate between *M. malmoense* and *M. shimoidei* on the one hand, and *M.*

*avium* on the other. It is positive for *M. malmoense* and *M. shimoidei* but negative for *M. avium.* Some other tests are unique for some species of this group; for example, a positive urease test is typical for *M. gastri* only, and a positive catalase semiquantitative test is typical for *M. triviale* and *M. nonchromogenicum.* A distinctive feature to differentiate between *M. malmoense* and *M. shimoidei* is the temperature tolerance: it is 22–32°C for *M. malmoense* and 32–42°C for *M. shimoidei.* Differentiation within this group of organisms can be based on the test results shown in Fig. 11.4.

The *M. avium* complex (*M. avium* and *M. intracellulare*) is, after the *M. tuberculosis* complex, the second most important isolate in developed countries. The most frequent source of isolation used to be sputum, but with the spread of AIDS, blood became the most frequent specimen submitted for the detection of disseminated *M. avium* infection. These organisms usually grow on 7H10/7H11 agar medium within 2 weeks, but they grow poorly and much more slowly on L-J and other egg-based media. The colony types most often found on primary isolation agar plates are smooth flat transparent (SmT) with an elevated center. If the plates are incubated for a prolonged period of time, the elevated center becomes extended, and then it is hard to distinguish this type from another type of original colony, opaque smooth domed (SmO).

The opaque colonies (SmO) are quite rare in primary isolates when the agar

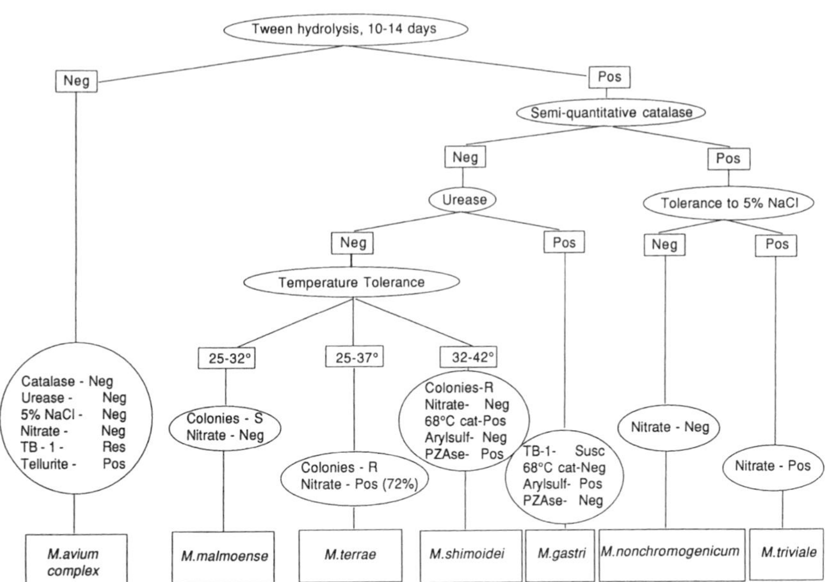


Figure 11.4. Workflow in speciation of nonchromogenic nontuberculous mycobacteria.

plates are examined no later than 2–3 weeks after incubation. They appear with higher frequency on L-J, a medium which can inhibit growth of SmT colonies. Prolonged cultivation of L-J cultures leads to a selective growth of the SmO colonies. During subcultures, especially in a liquid medium or on L-J, transformation from SmT to SmO takes place. Because the organisms forming SmO colonies grow and multiply much faster than those forming SmT colonies, the culture eventually is dominated by the SmO subpopulation, which is not quite representative of the patient's isolate. The SmO subpopulation is often less drug resistant than the original SmT isolate, and the SmO colonies can develop a yellow color, especially when the cultures are kept on the bench at room temperature for a prolonged period time. Some institutions that started dealing with the *M. avium* complex isolates from AIDS patients without having any previous experience with the same organisms from non-AIDS patients suddenly "discovered" that their cultures of *M. avium,* known as nonchromogenic from the literature, turned yellow on L-J slants. The authors of this "discovery" even stated that pigmentation is a distinctive feature of *M. avium* strains isolated from AIDS patients (14). It is interesting to mention that all initial isolates analyzed in this report were submitted to our laboratory in Denver for susceptibility testing and we found nothing special about these cultures compared with isolates from non-AIDS patients when they were subcultivated on agar medium.

The third major colony type is rough (R), morphologically often indistinguishable from the colonies of *M. tuberculosis.* They can be very susceptible to many of the conventional antituberculosis drugs, and this fact can create even more confusion. The important difference is that *M. avium* can grow not only at 35–37°C but also at 25°, and sometimes at 42°C.

The most important among the biochemical tests is the 10-day Tween-80 hydrolysis test which is negative for the *M. avium* complex and usually positive for all other mycobacteria of this group. Other test results that can be used for differentiation are negative semiquantitative and positive (76%) heat-stable (68°C) catalase tests, positive tellurite reduction test (82%), positive pyrazinamidase and negative results with all other tests, including nitrate reduction, arylsulfatase, and urease.

*Mycobacterium malmoense* is very similar to the *M. avium* complex. It usually produces smooth colonies, sometimes opaque and domed. They grow better at 25–32°C than at 37°C, and no growth is detected on agar plates incubated for 3 weeks at 42°C. An acidic environment is favorable, as it is for the *M. avium* complex, but some isolates of *M. malmoense* might not grow at all in the standard pH 12B BACTEC broth (pH 6.8), but they grow well in this medium at pH 6.0, manufactured for the pyrazinamide susceptibility testing of *M. tuberculosis* by Becton Dickinson Diagnostic Instrument Systems. The distinction from *M. avium* can be made by the Tween hydrolysis test, which is positive for *M. malmoense.* If results are questionable at 10 days, the test should be repeated at 14 days.

*Mycobacterium shimoidei* has broad temperature tolerance, and the appearance of growth at 42°C helps to differentiate this species from the *M. terrae* complex. The colonies are rough, elevated, with irregular edges, appearing on agar plates no sooner than after 3 weeks of cultivation. The negative nitrate reduction test is helpful in differentiation from *M. terrae,* whereas negative semiquantitative catalase test and negative 3-day arylsulfatase test help in differentiation from *M. nonchromogenicum.* A positive Tween hydrolysis test differentiates *M. shimoidei* from the *M. avium* complex. Negative urease and 14-day arylsulfatase, as well as positive heat-stable catalase and 4-day pyrazinamidase tests differentiate from *M. gastri.*

*Mycobacterium terrae* produces mostly rough, nonpigmented colonies that can grow on agar medium at 25°C and 37°C, but not at 42°C. The Tween hydrolysis test is positive, whereas urease and semiquantitative catalase tests are negative. Nitrate reduction, 14-day arylsulfatase, and pyrazinamidase tests can be either positive or negative; the 3-day arylsulfatase test is always negative.

*Mycobacterium nonchromogenicum,* along with *M. terrae,* belong to the *M. terrae* complex. Colonies grow at 25–37°C, are rough or smooth, often domed, and sometimes pale pink or beige. It differs from *M. terrae* by positive semiquantitative catalase and 3-day arylsulfatase tests.

*Mycobacterium triviale* is often also included in the *M. terrae* complex. Variable types of nonpigmented colonies grow on agar plates at the temperature range 25–37°C. The nitrate reduction test is positive. An important distinctive feature is growth on L-J slants in the presence of 5% NaCl. Other test results are similar to those of *M. nonchromogenicum.*

*Mycobacterium gastri* is similar to the *M. terrae* complex. The nonpigmented colonies are rough or smooth, elevated, and grow at 25°C and 37°C. Separating these species from other slowly growing nonphotochromogens are a positive urease test (with 79% probability) and susceptibility to thiacetazone. Positive 14-day arylsulfatase and negative 4-day pyrazinamidase and heat-stable catalase tests are also important for differentiation from *M. shimoidei.* Other test results are positive Tween hydrolysis test and negative nitrate reduction and semiquantitative catalase test results.

*Mycobacterium haemophilum* is usually isolated from skin lesions but also has been found in other specimens from HIV-positive individuals. The most distinguished feature is that it can only grow on media supplemented with hemoglobin or hemin. Therefore, isolation of these organisms is possible only on special media: chocolate agar, and either L-J or 7H10/7H11 agar supplemented with 2% ferric ammonium citrate, 0.4% hemoglobin, or 40 $\mu$g/ml of hemin. At the National Jewish Center in Denver we use two media supplemented with 40 $\mu$g/ml of hemin: 7H11 agar and BACTEC 12B broth. The optimal temperature for cultivation is 30–32°C. The nonpigmented colonies grow very slowly and usually are rough, flat, with irregular edges. It is pyrazinamidase positive but negative in all other biochemical tests.

*Mycobacterium ulcerans* can be found in skin lesions only from patients with Buruli disease. The specific feature of this organism is that it grows very slowly, only at 30–32°C, and very small colonies became visible only after 6–8 weeks of incubation.

### 3.7. Speciation Among the Rapidly Growing Mycobacteria

The appearance of mature colonies of acid-fast bacteria within the first week of incubation on any solid medium justifies use of a pathway of speciation designed for the rapidly growing mycobacteria. The first step, especially if the colonies are nonpigmented, is to perform two tests that distinguish the *M. fortuitum*–*M. chelonae* complex from other rapid growers: 3-day arylsulfatase and the ability to grow on crystal violet-free McConkey agar. If the results are positive, *M. fortuitum* can then be distinguished from *M. chelonae* by positive nitrate reduction and iron uptake tests. Further speciation of the biovars within the *M. fortuitum* group is based on the ability to grow at 42°C, tolerance of 5% NaCl in the medium, and fermentation of inositol, mannitol, and sodium citrate. If the results of the 3-day arylsulfatase test and growth on the McConkey agar are negative, further speciation is based on temperature tolerance (42°C and 52°C), pigmentation, 14-day arylsulfatase, iron uptake, and tellurite reduction tests. A possible pathway of differentiation among the rapid growers is shown in Fig. 11.5.

*Mycobacterium fortuitum* produces nonpigmented smooth or rough colonies, with wet surface and irregular edges. Temperature tolerance ranges from 25°C to 42°C. Identification is based on positive results in four tests: 3-day arylsulfatase, growth on McConkey agar (crystal violet-free), nitrate reduction, and iron uptake. All three known biovars can grow in the presence of 5% NaCl. Biovar "fortuitum" can be distinguished by the ability to grow at 42°C, and no fermentation of inositol, sodium citrate, and mannitol. Biovar "peregrinum" does not grow at 42°C and can ferment mannitol. The so-called "3rd biovar" also does not grow at 42°C but can use not only mannitol but also inositol and/or sodium citrate as well as a carbon source.

*Mycobacterium chelonae* produces colonies indistinguishable from those of *M. fortuitum.* It is positive in the 3-day arylsulfatase test and grows on the McConkey agar, but it is negative in nitrate reduction and iron uptake tests. Subspecies "*abscessus*" can grow at 42°C, is tolerant to 5% NaCl, and does not utilize sodium citrate. Subspecies "*chelonae*" does not grow at 42°C, does not grow in the presence of 5% NaCl, and can utilize sodium citrate.

*Mycobacterium phlei* and *M. thermoresistable* produce rough or smooth nonpigmented colonies and can grow at 42°C and 52°C and in the presence of 5% NaCl. These and all other rapid growers that do not belong to the *M. fortuitum*–

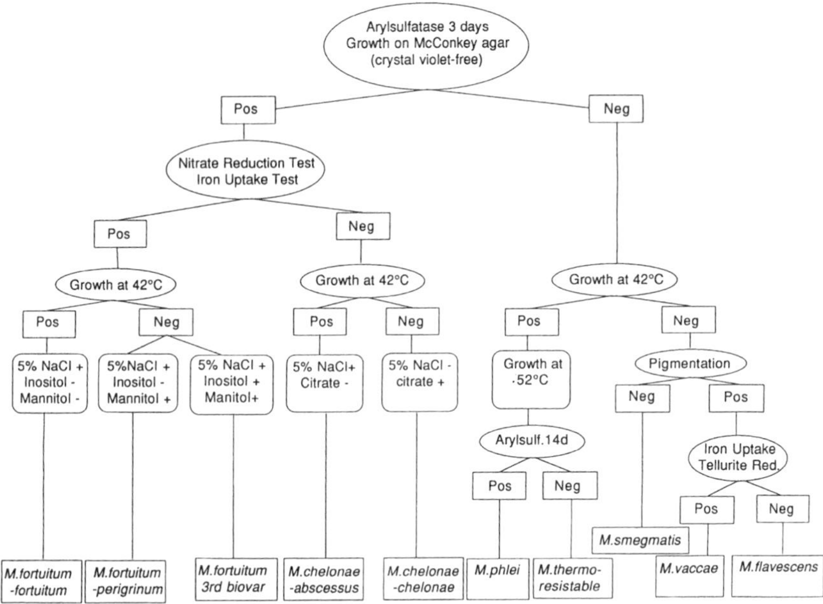


Figure 11.5. Workflow in speciation of rapidly growing mycobacteria.

*M. chelonae* complex have a negative 3-day arylsulfatase test and do not grow on the McConkey agar, but they are positive in the nitrate reduction test. In a suspension, *M. phlei* and *M. thermoresistable* can survive exposure to 64°C for 4 h. *M. thermoresistable* is usually negative in the 14-day arylsulfatase test, whereas *M. phlei* and other rapidly growing mycobacteria are positive (15).

*Mycobacterium flavescens* and *M. vaccae* are scotochromogenic and do not grow at 42°C. *M. flavescens* is negative and *M. vaccae* is positive in iron uptake and tellurite reduction tests.

*Mycobacterium smegmatis* produces nonpigmented colonies, sometimes light gray, with wet rough surfaces. It is positive in the nitrate reduction and iron uptake tests and grows in the presence of 5% NaCl.

## 4. Conventional Biochemical Test Techniques

### *4.1. Arylsulfatase*

Some mycobacterial species have and some do not have an enzyme that can hydrolyze substraces like potassium phenolphathalein disulfate by splitting the sulfate group from the aromatic ring of the substrate. The liberated phenolphthalein can be detected by the appearance of a red color when the culture is alkalized (16–18). The 3-day arylsulfatase test is essential for separation of *M. fortuitum* and *M. chelonae* from other rapidly growing mycobacteria, as well as for identification of *M. xenopi.* The 14-day test is useful in differentiation between *M. marinum* (positive) and *M. asiaticum* (mostly negative), between *M. gastri* (positive) and *M. shimoidei* (negative), as well as for identification of *M. triviale* (positive) and *M. flavescens* (positive for 78% of isolates). It is practical to set up two arylsulfatase tests: the 3-day and the 14-day tests, the application of which was discussed in the previous section. For this purpose, two tubes of media should be inoculated with the culture, one to be tested after 3 days of cultivation, the other after 14 days.

### THREE-DAY TEST IN AGAR SUBSTRATE MEDIUM (19)

It is convenient to use the agar-based medium available commercially in powder form under the label Wayne Sulfatase Agar (BBL Systems, Becton Dickinson and Co., Cockeysville, MD) for the 3-day test to differentiate among the rapid growers. The substrate, 20.6 g of powder, is dissolved in 1000 ml of water containing 10 ml of glycerol. After the powder is dissolved by heat and agitation, it is dispensed in 13 × 100-mm screw-cap tubes, 2.0 ml per tube, and autoclaved at 121°C for 10–15 min.

The butt surface of the agar is inoculated with 0.1 ml of a bacterial suspension or broth culture. After 3 days of incubation at 37°C (or 42°C if *M. xenopi* is suspected), 0.5 ml of 2 *N* sodium carbonate solution (10.6 g of anhydrous $Na_2CO_3$ in 100 ml of distilled water) is added. The test is positive if a pink or red band appears at the surface of the medium.

### THREE-DAY AND 14-DAY TESTS IN LIQUID SUBSTRATE MEDIUM (16)

The substrate medium can be made on the basis of the 7H9 broth to which the tripotassium salt of phenolphthalein disulfate 0.08 *M* solution is added. This substrate, 5.2 g, is dissolved in 100 ml of distilled water and filter sterilized. For the 3-day test medium, 2.5 ml of this solution is added to 200 ml of complete sterile 7H9 broth, providing a 0.001 *M* final concentration of the substrate. For the 14-day test medium, 7.5 ml of the substrate solution should be added to provide final contents of 0.003 *M*. Either medium is dispensed in 13 × 100-mm screw-cap tubes, 2 ml per tube.

Both tubes, properly labeled, are inoculated with 0.1 ml of a bacterial suspension or broth culture. 0.3 ml of the 2 *N* sodium carbonate solution is added after 3 days of cultivation in the 3-day test tube and after 14 days to the 14-day tube. Positive reaction is reported if pink to red color appears immediately after addition of the substrate solution. The intensity of the reaction can be reported (1+ to 5+) based on comparison with the color plate standard.

Quality controls include (1) preliminary test of substrate media on the presence of free phenolphthalein, by adding the carbonate solution (no color change), (2) positive control—*M. fortuitum,* and (3) negative control—*M. avium.*

### *4.2. Catalase Tests*

The presence of enzymes that can split hydrogen peroxide into water and oxygen can be tested by two procedures: semiquantitative test and heat-stable catalase test.

#### SEMIQUANTITATIVE TEST (20)

This test is useful in identification of the *M. tuberculosis* complex, the *M. avium* complex, *M. malmoense, M. gastri, M. shimoidei,* and *M. xenopi,* which have negative results, whereas other mycobacteria can have positive test results. The surface of an L-J medium is inoculated with 0.1 ml of a bacterial suspension or broth culture and tubes are incubated with caps loose for 14 days at 37°C. After incubation, a freshly made mixture of equal parts of 10% Tween-80 solution and 30% hydrogen peroxide is added, 1.0 ml per tube. The heights of the bubble column appearing above the medium surface after 5 min of incubation at room temperature is measured in millimeters. The test is positive if this column is greater than 45 mm.

#### HEAT-STABLE CATALASE TEST (21)

The test is designed to determine whether the isolate contains the heat-stable catalase, and negative results are important for identifying the *M. tuberculosis* complex, *M. gastri,* and *M. malmoense.* A harvest from colonies grown on a solid medium, preferably L-J, are suspended in a phosphate buffer at pH 7.0. The bacterial suspension, approximately 0.5 ml in a 13 × 100-mm tube, is then exposed to 68°C for 20 min in a water bath. After cooling at room temperature, an equal volume of the Tween–peroxide mixture (see above) is added. The results are recorded within 20 min. The appearance of bubbles in any amount is considered positive. Quality control: positive—*M. kansasii,* negative—*M. tuberculosis.*

### *4.3. Niacin Production (22,23,24)*

All mycobacteria produce niacin, but most of the species have an enzyme capable of converting it to niacin ribonucleotide. Lack of this enzyme, for example, in *M. tuberculosis* and *M. simiae,* leads to accumulation of niacin in the medium. The test is important for differentiation between *M. tuberculosis* and *M. bovis.* Niacin extracted from the solid medium can be detected by a reaction with cyanogen bromide in the presence of alanine, resulting in yellow colorization (24). Many laboratories in the United States use 3- or 4-week-old cultures grown on 7H10 or 7H11 agar. In some laboratories, 3-week-old 7H9 broth cultures have been used, in which case the supernate (after centrifugation, if necessary) is transferred to another tube. Unfortunately, not all *M. tuberculosis* strains produce sufficient amounts of niacin in this medium within this period of time, and results can be false negative. The niacin production can be increased by incorporating L-asparagine (0.25%) or its potassium salt (0.1%) into the medium (22). The best results with *M. tuberculosis* can be achieved on 6-week-old cultures grown on L-asparagine-supplemented L-J medium. In addition to timing, the intensity of growth is also important. Usually, a minimum of 50 well-grown colonies per slant are needed. On the other hand, confluent growth can block the extraction of niacin, and with these cultures, it is necessary to pierce through the bacterial growth with a pipette or loop to expose the medium. Saline or distilled water is added to the slant culture for a period of 30 min to 2 h to extract niacin. This extraction is usually done at room temperature, but incubation at 37°C can improve the results. About 0.6 ml of the extract is transferred to a sterile 13 × 100-mm screw-cap tube. The classical test with cyanogen bromide has been replaced in most laboratories by commercially available niacin paper strips (22,25). The strip is inserted into the tube containing the extract and immersed in the fluid with the end indicating an arrow. The cap should be tightened. The test is positive if the *liquid* turns yellow. If only the paper strip turns yellow, the test should be repeated with another subculture. Quality controls: positive—*M. tuberculosis,* negative—*M. avium.* Precaution: Purity of the tested culture should be assured, as some contaminants can give false-positive results.

### 4.4. Nitrate Reduction Tests

Different mycobacterial species and even cultures of the same species, depending on the age of the culture and other conditions, vary quantitatively in their ability to reduce nitrate ($NO_3$) to nitrite ($NO_2$) (26). The presence of nitrite is determined in a semiquantitative manner by adding sulfanilamide and N-napthylethylenediamine at acidic pH to a heavy bacterial suspension, resulting in the formation of a red diazonium dye (27). The test is important in differentiating *M. tuberculosis* (positive) from *M. bovis* (negative), *M. fortuitum* (positive) from *M. chelonae* (negative), *M. kansasii* (positive) from *M. marinum* (negative), as well as for identifying of *M. szulgai, M. triviale,* and *M. flavescens* (all positive).

For the slowly growing mycobacteria, the best results can be obtained with a culture well grown on 4-week-old egg-based medium. A 2-week-old culture can be used for rapidly growing mycobacteria. Two techniques are most commonly used: one with liquid reagents and another with commercially available paper strips. For either method, a heavy suspension of bacterial cells should be made from growth on solid medium or from a 7H9 broth culture, 0.2 ml in a 13 × 100-mm screw-cap tube, or 0.5 ml in a 16 × 125-mm tube.

For the classical procedure with liquid reagents, 0.2 ml of a 0.01 *M* sodium nitrate solution made in 0.022 *M* phosphate buffer (pH 7.0) should be added to the bacterial suspension, followed by incubation of the mixture for 2 h at 37°C 0in a water bath.* Three reagents are then added in the following strict order: No. 1—two drops of diluted 1 : 1 hydrochloric acid (made by adding concentrated HCl to an equal volume of distilled water); No. 2—two drops of 0.2% aqueous solution of sulfanilamide; No. 3—two drops of 0.1% *N*-naphthylethylenediamine dihydrochloride aqueous solution. The developed color ranges from pale pink (±) to deep red (+ + + + +) and it should be compared with color standards for proper judgment. The test is positive if the color is graded + + + or greater. Quality controls: positive—*M. tuberculosis,* negative—*M. avium.* To exclude false-negative results, a pinch of zinc dust should be added to negative test tubes: the appearance of a red color indicates the presence of unreacted nitrate. If color does not develop, the test should be repeated after checking the quality of all reagents.

If nitrate paper strips are used instead of the classical method (28), the strip is inserted arrow down into a tube containing the bacterial suspension, avoiding contact with fluid on the side of the tube. The tube is incubated for 2 h at 37°C, after which it is tilted to wet the entire strip. The test is positive if after 10 min, a blue area appears at the top of the strip.

### *4.5. Pyrazinamidase Test (29)*

The test detects the presence of the enzyme that converts pyrazinamide (PZA) to pyrazinoic acid (POA). This test is important for differentiation between *M. tuberculosis* (positive) and *M. bovis* (negative), especially after 7 days of incubation. It also can be useful in differentiation between *M. marinum* (positive) and *M. kansasii* (negative) at 4 days. Special semisolid agar medium should be prepared for this test:

| | |
|---|---|
| Dubos broth base | 6.5 g |
| Pyrazinamide | 0.1 g |
| Pyruvic acid, sodium salt | 0.2 g |
| Agar | 15.0 g |
| Distilled water | 1000.0 g |

*Sodium nitrate solution is made by dissolving the following reagents in 100 ml of distilled water: $NaNO_3$—0.08 g, $KH_2PO_4$—0.117 g, $Na_2HPO_4 \cdot 12H_2O$—0.485 g.

The mixture should be stirred with heat to dissolve the ingredients. Dispensed in 13 × 100-mm screw-cap tubes, 2 ml per tube, the medium is autoclaved at 121°C for 15 min. The tubes are left in an upright position to form a butt. The medium can be kept refrigerated for 3 months.

The surface of the medium in two tubes is heavily inoculated with a bacterial suspension made from colonies scraped from the surface of an actively growing 2–3-week-old culture on solid medium (L-J or 7H10/7H11 agar). The inoculum should be visible to the naked eye. To ensure good interaction of the inoculum with pyraziamide, the surface of the medium can be jabbed repeatedly with the loop.

After 4 days of incubation at 37°C, 1.0 ml of a freshly prepared 1% ferrous ammonium sulfate solution is added to one of the two tubes. A positive reaction is indicated by a pink band in the subsurface of the medium appearing within 30 min after addition of the reagent. If there are no reactions or results are doubtful, the tube can be examined again after 4 h of refrigeration. If the test is negative, it can be repeated with the second tube after additional incubation at 37°C for 3 more days. Quality controls: positive—*M. tuberculosis*-susceptible strains, negative—*M. bovis* and a tube with an uninoculated medium.

### 4.6. Susceptibility to Thiophen-2-carboxylic Acid Hydrazide (TCH)

*Mycobacterium bovis* is the only mycobacterial species susceptible to TCH, and therefore this test is important for differentiation between this organism and *M. tuberculosis* (30,31) The most common procedure is to use 7H10 or 7H11 agar incorporating 1 $\mu$g/ml and 5 $\mu$g/ml of TCH. TCH aqueous solution should be filter sterilized. It is added to the agar along with the oleic acid albumin dextrose complex (OADC) and then dispensed into tubes to make slants or into four-sectioned petri dishes. The same batch of the agar without TCH should be used as a growth control. Two sets of plates or tubes should be used for two inocula made from a 7–9-day-old 7H9 broth culture diluted $10^{-3}$ and $10^{-5}$. After incubation at 37°C in the presence of 5–10% $CO_2$, the growth on TCH-containing and TCH-free media is compared.

### 4.7. Tween Hydrolysis Test

This test is based on the fact that some mycobacteria have a lipase that splits Tween-80 (polyoxyethelene sorbitan monooleate) into oleic acid and polyoxyethilated sorbitol (32). Hydrolysis leads to the change of the neutral red indicator from yellow to pink. This test is important for differentiation between the *M. avium* complex (negative) and other slowly growing nonchromogens (positive), and between *M. scrofulaceum* (negative) and *M. gordonae* (positive). It can also be helpful in differentiation between *M. simiae* (negative) and other photochromogens (positive).

A substrate medium for this test is made by mixing three reagents: (1) 0.067 *M* phosphate buffer (pH 7.0)—100.0 ml, (2) 10% Tween-80 solution—5.0 ml, and (3) 0.1% aqueous solution of neutral red (adjusted to the actual dye content)—2.0 ml. This mixture is dispensed in 13 × 100-mm screw-cap tubes, 2 ml per tube. Immediately after autoclaving at 121°C for 15 min, each tube should be shaken to resuspend the Tween that might settle to the bottom of the tube. The medium should be kept refrigerated for no more than 2 weeks, protected from light. Sterility control of each batch is essential.

The test can be performed with either a 7H9 broth culture or with a heavy suspension made from colonies grown on solid media, adding 0.4 ml of the inoculum per tube. Formation of a pink color should be observed after incubation at 37°C at 5 and 10 days. If *M. malmoense* is suspected, the reaction should be also be checked after 14 days. Quality controls: positive—*M. kansasii,* negative—*M. avium* and uninoculated medium.

### *4.8. Urease Tests*

The ability of bacteria to hydrolyze urea can be helpful in differentiation between *M. scrofulaceum* (positive) and the *M. avium* complex (negative), between *M. gastri* (positive) and *M. terrae* (negative), and between *M. gordonae* (negative) on the one hand and *M. szulgai* and *M. flavescens* (positive) on the other. Two common techniques to perform the test use either urea broth prepared in the laboratory (33) or commercially available urea-impregnated disk (34).

Urea broth is prepared by dissolving the following ingredients in 100 ml of distilled water: peptone—1 g, dextrose—1 g, sodium chloride—5 g, $KH_2PO_4$—0.4 g, and urea—20 g. Two more ingredients should be added: (1) 1% sodium phenol red solution—1.0 ml and (2) 2% Tween-80 solution—1.0 ml. The pH is adjusted to 5.8, and the medium is filter sterilized and dispensed into 13 × 100-mm sterile screw-cap tubes, 1.5 ml per tube. The medium can be stored refrigerated up to 8 weeks.

A spadeful of growth from solid medium is suspended in broth and incubated at 37°C (without $CO_2$) for 3 and 7 days. Positive results are shown by a color change of the medium from yellow to dark pink or red.

For the disk method, 0.6 ml of a heavy 7-day-old 7H9 broth culture is transferred to a 13 × 100-mm screw-cap tube. A urea disk is placed in the tube, and the results are read after 3 and 7 days of incubation at 37°C without $CO_2$. Quality controls: positive—*M. scrofulaceum,* negative—*M. avium* and uninoculated medium.

### *4.9. Sodium Chloride Tolerance*

This test is important for identification of the few species (e.g., *M. triviale*) that will grow on L-J medium containing 5% NaCl. About 70% of *M. flavescens* strains also grow on this medium. The test is useful for differentiation among the members of the *M. fortuitum–M. chelonae* complex: *M. chelonae* biovar *abscessus* grows on this medium, whereas biovar *chelonae* does not; about 75% of *M. fortuitum* isolates can also grow in the presence of 5% NaCl. Commercially available slants of L-J medium containing 5% NaCl are inoculated, along with the sodium chloride-free L-J medium, 0.1 ml per slant. The growth or its absence is recorded after 4 weeks of cultivation at 37°C. Quality controls: positive—*M. chelonae* (biovar *abscessus*) or *M. fortuitum*, negative—*M. avium* or *M. tuberculosis.*

### *4.10. Susceptibility to Thiacetazone (TB-1)*

This test can be useful in differentiation between *M. marinum,* which is resistant to TB-1, and *M. kansasii,* which is susceptible to 10 $\mu$g/ml of this drug incorporated into 7H11 agar (35). The appearance of growth is compared on TB-1-containing and drug-free agar after 3 weeks of cultivation at 37°C. It is rational to combine this test with the susceptibility to TCH described above if four-segment petri dishes are used. In this case, one segment is filled with the drug-free agar, one with medium containing TB-1, and the remaining two with the medium containing two concentrations of TCH. Quality controls: susceptible—*M. kansasii,* resistant—*M. marinum.*

### *4.11. Iron Uptake*

This test is useful in differentiation between *M. chelonae* (negative) and *M. fortuitum* (positive). *M. phlei, M. smegmatis,* and *M. vaccae* can also be positive in this test. Some *M. szulgai* strains are also capable of taking up iron. The test is based on determining whether the isolate is capable of converting ferric ammonium citrate to an iron oxide. The technique most often used for this test (20) include cultivation, in a slanted position, of standard L-J medium inoculated with 0.1 ml of a suspension or broth culture diluted approximately to the optical density of the McFarland standard No. 1. After visible growth appears, 5–7 drops of a sterile aqueous 20% ferric ammonium citrate solution are added to one of the two slants. The slants are reincubated for an additional 3 weeks at 30–32°C if the test is performed to differentiate between *M. fortuitum* and *M. chelonae.* The slants should be examined weekly for rusty brown colonies if they are taking up the iron. Quality controls: positive—*M. fortuitum,* negative—*M. chelonae.*

### *4.12. Growth on McConkey Agar*

Among the rapidly growing mycobacteria, only *M. fortuitum* and *M. chelonae* can grow on McConkey agar that does not contain crystal violet (36). The medium for this test is made from commercially available base. Petri dishes, prepared according to the manufacturer's instructions, should be inoculated by streaking a loopful of a bacterial suspension or broth culture to obtain isolated colonies. After 5 and 10 days of incubation at 30–32°C, without $CO_2$, the appearance of growth should be checked at the end of the streaks, as some growth of mycobacteria other than *M. fortuitum–M. chelonae* can appear in the heavily inoculated areas. Quality controls: positive—*M. fortuitum,* negative—*M. smegmatis* or *M. phlei* and an uninoculated plate.

### *4.13. Tellurite Reduction*

Among the slowly growing mycobacteria, only the *M. avium* complex and *M. malmoense* give positive results (with 82% and 71% probability, respectively) within 3–4 days, whereas it takes about 7 days for other species. Tellurite reductase of the bacteria reduces potassium tellurite to metallic tellerium, which appears as a black precipitate on the bottom of the tube with broth cultures (37). The 7H9 broth in 20 × 125-mm screw-cap tubes, 4.5 ml in each, is inoculated with a heavy bacterial suspension, followed by a 7-day incubation at 37°C. The tubes are placed on a rotating wheel for better oxygenation, or hand-shaken daily. Two drops of an aqueous 0.2% potassium tellurite solution, sterilized at 121°C for 10 min, is added to each broth culture. After shaking the tubes, they are incubated in an upright stationary position for 3 days at 37°C. The tubes should be examined daily for the appearance of the black precipitate that can be seen on the bottom of the tubes along with the bacterial cell sediment. The test is positive if the black precipitate appears in no more than 4 days of incubation.

### *4.14. Utilization of Different Carbon Sources*

This test is performed for identification of the biovars of *M. fortuitum* and subspecies of *M. chelonae* and is designed to determine the ability of the isolate to utilize sodium citrate, mannitol, and inositol, each as the only source of carbon (38). The basal agar medium for this procedure is made by dissolving the following ingredients in 950.0 ml of distilled water: $(NH_4)_2\ SO_4$—2.4 g, $KH_2PO_4$—0.5 g, and $MgSO_4 \cdot 7H_2O$—0.5 g. After adjustment of pH to 7.0 for citrate medium and 7.2 for controls and mannitol and inositol media, 20 g of purified agar is added. After autoclaving for 20 min at 121°C, the medium is cooled to 56°C in a water bath and filter-sterilized substrate solutions are added to each: for citrate medium, 50.0 ml containing 5.6 g of sodium citrate; for mannitol and inositol media, 50.0 ml containing 5.0 g of either substrate; and 50.0 ml of distilled water for controls. Each of four media is dispensed into 20 × 150-mm screw-cap tubes, 8.0 ml per tube, and solidified in a slanted position. A set of four slants is inoculated with a 7-day-old 7H9 broth culture or bacterial suspension made from colonies on solid medium, 0.1 ml per slant. After cultivation at 30°C for 2 weeks, no growth should appear on controls, whereas growth on any of the substrate-containing media is interpreted as a positive test result. Quality controls for inositol and mannitol: positive—third biovar of *M. fortuitum,* negative—biovar *fortuitum;* for citrate: positive—subspecies *chelonae,* negative—subspecies *abscessus.*

### 4.15. Tween Opacity Test

This test was recommended to help identify *M. flavescens* if it grows slowly and has to be differentiated from other slow growers (39). The medium for this test is made from commercially available Dubos oleic agar base: 130 ml of distilled water, 4 g of agar base, and 50.0 ml of 10% Tween-80 solution. After autoclaving for 10 min at 121°C, the medium is cooled to 56°C in a waterbath, and 20 ml of commercially available Dubos albumin supplement is added. After dispersion of 5.0 ml in the 16 × 125-mm screw-cap tubes, the medium is solidified in an upright position. This medium, along with the control medium containing only 0.4 ml of the 10% Tween solution in 200.0 ml, should be inoculated with 0.1 ml of a 7-day-old 7H9 broth culture diluted 1 : 100. If the test is positive (*M. scrofulaceum*), an opaque ring will appear just below the surface. This should be detected by a once-a-week observation during a period of up to 6 weeks of cultivation at 37°C.

## 5. Nucleic Acid Probes

A hybridization protection assay (40–43) is used in the commercially available AccuProbe technology (Gen-Probe, San Diego, CA). A single-stranded DNA probe used for this test is complementary to the rRNA of the organism. When the probe is added to the bacterial lysate, it reacts with the released rRNA and forms a stable DNA–RNA hybrid. A chemiluminescent acridinium ester label is attached to the probe in such a way that the DNA–RNA hybrid is protected from the effect of the hydrolysis reagent that eliminates the chemiluminescence of the unbound acridinium. The acridinium protected in the hybrid preserves its chemiluminescence, which is developed by interaction with hydrogen peroxide and is measured in a luminometer. The following probes are currently available from the manufacturer: *M. tuberculosis* complex, *M. avium* complex (complementary to *M. avium, M. intracellulare,* and "x"-strains), separate *M. avium* and *M. intracellulare* probes, *M. kansasii,* and *M. gordonae.*

The following are the basic steps of the procedure, as described in the manufacturer's product manual.

1. *Sample preparation.* Two reagents, No. 1 (Lysis Reagent) and No. 2 (Hybridization Buffer), are added to a Lysing Reagent Tube with glass beads (provided by the manufacturer), 100 $\mu$l each, after which a loopful of growth from solid medium is transferred into this tube and suspended. If the test is done with a broth culture, without blood present, it should be centrifuged first at 8000 $\times$ **g** for 15 min. After removal of the supernate, 100 $\mu$l each of reagents Nos. 1 and 2 are added to the microfuge tube and the pellet suspended by vortexing. This bacterial suspension is then transferred into the Lysing Reagent Tube, 200 $\mu$l per sample.
2. *Lysis.* The Lysing Reagent Tubes containing 200 $\mu$l each of the prepared culture samples are placed in a sonicator bath so that the tubes with their contents are submerged but the caps are above the water. The tubes, placed through the sonicator rack, should not touch the bottom or sides of the sonicator. The water temperature in the sonicator should be adjusted to 60°C in advance, after the water has been thoroughly degassed. After 15–20 min of sonication, the tubes are transferred to a heating block for 10 min at 95 $\pm$ 5°C. This operation leads to the lysis of the bacterial cells with the release of RNA in the solution.
3. *Hybridization.* Probe Reagent Tubes, provided by the manufacturer, contain one of the DNA probes. Lysed samples are transferred in properly labeled Probe Reagent Tubes, 100 $\mu$l per tube. Tubes are recapped and placed in a water bath at 60°C for 15 min.
4. *Selection.* A selection reagent (Reagent No. 3) is added to each of the probe Reagent Tubes, 300 $\mu$l per tube. The tubes are recapped and vortexed, after which they are incubated in a water bath at 60°C for 5 min (with *M. tuberculosis* probe—10 min) and then remain for 5 min more at the room temperature. The reading in the luminometer should be done within 1 h, with the caps removed from the tubes.
5. *Detection.* The surface of each tube should be wiped before it is placed in the luminometer. There are two types of luminometers recommended by the AccuProbe manufacturers: AccuLDR™ and LEADER.™ Use of these instruments should comply with their respective specifications. The results are expressed in relative luminescence units (RLU). The cutoff values in RLU are different for each of the two instruments. The manufacturer suggests that the results are positive if the reading is 900 or greater in the AccuLDR™ and 30,000 or greater in the LEADER™. The repeat ranges suggested by the manufacturer are 600–999 and 20,000–29,999, correspondingly, and the reading below that is negative.

At National Jewish Center in Denver, we have realized that these interpretations are quite reasonable only for cultures grown on solid media, but when the BAC-

TEC cultures were used, the readings were much lower, especially with *M. avium* cultures even at maximum growth of GI 999. We found that 10,000 RLU in the LEADER™ instrument can be safely used as a cutoff for positive results. The repeat zone is 8,000–10,000 RLU, and negative and positive quality controls must give satisfactory results. It is reasonable to suggest that the cutoff standards should be established in each laboratory, after proper multiple testing with the various QC strains.

Special attention should be given to the BACTEC cultures inoculated with blood, even if such specimens were lysed and the bacterial contents concentrated before inoculation. The BACTEC 12B broth cultures with blood processed or cultured not processed in 13A medium always cause false-positive results in the AccuProbe test. To prevent this, we inoculate 0.1 ml of the original culture into a new 12B vial for subcultivation. Small amounts of blood still can cause false-positive results, so these subcultures are processed twice with sodium dodecyl sulfate (SDS). The first process consists of mixing 0.8–1.0 ml of the culture with 100 $\mu$l of 10% SDS solution containing 50 m*M* EDTA (pH 7.2). The second process is done with 1% SDS solution. After centrifugation in a microcentrifuge tube at 11,000 $\times$ **g** for 15 min, the supernatant is removed, and the pellet is resuspend in 0.8 ml of sterile distilled water. The next centrifugation at 11,000 $\times$ **g** for 15 min is the first step of the procedure described above for blood-free BACTEC broth cultures. An alternative to this procedure is to wait until the colonies on 7H11 agar plates, inoculated along with the 12B vials, are mature enough to be used for the AccuProbe test. Bacterial suspensions made from these colonies do not cause false-positive results if the blood specimens used to inoculate these plates were processed by blood cell lysis and centrifugation (5).

## 6. Polymerase Chain Reaction and Other Amplification Techniques*

Amplification techniques recently became the subject of research of many scientific groups and industry. It is hard to predict at this moment which of these approaches will become the most favorable tools for rapid mycobacterial identification in clinical laboratories. Automation, cost efficiency, specificity, and reliability are the most important features that will be taken into account. We have limited our description in this chapter to three methodologies which at this moment seem to be the most promising: (1) polymerase chain reaction by Roche Molecular Systems, (2) Gen-Probe amplification assay by Gen-Probe Inc., and (3) strand displacement amplification (SDA) by Becton Dickinson DNA Diagnostics.

---

*Some recent developments of these techniques and the FDA approval might have taken place after this chapter was submitted.

### *6.1. PCR*

Polymerase chain reaction allows direct detection and identification of the mycobacterial species by targeting their specific DNA sequences and various technical approaches have been suggested for this rapid diagnostic technique (44–54).

The Roche Amplicor™ *Mycobacterium tuberculosis* test is designed to amplify a segment of the 16S ribosomal RNA gene specific for mycobacteria and detect the amplified DNA using a DNA probe specific for the *M. tuberculosis* complex. The assay is performed on raw specimens (expectorated or induced sputum, BALs, bronchial washings) obtained from patients suspected of having pulmonary tuberculosis. The procedure includes four phases: (1) sample preparation to release the DNA from the organism, (2) amplification, (3) hybridization of the amplified DNA sequence to the specific *M. tuberculosis* probe, and (4) detection.

The test is performed in three areas assigned by the following procedures: (1) reagent preparation, (2) specimen preparation, and (3) amplification–hybridization–detection.

Specimen preparation starts with the conventional digestion–decontamination procedure, for example, with NaOH-NALC, resulting in a liquefied concentrated specimen. Special attention should be given to avoid any possible cross-contamination. One hundred microliters of the decontaminated specimen is required for PCR. The sample can be used for PCR the same day or kept refrigerated for the next day. Any remaining portion of the aliquot can be stored at $-70°C$ for possible future retesting. QC samples should be arranged at the same time: one positive—with *M. tuberculosis*—and three negative controls supplied in the test kit.

Amplification reagents are prepared in the reagent preparation area. To each specimen (0.1 ml) prepared in the Specimen Preparation area, 0.5 ml of Sputum Wash Solution is added. After centrifugation at $12{,}500 \times \mathbf{g}$ in a microfuge for 10 min, 100 $\mu$l of Specimen Lysis Reagent is added to the pellet. The tubes are vortexed and incubated at 60°C for 45 min, after which the tubes are centrifuged for about 5 s, and 100 $\mu$l of Specimen Neutralization Reagent is added. After vortexing, each specimen is transferred to the special Microamp PCR tubes positioned in the proper order in the tube rack, each tube containing 50 $\mu$l of PCR Master Mix Reagent. This reagent contains the mycobacteria-specific biotinylated primers for amplification, TAQ polymerase, and deoxynucleotides. The tubes are then tightly capped and transferred to the amplification area.

The amplification–detection area should be completely isolated from areas where any manipulations with mycobacteria are performed, best of all, far away from the mycobacteriology laboratory. Glassware, plastic, and other supplies should not come from the mycobacteriology laboratory. Personnel should use laboratory coats and gloves designated especially for this area. On the other hand, no materials, especially amplified samples, should be brought to the mycobacteriology laboratory, to avoid DNA contamination of the specimens arriving at the laboratory.

For amplification, the sample tray containing Micramp tubes is placed into the thermal cycler, which is programmed according to the manufacturer's instructions. Heating at 98°C for the first 3 cycles and 94°C for the remaining 36 cycles, each time for 20 s, denatures the double-stranded DNA and exposes its target sequence. Cooling to 62°C, also for 20 s in each cycle, allows the primers to anneal to their targets. The thermo-stable TAQ polymerase present in the PCR reaction mixture extends the annealed primers along the target templates to produce amplicons. This takes place by using the excess deoxynucleotides–ademine, guanine, cytosine, and uracil (instead of thymine) present in the reaction mixture. Each cycle is effectively doubling the amount of target DNA, resulting in approximately $10^9$-fold amplification from each copy of a target DNA after 36 cycles.

The PCR amplification reaction is so sensitive that fewer than 10 input molecules of specific target DNA can lead to a positive signal in this system. To avoid an aerosol contamination of a new clinical specimen with previously amplified DNA, the Roche Amplicor MTB Test has incorporated the use of AmpErase™. AmpErase contains the enzyme Uracil-N-Glycosylase (UNG), which recognizes and catalyzes the destruction of uracil-containing DNA but not thymine-containing DNA. Uracil is not present in microbial DNA but is always present in amplicons due to the use of uracil (in place of thymine) as one of the dNTPs in the reaction mixture; thus, only amplicons will contain uracil. The presence of uracil in amplicons renders contaminating amplicons susceptible to destruction by UNG prior to the amplification of the target DNA. UNG catalyzes the cleavage of an oligonucleotide at a uracil base by opening the deoxyribose chain at the 1 position. The opened chain, when heated in the first thermal cycling step (at the alkaline pH of Master Mix), causes the amplicon's DNA chain to break at the position of the uracil, thereby rendering the DNA nonamplifiable. The UNG enzyme, which is inactive at temperatures above 55°C (i.e., throughout the thermal cycling steps), is then itself denatured by the addition to the Denaturation Solution immediately after the amplification step is completed, thereby preventing it from destroying any "true" amplified products of the test.

After amplification, the amplicons themselves are also denatured by adding the Denaturation Solution, 0.1 ml per tube, to form single strands that can hybridize to the *M. tuberculosis*-specific probe bound to 96-well microtiter plates provided by the manufacturer. Twenty-five microliters of denatured amplicon is added to the wells of the microtiter plates, each containing 100 $\mu$l of the Hybridization Solution. After incubation for 90 min at 37°C, the amplicons hybridize (bind) to the plate.

Detection is performed after the plates are washed five times with a buffer to remove the unbound material. A horseradish peroxidase–avidin conjugate is added, 0.1 ml per well, which then reacts with the biotinylated amplicons during incubation for 15 min at 37°C. The plate is again washed five times, and 100 $\mu$l of a substrate reagent containing peroxide and tetramethylbenzidine is then added, 0.1 ml per well. The color is developed after the plate is incubated for 10 min at room temperature in the dark. After that, 0.1 ml of the Stop Solution is added, and the intensity of color is read in the microwell plate reader at 450 nm. More details on this test can be found in the manufacturer's manual.

### *6.2. Gen-Probe Amplification Mycobacterium Tuberculosis Direct (MTD) Test*

This test differs from the original hybridization protection assay (HPA) described in Section 4 by adding the amplification procedure to it. The test developed by Gen-Probe, Inc. (San Diego, CA) is an isothermal transcription-mediated amplification (TMA) that provides about $10^9$-fold amplification of the rRNA targets in *M. tuberculosis* (55,56). It differs from the above-described PCR test by using RNA instead of DNA as a target, it excludes the temperature cycling, and the whole assay is run in a single tube. A concentrated sputum specimen, after digestion–decontamination procedure, is subjected to sonication for disruption of the bacterial cell wall and release of the rRNA, which serves as a template for the following replication.

This lysis can be performed with 0.05 ml of the homogenized sputum added to 0.2 ml of a buffer. The following are the procedures suggested by the manufacturer. The lysate (0.05 ml) is transferred to a tube containing Amplification Reagent and heated for 15 min in a heat block at 95°C, followed by cooling at 42°C for 5 min. After this, 0.025 ml of the Enzyme Mix is added and the tube is incubated for 2 h at 42°C. After addition of 0.02 ml of the Termination Reagent, the sample is ready for the previously described HPA assay.

Specific primers (small pieces of nucleic acid that initiate DNA synthesis) anneal to the rRNA template at defined sites. The Primer (rRNA hybrid) is converted to a transcription complex by reverse transcriptase. Numerous RNA transcripts are then synthesized from the transcription complex via RNA polymerase. The process then repeats automatically. The Termination Reagent destroys selected products, templates, and the product RNA. The single assay tube format, with no wash steps, combined with the use of the Termination Reagent and HPA Detection Reagent minimizes the chance of carryover contamination.

### *6.3. Strand Displacement Amplification (SDA)*

This method developed by Becton Dickinson Diagnostic Instrument Systems (Sparks, MD) is an isothermal DNA amplification method that utilizes a restriction enzyme (Hinc II) and a specific polymerase (exoklenow) (42,43). It was designed to amplify IS6110 and 16S rDNA. The SDA products are detected using a microwell chemiluminescent assay that includes simultaneous hybridization and capture of the amplified product. By incorporating a modified Hinc II restriction site into the target specific probe (i.e., IS6110), the target-specific assay is created. Upon a heat denaturing of the double-stranded DNA into single-stranded DNA, the modified target-specific probe attaches, if the target sequence is present in the sample. The assay is run at 41°C for 2 h while the polymerase makes $10^7$–$10^8$ copies of the target. The amplified materials are transferred to a microwell solid phase for capture and hybridization. Finally, a chemiluminescent substrate is added for reading.

Assay Protocol:

1. Perform standard NALC procedure
2. Transfer 100 $\mu$l of the pellet to microfuge tube
3. Wash 2× with a buffer
4. Place washed pellet in dry air oven (30 min)
5. Place on instrument (about 3.5 h):
   Instrument performs:
   - Decontamination
   - Amplification
   - Hybridize to probes
   - Chemiluminescence reaction
6. Transfer to luminometer (minute/plate)

The test is still under development for identification of the *M. tuberculosis* complex, the *M. avium* complex, *M. kansasii,* and *Mycobacterium* genus.

## 7. Cell-Wall Lipid Analysis

Cell-wall lipid analysis of mycobacteria by chromatographic methods is based on separation through varying extraction or partitioning and absorption of sample components between the mobile and stationary phase (57). Three chromatographic methods were used for mycobacterial speciation: gas–liquid chromatography (GLC) uses gas (hydrogen) in the mobile phase and liquid for the stationary phase (58,59). High-pressure (or high-performance) liquid chromatography (HPLC) uses a liquid mobile phase at high pressure to carry a sample through a column packed with particulate material for stationary phase, where the separation into components takes place (60,61). Thin-layer chromatography (TLC) utilizes liquid as the mobile phase and a solid stationary phase (62). From among these three approaches, GLC and HPLC are the most popular in clinical mycobacteriology.

### *7.1. GLC*

This method was designed for analyzing the methyl esters of relatively short-chain fatty acids in a range of 9–20 carbons in length of the bacterial wall. It has been successfully used in several laboratories (58,63,64) by employing commercially available chromatograph (Hewlett-Packard CO, Palo Alto, CA). The system is computerized by the manufacturer and the software, the Microbial Identification System, is also commercially available (Microbial ID, Inc., Newark, DE). The computerized library includes the lipid patterns of about 30 mycobacterial species and is systematically updated. Profiles of the tested cultures are compared with those in the library, and the results are presented in an actual chromatogram, as well as in a probability (%%) that the isolate belongs to a certain species.

The bacterial suspension (a harvest from solid medium) is subjected to saponification and methylation to extract fatty acids and to form their methyl esters. The components of the samples injected into the system are separated into individual compounds by partitioning between the stationary and mobile phases. Those that have the lowest affinity for the liquid stationary phase are eluted from the column first, and those that are absorbed by the liquid phase are eluted later. The results are analyzed by comparison of retention time (RT) and peak heights of individual components of the sample with the corresponding data for the known species accumulated in the computerized library. RT is the time between sample injection and elution from the column, and it characterized the sample qualitatively. Quantitative assessment is based on comparison of the peak heights and areas produced by the specific components of the tested sample in comparison with the known species.

### *7.2. HPLC*

This method can be used for the analysis of very large fatty acid molecules such as mycolic acids of mycobacteria (61,65–67). Mycolic acid extracted from saponified mycobacteria are converted to the *p*-bromophenacyl esters. The samples in a liquid mobile phase are carried at high pressure through a column packed with particulate material that represent a stationary phase, where they separate into the individual components. The analysis of the chromatogram includes qualitative analysis by comparing RT to the RT obtained with the known species, and quantitative analysis by the heights and areas of individual compounds as described above for GLC.

The method has been used for accurate identification of *M. tuberculosis, M. bovis,* the *M. avium* complex, *M. kansasii, M. marinum, M. szulgai, M. asiaticum, M. gordonae,* and *M. gastri* (61,67). The method has been recently upgraded with a fluorescens dtection system, which made sensitive enough for using the BACTEC cultures (Roberts GD et al, Clinics in Lab Med (1996) 16:603–616).

## References

1. Runyon EH (1970) Identification of mycobacterial pathogens using colony characteristics. Am J Clin Pathol 54:578–586.
2. Gangadharam PRJ, Droubi AJ (1981) Identification of mycobacteria by smear examination of the culture. Tubercle 62:123–127.
3. Dixon D, Gutherberg EH (1967) Isolation of tubercle bacilli from uncentrifuged sputum on pyruvic acid medium. Am Rev Respir Dis 96:119–122.
4. Heifets LB, Good RC (1994) Current laboratory methods for the diagnosis of tuberculosis. In: Bloom BR, ed. Tuberculosis: Pathogenesis, Protection, and Control, pp. 85–110. Washington, DC: ASM Press.
5. Heifets LB (1994) Quantitative cultures and drug susceptibility testing of *M. avium* clinical isolates before and during the antimicrobial therapy. Res Microbiol 145:188–196.
6. Kent PT, Kubica GP (1985) Public Health Mycobacteriology. A guide for the Level III Laboratory. Atlanta, GA: Centers for Disease Control.
7. Hirschel B, Chang HR, Mach N (1990) Fatal infection with a novel unidentified mycobacterium in a man with acquired immunodeficiency syndrome. New Engl J Med 323:109–113.
8. Böttger EC, Teske A, Kirschner P, Bost S, Chang HR, Beer V, Hirschel B (1992) Disseminated "*Mycobacterium genavense*" infections in patients with AIDS. Lancet 340:76.
9. Coyle MB, Carlson LC, Wallis CK, Leonard RB, Raisys VA, Kilburn JO, Samadpour M, Bottger EC (1992) Laboratory aspects of "*Mycobacterium genavense,*" a proposed species isolated from AIDS patients. J Clin Microbiol 30:3206.
10. Jackson K, Sievers A, Ross BC, Dwyer B (1992) Isolation of a fastidious *Mycobacterium* species from two AIDS patients. J Clin Microbiol 30:11.
11. Butler WR, Thibert L, Kilburn JO (1992) Identification of *M. avium* complex strains and some similar species by high-performance liquid chromatography. J Clin Microbiol 30:2698–2704.
12. Butler WR, O'Connor PO, Yakrus MA, Smithwick RW, Plikaytis BB, Moss CW, Floyd MM, Woodley CL, Kilburn JO, Vadney FS, Gross WM (1993) *Mycobacterium celatum* sp. nov. Int J System Bacteriol 43:1540–1550.
13. Butler WR, O'Connor SP, Yakrus MA, Gross WM (1994) Cross-reactivity of genetic probe for detection of *Mycobacterium tuberculosis* with newly described species *Mycobacterium selatum.* J Clin Microbiol 32:536–538.
14. Kiehn TE, Edwards FF, Brennan P, Tsang AY, Maio M, Gold JWM, Whimbly E, Wong B, McClatchy JK, Armstrong D (1985) Infections caused by *M. avium* complex in immunocompromised patients: Diagnosis by blood culture and fecal examination, antimicrobial susceptibility tests, and morphological and seroagglutination characteristics. J Clin Microbiol 21:168–173.
15. Kubica GP, Vestal AL (1961) The arylsulfatase activity of acid-fast bacilli.

Investigation of the activity of stock cultures of acid-fast bacilli. Am Rev Respir Dis 83:728–732.

16. Kubica GP, Ridgon AL (1961) The arylsulfatase activity of acid-fast bacilli. III. Preliminary investigation of rapidly growing acid-fast bacilli. Am Rev Respir Dis 83:737–740.
17. Whitehead JEM, Morrison HR, Young L (1952) Bacterial arylsulfatase. Biochem J 51:585–593.
18. Whitehead JEM, Wildy P, Engback HC (1953) Arylsulfatase activity of mycobacteria. J Pathol Bacteriol 65:451–460.
19. Wayne LG (1961) Recognition of *M. fortuitum* by means of 3-day phenolphthalein sulfatase test. Am J Clin Pathol 36:185–187.
20. Wayne GL, Dubek JR (1968) Diagnostic key to mycobacteria encountered in clinical laboratories. Appl Microbiol 16:925–931.
21. Kubica GP, Pool GL (1960) Studies on the catalase activity of acid-fast bacilli (I). Am Rev Respir Dis 81:387–391.
22. Kilburn JO, Kubica GP (1968) Reagent impregnated paper strips for detection of niacin. Am J Clin Pathol 50:530–532.
23. Konno K (1956) New chemical method to differentiate human-type tubercle bacill from other mycobacteria. Science 124:985.
24. Runyon EH, Selin MJ, Harris HW (1959) Distinguishing mycobacter by the niacin test. Am Rev Respir Dis 79:663–665.
25. Young Jr. WD, Maslansky A, Lefar MS, Kronish DP (1970) Development of a paper strip test for detection of niacin produced by mycobacteria. Appl Microbiol 20:939–945.
26. Virtanen S (1960) A study of nitrate reduction by mycobacteria. Acta Tuberc Scand 48(Suppl):1–119.
27. Wayne LD, Doubek SR (1965) Classification and identification of nitrite as a substrate. Am Rev Respir Dis 91:738.
28. Quigley HS, Elston HR (1970) Nitrite strips for detection of nitrite reduction by mycobacteria. Am J Clin Pathol 53:663–665.
29. Wayne LG (1974) Simple pyrazinamide and urease tests for routine identification of mycobacteria. Am Rev Respir Dis 109:147–151.
30. Bönike R (1958) Die differenzierung humaner and boviner tuberkelterien mit hilfe von thiophen-2-carbonsaure-hydrazid. Naturwissenschaften 46:392–393.
31. Harrington R, Karlson AG (1967) Differentiation between *M. tuberculosis* and *M. bovis* by *in vitro* procedures. Am J Vetin Res 27:1193.
32. Wayne LG, Doubek JR, Russell RL (1964) Tests employing Tween-80 as substrate. Am Rev Respir Dis 90:588–597.
33. Steadham JE (1979) Reliable urease test for identification of mycobacteria. J Clin Microbiol 10:134–137.

34. Murphy DB, Hawkins JE (1975) Use of urease test disks in the identification of mycobacteria. J Clin Microbiol 1:465–468.

35. Silcox VA, David HL (1971) Differential identification of *M. kansasii* and *M. marinum.* Appl Microbiol 21:327–334.

36. Jones WD, Kubica GP (1964) The use of McConkey agar for differential typing of *M. fortuitum.* Am J Med Technol 30:187–195.

37. Kilburn JO, Silcox VA, Kubica GP (1969) Differential identification of mycobacteria. V. The tellurite reduction test. Am Rev Respir Dis 99:94–100.

38. Silcox VA, Good RC, Floyd MM (1981) Identification of clinically significant *M. fortuitum* complex isolates. J Clin Microbiol 14:686–691.

39. Wayne LG, Doubek JR, Russell RL (1964) Classification and identification of mycobacteria tests employing Tween-80 as substrate. Am Rev Respir Dis 30:588–597.

40. Arnold LJ, Hammond PW, Wiese WA, Nelson NC (1989) Assay formats involving acridinium–ester-labeled DNA probes. Clin Chem 35:1588–1594.

41. Tenover FC (1991) Molecular methods for the clinical microbiology laboratory. In: Ballows A, Hausler WJ Jr, Herrmann KL, Isenberg HD, Shadomy HJ, eds. Manual of Clinical Microbiology, 5th ed., pp. 119–127. Washington, DC: American Society of Microbiology.

42. Spargo CA, Haaland PD, Jurgensen SR, Shank DD, Walker GT (1993) Chemiluminescent detection of strand displacement amplified DNA from species comprising the *Mycobacterium tuberculosis* complex. Molec Cell Probes 7:395–404.

43. Walker GT, Little MC, Nadeau JG, Shank DD (1992) Isothermal *in vitro* amplification of DNA by a restriction enzyme/DNA polymerase system. Proc Natl Acad Sci USA 89:392–396.

44. Brisson-Noel A, Aznar C (1991) Diagnosis of tuberculosis by DNA amplification in clinical practice evaluation. Lancet 338:364–366.

45. Cousins DV, Wilton SD, Francis BR, Gow BL (1992) Use of polymerase chain reaction for rapid diagnosis of tuberculosis. J Clin Microbiol 30:255–258.

46. De Wit D, Steyn L, Shoemaker S, Sogin M (1990) Direct detection of *Mycobacterium tuberculosis* in clinical specimens by DNA amplification. J Clin Microbiol 28:2437–2441.

47. Eisenach KD, Sifford MD, Cave MD, Bates JH, Crawford JT (1991) Detection of *Mycobacterium tuberculosis* in sputum samples using polymerase chain reaction. Am Rev Respir Dis 144:1160–1163.

48. Sjobring U, Mecklenburg M, Andersen AB, Miorner H (1990) Polymerase chain reaction for detection of *Mycobacterium tuberculosis.* J Clin Microbiol 28:2200–2204.

49. Sritharan V, Barker RH Jr (1991) A simple method for diagnosing *M. tuberculosis* infection in clinical samples using PCR. Molec Cell Probes 5:385–395.

50. Forbes BA, Hicks KE (1993) Direct detection of *Mycobacterium tuberculosis* in respiratory specimens in a clinical laboratory by polymerase chain reaction. J Clin Microbiol 31:1688–1694.

51. Boddinghaus B, Rogall T, Blocker H, Bottger EC (1990) Detection and identification of mycobacteria by amplification of rRNA. J Clin Microbiol 28:1751–1759.

52. Buck GE, O'Hara LC, Summersgill JT (1992) Rapid, simple method for treating clinical specimens containing *Mycobacterium tuberculosis* to remove DNA for polymerase chain reaction. J Clin Microbiol 30:1331–1334.

53. Portillo PD, Murillo LA, Patarroyo ME (1991) Amplification of a species specific DNA fragment of *Mycobacterium tuberculosis* and its possible use in diagnosis. J Clin Microbiol 29:2163–2168.

54. Soini H, Skurnik M, Liippo K, Tala E, Vilganen MK (1992) Detection and identification of mycobacteria by amplification of a segment of the gene coding for the 32-kilodalton protein. J Clin Microbiol 30:2025–2028.

55. Jonas V, Alden MJ, Curri JI, Kamisango K, Knott CA, Lankford R, Wolfe JM, Moore DF (1993) Detection and identification of *Mycobacterium tuberculosis* directly from sputum sediments by amplification of rRNA. J Clin Microbiol 31:2410–2416.

56. Pfyffer GE, Kissling P, Wirth R, Weber R (1994) Direct detection of *Mycobacterium tuberculosis* complex in respiratory specimens by a target-amplified test system. J Clin Microbiol 32:918–923.

57. Susser M, Wichman MD (1991) Identification of microorganisms through use of gas chromatography and high-performance liquid chromatography. In: Ballows A, Hausler WJ Jr, Herrmann KL, Isenberg HD, Shadomy HJ, eds, Manual of Clinical Microbiology, 5th ed., pp. 111–118. Washington, DC: American Society of Microbiology.

58. Tisdall PA, DeYoung DR, Roberts GD, Anhalt JR (1982) Identification of clinical isolates of mycobacteria with gas-liquid chromatography: a 10-month follow-up study. J Clin Microbiol 16:400–402.

59. Jiminez J, Larsson L (1986) Heating cells in acid methanol for 30 minutes without freeze-drying provides adequate yields of fatty acids and alcohols for gas chromatographic characterization of mycobacteria. J Clin Microbiol 24:844–845.

60. Butler WR, Ahearn DG, Kilburn JO (1986) High-performance liquid chromatography of mycolic acids as a tool in the identification of *Corynebacterium, Nocardia, Rhodococcus,* and *Mycobacterium* species. J Clin Microbiol 23:182–185.

61. Butler WR, Kilburn JO (1988) Identification of major slowly growing pathogenic mycobacteria and *Mycobacterium gordonae* by high-performance liquid chromatography of their mycolic acids. J Clin Microbiol 26:50–53.

62. Brennan PJ, Heifets M, Ullom BP (1982) Thin layer chromatography of lipid antigens as a means of identifying nontuberculous mycobacteria. J Clin Microbiol 15:447–455.

63. Larsson LJ, Jiminez J, Sonesson A, Portaels F (1989) Two-dimensional gas chromatography with electron capture detection for the sensitive determination of specific mycobacterial lipid constituents. J Clin Microbiol 27:2230–2231.

64. Maliwan NR, Reid W, Pliska SR, Bird TJ, Zvetina JR (1988) Identifying *Mycobacterium tuberculosis* cultures by gas–liquid chromatography and a computer-aided pattern recognition model. J Clin Microbiol 26:182–187.

65. Lévy-Frébault V, Daffé M, Restrepo E, Grimont F, Grimont PA, David HL (1986) Differentiation of *Mycobacterium thermoresistable* from *Mycobacterium phlei* and other rapidly growing mycobacteria. Ann Pasteur Microbiol 137A:143–151.

66. Butler WR, Kilburn JO, Kubica GP (1987) High-performance liquid chromatography analysis of mycolic acids as an aid in laboratory identification of *Rhodococcus* and *Nocardia* species. J Clin Microbiol 25:2126–2131.

67. Butler WR, Jost KC, Kilburn JO (1991) Identification of mycobacteria by high-performance liquid chromatography. J Clin Microbiol 29:2468–2472.

# 12

# Tuberculosis Laboratory Procedures for Developing Countries

*Isabel N. de Kantor and Adalbert Laszlo*

## 1. Introduction

Two of the main requirements for a successful National Tuberculosis Program (NTP) in developing countries have been clearly defined (1):

1. Government commitment to a tuberculosis (TB) program aiming at nationwide coverage, conceived as a permanent health system activity, integrated into the existing health structure with effective technical leadership from a central unit
2. Case detection and treatment follow-up based on bacteriological examinations

Tuberculosis bacteriology—a fundamental element of the NTP—must interrelate closely with the administrative, epidemiological, and clinical components of the program. Bacteriological diagnostic services must be developed concurrently with the other activities of the NTP and should be integrated in the general health services of the country to attain maximum coverage (2).

Tuberculosis case detection in a developing country, especially in rural and poor urban areas, is based on sputum smear examination performed on any person presenting to a health facility with respiratory symptoms, henceforth referred to as suspect of tuberculosis or ST. A ST is defined as a person who presents productive persistent cough for 3 consecutive weeks or more. This operational definition is very useful for TB case detection in a developing country because these individuals are, without further selection, referred to the appropriate health service for confirmation of diagnosis primarily by sputum smear microscopy. This strategy is called "passive case finding" (3).

Other coincident symptoms like fatigue, loss of appetite and weight, night sweats, fever, shortness of breath, and chest pain can also be present. These symptoms are easily detectable but not specific of TB. The detection of acid fast bacilli (AFB) by smear examination of sputum specimens is a highly specific diagnostic tool for pulmonary tuberculosis.

In some health institutions, case-finding among STs usually begins with a chest x-ray and is followed by clinical examination. Bacteriological tests are then proposed to those individuals presenting with a chest x-ray suggestive of TB. There is, therefore, preselection prior to sputum smear examination. However, without bacteriological confirmation, x-ray examination and clinical assessment alone cannot be the basis for definitive diagnosis of pulmonary TB. Only by finding the sources of infection in the community, mainly through sputum smear examination, and by rendering these patients noninfectious by adequate treatment can illness be shortened and the chain of TB transmission interrupted. So, for the purposes of a NTP, a case of tuberculosis is any individual discharging tubercle bacilli, especially if the bacilli in the sputum can be seen by direct microscopy (smear positive patients) (3).

## 2. Structure of Laboratory TB Diagnostic Services

General health services in developing countries should be organized into three main levels:

The national or central level
The intermediate or regional level
The peripheral or local level

Members of the managerial team at the national and intermediate levels should be capable of supervising and monitoring all the activities of the NTP. This means that laboratory workers should be involved—at their level—in practical epidemiology and in the program decision-making process (3).

It is at the peripheral level that the most direct contact between patients and health workers takes place. Basic information on the extent of the local TB problem and on the efficacy of the program activities is also obtained at this level. This information flows upward, to the regional and central levels for analysis and evaluation, and therefore nurtures the whole system.

### *2.1. The Laboratory Network*

The laboratory network organization takes into account the degree of technical complexity of the TB diagnostic procedures required and reflects the overall structure of the general health services in the country (2,3–6). It ideally comprises the following levels:

Central or reference Type I laboratory
Intermediate or regional Type II laboratories
Peripheral Type III laboratories—including specimen collection units

Type I laboratories are usually located in the capital city of the country. They can be highly sophisticated and are often closely associated with university teaching hospitals. Sometimes, especially in large developing countries, two or more central laboratories become established in regions reaching concurrent, albeit separate, socioeconomic development. Each central laboratory has its own network of regional and peripheral laboratories; good coordination between networks is essential.

Central laboratories perform all types of bacteriological analyses: smear microscopy, culture, and drug-susceptibility tests, as well as basic differentiation of mycobacteria. In addition, they supervise the laboratory network and conduct operational research (i.e., coordination of surveys on drug resistance, quality assurance programs, train laboratory personnel, etc.).

Type II regional laboratories perform smear microscopy and culture, supervise the peripheral laboratory network, and facilitate training of regional staff.

Type III peripheral laboratories perform smear examination and select samples in accordance to national guidelines to be referred to the regional laboratory for culturing. There should ideally be one smear microscopy center per 50,000–70,000 population. Specimens collection units are usually situated in primary health care clinics where clinical specimens are collected and referred as soon as possible to the nearest Type III laboratory for smear microscopy. The practice of preparing smears at these clinics and referring slides to the local laboratory for staining and for microscopic examination should be avoided. Tubercle bacilli can remain viable on unfixed and even on fixed smears. On the other hand, preparation of smears by personnel who are not adequately trained in this procedure and who work in unsafe conditions, can lead to unreliable test results, to increased risk of infection of laboratory personnel, and to contamination of clinical specimens (2).

## 3. Basic Bacteriological Diagnostic Procedures

### *3.1. The Specimen (2,5–7,10,11,19).*

"A laboratory test is no better than the specimen, and the specimen no better than the manner in which it was collected. A good specimen is one taken from the lesion being investigated, obtained in sufficient quantity, placed in a suitable container, well identified, preserved, and correctly transported" (2).

A good sputum sample is the one raised from the bronchial tree following a coughing effort, not one that is discharged from the pharynx or obtained by aspiration of nasal secretions or from saliva. It is recommended that two or three sputum specimens be taken from a ST individual, one at the time of consultation (spot specimen) and a second one to be collected by the patient at home (overnight or early morning specimen). The third sample might be collected at the time the second specimen is brought to the clinic.

### 3.2. Quality of the Sputum Specimen

Enough volume (3–5 ml) of a specimen containing mucopurulent particles, not merely saliva, should be obtained. This type of sputum sample is not often produced in a single coughing effort. Clear instructions and sufficient time should be given to the patient to produce expectoration through deep coughing. The risk of infection is high in the immediate surroundings of a coughing TB patient. A suitable, secluded, and well-ventilated area should be set aside for sputum collection. The patient should be instructed to expectorate carefully so as to not contaminate the outside surface of the container (7).

### 3.3. The Specimen Container

A widemouthed (50-mm diameter) 50-ml capacity container is recommended for sputum collection. The container, preferably made of a translucent, disposable and combustible plastic material, should have a screw cap to allow an airtight seal. The patient's identification must be written indelibly on the container, never on the lid.

### 3.4. Specimen Collection

Sputum is the best clinical sample for the diagnosis of pulmonary tuberculosis and is relatively easy to obtain. Microscopic smear examination of sputum is a relatively simple, rapid, and inexpensive procedure. When AFB are detected in a sputum specimen by smear examination, the diagnosis of active infectious pulmonary tuberculosis is made and chemotherapy must be started at once.

In some patients—children or adults who swallow their bronchopulmonary secretions—sputum is difficult to obtain. In these cases, other, rather simple procedures should be used to obtain these secretions: laryngeal swab collection, sputum induction, or gastric lavage. When bronchoscopy is indicated due to other medical reasons, a bronchial washing or aspiration sample should be collected at the same time for smear microscopy. The culture of these clinical specimens must be performed in a Type II or I laboratory.

#### LARYNGEAL SWABS

The purpose of their use is to collect mucous material originating in the bronchi and gathering on the larynx. The amount collected is always small and should be entirely devoted to culture. Because cotton fibers can cause false positivity in smears, the microscopic examination of laryngeal swabs is not recommended.

### INDUCED SPUTUM

It is induced by a warm, sterile hypertonic saline solution which is aerosolized by a nebulizer into the upper respiratory tract. A first sputum is collected immediately after aerosolization. The patient is instructed to collect a second sample produced at home. Because induced specimens resemble saliva, they should be appropriately labeled as such. The risk entailed in the production of aerosols during this procedure is high. Therefore, this procedure should only be attempted when the appropriate conditions of containment, ventilation, and protection of health care workers exist.

### GASTRIC LAVAGE

As in the case of laryngeal swabs, this specimen contains usually only a scant number of bacilli. In addition, food-derived mycobacteria could be present, yielding false-positive smear microscopy results. Therefore, direct microscopic examination of sediment should be avoided. The specimen should be obtained in the morning, before the patient eats or gets out of bed if possible, in order to collect respiratory secretions swallowed during the night. In breast-fed babies, the specimen should be collected between feedings.

### BRONCHIAL LAVAGE

Bronchoscopy is often used to facilitate rapid, differential diagnosis of pulmonary tuberculosis. Its benefits must be weighed against the cost of the procedure, the inconvenience to the patient due to the invasiveness of the procedure, the anesthesia, and the risks of TB transmission by aerosols.

This procedure should only be performed by trained personnel. Bronchial secretion aspirates are sent to the laboratory for bacteriological examination. The search for mycobacteria must always include the culturing of the specimen. In addition to the secretions collected during bronchoscopy, the patient should be instructed to collect a sputum sample the following morning.

### EXTRAPULMONARY SPECIMENS

Most of these clinical samples contain only small amounts of bacilli (paucibacillary specimens). Smear examination is therefore rarely positive and culturing should always be attempted. Other mycobacteria, mostly saprophytes, might be present in the specimen and yield false-positive results in smear microscopy.

These samples are grouped into two categories: nonaseptically collected and presumed contaminated (urine, pus from fistulized abscesses) and aseptically collected presumed not contaminated (pleural, ascitic, pericardial, synovial and spinal fluids, tissue biopsies). Blood samples from AIDS patients are often submitted for the culturing of mycobacteria.

Specimens presumed contaminated should be processed for culture following decontamination.

### URINE

A minimum of three and a maximum of six specimens should be collected on different days, because the elimination of bacilli could be a discontinuous process. Early morning midstream urine is collected and referred immediately to the laboratory.

### PUS FROM FISTULIZED ABSCESSES

It is obtained by aspiration of a suitable volume. Pus swabs should be avoided.

### PLEURAL, ASCITIC, AND OTHER FLUIDS

These specimens often contain very few bacilli. They are collected in sterile containers under aseptic conditions by the medical staff. An anticoagulant, like a 10% solution of sodium citrate (three drops per 10-ml sample volume) or heparin, could be added.

### CEREBROSPINAL FLUID (CSF)

This specimen is obtained by punction, carried out by the medical staff. The fluid should be aseptically collected into a sterile container, without the addition of anticoagulant. The centrifuged sediment is examined by smear microscopy—carefully and painstakingly—considering the urgency of diagnosis and the often paucibacillary character of the sample.

### BLOOD

Isolation of mycobacteria from blood has been reported in cases of disseminated disease associated with AIDS. A rather simple method for treatment of these specimens which can be performed in Type II and III laboratories is as follows.

About 8.5 ml of blood are aseptically collected in a tube containing 1.5 ml of a 0.35% sodium polyanetholsufonate (SPS) solution; the contents of the tube are then mixed gently. SPS is recommended because of its low toxicity for mycobacteria and its leukocytes lysing ability. If SPS is not available, a 10% solution of sodium citrate or heparin can be used instead, as described above.

The culture of specimens presumed uncontaminated, either CSF, other fluids, or blood, follows a similar procedure. The original specimen is split into two aliquots. Both aliquots are centrifuged, the supernatants are discarded, and one sediment is inoculated onto the culture media. The remaining one is kept under refrigeration. After 48 h of incubation at 37°C, the cultures are examined. If contamination is detected, these cultures are autoclaved and discarded. The remaining sediment is decontaminated and inoculated onto freshly prepared culture media.

If no contamination was detected in the first control at 48 h incubation, the remaining sediment is also directly inoculated onto other culture medium slants. These sediments should be examined by smear microscopy prior to culture.

### *3.5. Preservation and Transportation of Specimens*

Specimens collected in peripheral clinics or local collection units might need to be transported to a Type III laboratory for smear examination, and in some cases, to a Type II laboratory for culture. These laboratories are often situated in urban areas, sometimes relatively far from the collection units. Often, transportation might not be available on a daily basis, and in this case, the unit should have enough refrigerator space for the storage of specimens. Refrigeration inhibits the multiplication of saprophytic microorganisms, which are always present in sputum.

Where refrigeration facilities are not available, a cool place protected from light should be set aside for this purpose. The maximum allowable time lapse between collection of the specimen and its examination is 1 week (2–11). Saprophytic microorganisms thrive at room temperature (20–30°C) and their proteolytic activity liquefies specimens in a few days. As a result, the selection of purulent particles for smear examination becomes difficult. Liquefied sputum samples should be centrifuged to concentrate possible mycobacteria for smear and culture. The need for centrifugation is a complicating factor in performance of smear microscopy. Liquefaction—a sort of spontaneous homogenization—does not damage the acid-fastness of mycobacteria; however, prolonged periods at room temperature (i.e., more than 6 days) usually kill most of the mycobacteria present in the specimen. Therefore, culturing could yield false-negative results. In addition, the proportion of contaminated cultures increases because of the high content of saprophytic microorganisms in these specimens.

If sputum specimens are solely intended for microscopy, five drops of a 5% phenol solution or 2–3 ml of a commercial concentrated hypochlorite solution can be added to the specimen after collection to delay multiplication of saprophytes.

## TRANSPORTATION OF SPUTUM SPECIMENS

For specimens to be transported, they need to be protected from excessive heat and from sunlight, and the shipping containers must be tightly sealed to prevent leakage. A wooden, metal, or even thick cardboard box with partitions can be used to transport specimens. The space between containers should be filled with an absorbent material such as newspaper or other similar absorbing material to prevent breakage and the spilling of contents. Each container should also be enclosed in a watertight plastic bag sealed with tape. Each parcel should be accompanied by the appropriate request forms and a list of specimens.

Further information on the rules and regulations for the safe postal transportation of pathological specimens including cultures can be found elsewhere (4,7,8)

## CONTROL ON RECEIPT OF SPECIMENS

When a shipment of potentially infectious material such as sputum specimens is received in the laboratory, a careful inspection of containers is performed. Special attention should be given to matching the sample identification and the names on the attached list.

This procedure should be carried out in a biosafety hood; however, if such a hood is not available, a work bench can suffice, provided that its surface is carefully treated with an appropriate disinfectant after use.

The outside of each container should be disinfected with cotton or paper towels soaked in a 5% phenol solution. In case of massive leakage, the box should be autoclaved or burned.

The addition of a 10% trisodium phosphate solution (a 23% solution of trisodium phosphate · 10 $H_2O$ is used) or a 1% solution of cetyl pyridinium bromide or chloride equal to the volume of the specimen is an acceptable method to limit multiplication of saprophytes in sputum samples destined to be cultured. In this case, specimens are centrifuged upon arrival in the laboratory. The sediments could either be directly inoculated onto culture media or preferably submitted to a second round of decontamination (e.g., with a 4% solution of sodium hydroxide) according to Petroff's method (see below).

Nevertheless, the best strategy to avoid massive contamination without decreasing significantly the yield of culture is to refrigerate during transportation and to limit the time elapsed between collection and culture.

### 3.6. Direct Smear Microscopy Examination

Smear microscopy examination of a sputum specimen is probably the most efficient diagnostic test for pulmonary tuberculosis. As previously mentioned, it is a robust, simple, rapid, specific, and inexpensive test which can efficiently detect infectious TB cases in developing countries. It remains the cornerstone for case-finding in the framework of the NTP. It fulfills two important requirements: (a) It can be performed by staff with basic skills and training in very simple laboratory settings, with minimal facilities and (b) it detects the actual sources of TB infection in the community. Once these cases are detected and inactivated by chemotherapy, the main chain of TB transmission is interrupted.

Case-finding and treatment go together, the former has no meaning without the latter (3). Periodic smear examination of patients already under treatment is the basic method for monitoring progress of chemotherapy and cure.

#### PREPARATION OF SPECIMENS FOR MICROSCOPY (2,6–8,10,11)

This procedure involves the risk of aerosol formation and should be carried out, whenever possible, in a biosafety hood.

The following basic precautions should be followed when preparing smears, whether the safety hood or on the workbench. A metal tray placed on the work surface is covered with absorbent paper, dampened with a 5% phenol solution. If a tray is not available, a double sheet of full-size newspaper can be used. The tray or the sheet of paper delimits the contaminated area. The specimen containers are placed on this surface, in the order in which they will be processed. Each container is cautiously opened and the sputum specimens are carefully scrutinized. The technologist uses gloves or a paper soaked in a 5% phenol solution to hold the specimen container. The physical appearance of each sample is recorded (e.g., mucoid, mucopurulent, purulent, blood-stained sputum, or saliva). Only new slides should be used. If the slides are not precleaned, they can be immersed in 95% alcohol, individually removed and dried with a paper towel before use. Each slide is identified with a permanent marking device, diamond-pointed stylus, or grease pencil. In the latter case, the identifier is placed on the back of the slide to avoid erasure during the staining procedure. Whenever possible, purulent particles in the sputum are selected for transfer onto the slide. They are picked out of the container using a wooden applicator and smeared onto the slide. Once the smear is ready, the applicators are discarded into a receptacle containing a 5% phenol solution, which is then autoclaved or burned.

Wire loops could be used instead of wooden applicators, but the potential for creating aerosols is greater. If a wire loop is used, it should be cleaned in a sand flask containing alcohol before being flamed with a Bunsen burner.

Smears are dried in the "contaminated area," at room temperature, for 15–20 min, and then heat-fixed by passing the slide rapidly several times through the flame of a Bunsen burner or the alcohol lamp. Alternatively, slides could be fixed on an electric heater at 65°C for 2 h.

Smears, fixed or unfixed, are never left exposed overnight, as they could contain live bacilli. Preferably, staining should be carried out immediately after fixing. If the smears cannot be stained immediately, they should not be left on the work bench but kept in slide boxes. The Ziehl–Neelsen (ZN) method of staining is recommended. For the preparation of reagents needed in this procedure please, refer to Appendix I.

### STAINING TECHNIQUE (2,5,6,8–11)

Slides are positioned horizontally over parallel glass rods or a metallic slide rack placed over a wash basin. The whole surface of each slide is covered with filtered carbol fuchsin solution. The slides are gently heated from underneath with either the flame of a Bunsen burner or that of an alcohol lamp until emission of vapor occurs. A wad of cotton-wool soaked in alcohol fixed on the end of a metal rod or a strong stick of wood can also be used. Fuchsin should not boil or dry out on the slides. If the stain spills over, more fuchsin should be added and heated again.

The stain is allowed to act for 3–5 min, before it is removed by rinsing each slide gently with running tap water. The decolorizing solution is poured on the slides and allowed to stand for 2–3 min until no more pink color is observed. After rinsing again with running tap water and draining, the slides are covered with the counterstaining solution for 1–2 min and rinsed. Slides are placed on absorbent paper, smear face up, and left to dry at room temperature before examination by microscopy.

### EXAMINATION BY MICROSCOPY

The stained smear is examined with an oil-immersion objective (100×) and an eyepiece (8× or 10×). A drop of immersion oil is placed on the slide; care should be taken not to touch the slide with the oil applicator to avoid possible contamination of the immersion oil and subsequent carryover of AFBs to the next slide. Tubercle bacilli appear like fine little red rods, slightly curved, standing out clearly against a blue background.

Examination by microscopy should follow a uniform pattern; for example, reading from left to right of the smear a minimum of 100 effectual fields before calling the slide negative. An effectual microscopic field is one in which cellular elements of bronchial origin are observed.

The number of fields to be observed will vary according to the number of AFB seen on the smear:

(a) A minimum of 100 fields should be examined when no AFB or less than one AFB per field is observed on average.
(b) Only 50 fields need to be read when an average of 1–10 AFB per field are observed.
(c) Only 20 fields need to be read when an average of more than 10 AFB per field are observed.

Following the examination by microscopy, the immersion lens is cleaned with soft tissue paper or a piece of clean cotton to remove the remaining oil.

## REPORTING OF SMEAR EXAMINATION RESULTS

The number of bacilli in a specimen relates to the degree of infectivity of the patient, as well as to the severity of disease. In the case of patients undergoing treatment, a quantitative report is relevant to treatment follow-up.

The following scale is recommended for reporting results:

Negative (−): No AFB per 100 examined fields
Positive (+): An average of less than one AFB per field in 100 examined fields
Positive (+ +): An average of 1–10 AFB per field in 50 examined fields.
Positive (+ + +): An average of more than 10 AFB per field in 20 examined fields.

In cases where only 1–4 AFB are observed in 100 examined fields, it is recommended that 200 additional fields be read and/or prepare another smear from the same specimen. If the results remain unchanged, record the exact figure in the laboratory book; consider the specimen negative but request a new specimen from the patient. Whenever possible, a culture should be performed as well.

## CONSERVATION AND DISPOSAL OF EXAMINED SLIDES

All the positive and a percentage of negative slides should be retained for the period of time stipulated in the National Tuberculosis Control program manual for the purpose of quality control by the corresponding higher-level supervisory laboratory. The rereading of smears is a form of indirect supervision. Results might be questioned, and it is advisable that all positive readings be confirmed by a second reader. Only when these requirements are met can the slides be disposed of.

**EXAMINATION BY FLUORESCENCE MICROSCOPY**

In fluorescence microscopy, mycobacteria are stained by a fluorochrome, like auramine, which fluoresces (i.e., emits yellow light when exposed to ultraviolet (UV) or blue incident light (5,8,10). This technique is an alternative to ZN staining. Its main advantage is that a low-power objective can be used and, therefore, a larger area of the smear can be scanned in the same amount of time. Fluorescence microscopy would be the method of choice when more than 50 specimens are examined daily.

Nevertheless, fluorescence microscopy is scarcely used in developing countries because it requires a specially adapted microscope and a source of UV light, which is rather expensive and often unavailable. Its use in developing countries could be contemplated at the central laboratory level, sometimes at the intermediate laboratory level, but never at the peripheral level. Good laboratory practice requires that a positive fluorescence microscopy result be confirmed by restaining the slide by ZN.

### *3.7. Culturing of Clinical Specimens*

The culturing of clinical specimens for TB is a direct diagnostic method for TB and is more sensitive than smear microscopy. Its specificity is almost absolute and it is the gold standard of diagnostic tuberculosis bacteriology. Its cost and the complexity of this technique are also higher than those of the simple and inexpensive ZN smear microscopy. The time needed to obtain primary isolation by culture is between 3 and 6 weeks.

Sputum smear microscopy is well suited for case-finding in simple dispensaries and health centers. Culturing capacity in developing countries is usually limited to one or to a few laboratories. Specimens must be shipped from peripheral clinics to these laboratories, often located in main urban areas. As a consequence, in developing countries, culture for TB diagnosis is restricted to the following priority situations:

a. Extrapulmonary specimens: urine, pleural, cerebrospinal fluid (CSF), blood, pus, biopsies, and bronchopulmonary secretions other than sputum (i.e., laryngeal swabs, bronchial washing, and gastric lavage). Specimens collected from children, either pulmonary or extrapulmonary, are also considered priorities for culture.
b. Cases of clinically and radiologically suspected TB, where the sputum specimen's smear examination has proved repeatedly negative for AFB.
c. In retreatment, default, or treatment failure cases who continue eliminating bacilli while receiving chemotherapy or after completion of treatment. In these cases, drug-susceptibility tests (DST) are performed on mycobacteria isolated

by culture in order to determine whether these bacillary populations would present significant resistance to antituberculosis drugs used in treatment.

d. In epidemiological surveys on initial or acquired drug resistance in TB. In these surveys, cultures and DST are carried out on AFB-positive sputum specimens obtained from pulmonary TB before or after treatment.

## DECONTAMINATION PROCEDURES (2,7,8)

Specimen decontamination and culturing procedures should be performed in a microbiological biosafety cabinet or in an area meeting basic safety rules and recommendations given below (see Section 5.1). Sputum, urine, and most specimens submitted for culture usually contain microorganisms other than mycobacteria. These other microorganisms can grow on tuberculosis-dedicated culture media in 1 or 2 days. If these specimens were directly inoculated onto the culture medium, they would totally overgrow mycobacteria in a few days.

*Mycobacterium tuberculosis* complex bacilli are slow growers. This means that they usually need 2–3 weeks of incubation at 37°C to produce visible colonies on egg-containing or other solid media.

In order to isolate eventual mycobacteria present in the sample, it is necessary to eliminate contaminants before inoculating specimens on the medium. In addition, in the case of sputum, the specimen must be homogenized to free bacilli from mucus. Homogenization and decontamination prior to culture can be carried out either with "soft" reagents, which have minimal effect on mycobacteria but also weak action against other contaminant microorganisms, or with "hard" reagents, which kill almost all contaminants but also eliminate a proportion of the existing mycobacteria. Thus, specimen decontamination is a balancing act between maintaining mycobacterial viability and killing contaminating flora.

The requirement for centrifugation is one of the main obstacles for the culturing of mycobacteria in developing countries. Successful culturing of mycobacteria can be seen as a balancing act between centrifugal efficiency and digestant toxicity and the lethal effect of temperature created by friction during centrifugation.

There is a relationship among relative centrifugal force (RCF), revolutions per minute (rpm), and the radius of the centrifuge head. It is illustrated by a nomogram, where the rpm required to attain desired RCF can be determined (8). The relationship between RCF and rpm is given by the following equation: RCF (**g**) = 1.2 $R$(rpm/1000) $\times$ 2; where $R$ is the radius from the center of the rotating head of the centrifuge to the bottom of the spinning centrifuge tube expressed in millimeters (8,10). It is important to keep the spinning time low (15 min), the RCF high (3000 $\times$ **g**), and the temperature low (20°C) to obtain 95% sedimentation and good viability. The use of angle head rotors minimizes heat buildup due to air friction.

Laboratories in developing countries should select methods for decontamination of specimens taking into consideration efficacy, cost, simplicity of procedure, and availability of reagents. There are many different methods in use throughout the world, but only a few can be considered appropriate within the framework of a NTP in a developing country.

The most commonly used is the Petroff's sodium hydroxide (NaOH) method which consists of treating the specimen with a 4% (1 *N*) sodium hydroxide (NaOH) solution. Because of its simplicity and low cost, it remains the most common choice of laboratories in developing countries where a centrifuge capable of providing a relative centrifugal force (RCF) of 2000–3000 × **g** is available.

Equivalent volumes of the sample and 4% NaOH solution are mixed in a screw-cap tube. This mixture is incubated at 37°C for 15 min, with occasional shaking.

In strongly contaminated samples, incubation time could be extended for an additional 5–10 min; one should, however, keep in mind that the digestion–decontamination procedure should be as gentle as possible (i.e., compatible with a contamination rate not in excess of 5%). After that, the resulting suspension is centrifuged at 2000–3000 × **g** for 15 min. The supernatant is poured off, 4 ml of sterile distilled water are added, and the resultant suspension is centrifuged again at 2000–3000 × **g** for 10 min. The supernatant is poured off, and the sediment, diluted in 1 ml of water, is inoculated onto two slopes of Loewenstein–Jensen medium and onto at least one slope of 0.5% pyruvate-containing egg medium (Stonebrink medium). Pyruvate stimulates growth of *M. bovis.* Inoculated tubes are incubated at 37°C.

## DECONTAMINATION WITHOUT CENTRIFUGATION

If no adequate equipment is available for centrifugation, an alternative method for decontamination without centrifugation could be used. As a rule, flocculation methods are less sensitive than those using centrifugation. Differences in sensitivity as measured by percentage culture positivity obtained from paucibacillary specimens are evident when compared to decontamination followed by centrifugation. However, if the RCF obtained by centrifugation are lower than 2000 × **g,** the actual sensitivity of the Petroff's method is decreased and there is little difference between these methods. The solution used in the flocculation method as well as the procedure can be found in Appendix I.

## CULTURE MEDIA

Whereas a variety of growth media are available for growing mycobacteria, it is recommended that at least two be used for the purpose of primary isolation. The two most convenient ones for use in developing countries are Loewenstein–Jensen and Stonebrink media (see Appendix I for media formulation).

**CULTURE READING SCHEDULE**

Inoculated tubes must be examined after 48 h incubation to ensure that the inoculation liquid has completely evaporated. Stoppers are then firmly tightened to prevent desiccation of the medium during incubation. Slopes are checked for appearance and possible contamination. Subsequent readings are made at 1, 2, 4, and 8 weeks after inoculation. Positive cultures are reported without delay, whereas negative results might only be reported after 8 weeks of incubation.

**REPORTING OF CULTURE RESULTS**

The following scale is recommended:

| | |
|---|---|
| Positive (+ + +) | Confluent colonies |
| Positive (+ +) | Isolated colonies, over 100 |
| Positive (+) | 20–100 colonies |
| Positive (number) | 1–20 |
| Negative (−) | No colonies observed |
| Contaminated (C) | Contaminated culture |

The culturing of mycobacteria clears the way for further characterization of the isolate by drug-susceptibility testing and differentiation.

### *3.8 Drug Resistance in Tuberculosis*

The emergence of significant levels of drug resistance in tuberculosis can be a major obstacle for the implementation of an effective NTP. It is, therefore, imperative that high-prevalence countries possess the capability to measure drug resistance in the laboratory. This capability—because of the technical complexity of the testing procedures—must reside at the central TB laboratory level, except in very large countries where strategically located large regional labs under the direct and strict supervision of the central lab could also perform the task.

Within the framework of a developing country's NTP, the laboratory measurement of TB drug resistance has two main purposes: (1) epidemiological surveillance and (2) planning of wide-scale chemotherapy. Its use under program conditions in the management of the individual patient can only be justified in the case of treatment failure or relapse.

Standard drug regimens used in high-prevalence countries for the treatment of newly diagnosed cases include isoniazid (H), thiacetazone (Tb1), and streptomycin (S) with ethambutol (E) as a substitute drug in cases of streptomycin intolerance. The more effective short-course chemotherapy (SCC) regimens incorporate rifampicin (R) and pyrazinamide (Z) in addition to the above-mentioned drugs. There is a definite worldwide trend to implement short-course chemotherapy regimens because in the long term, it is more cost-effective than the initially less expensive standard drug regimens. Retreatment of smear-positive relapses and failure cases also relies on the same six drugs, except for those cases of multiple (three to four drugs) resistance which under program conditions prevailing in most developing countries are practically untreatable due to the high cost or the unavailability of retreatment drugs.

Resistance to H or R is a serious situation because these two drugs are the most potent antituberculosis weapons in our present armamentarium. The loss of one or both of these drugs curtails the effectivity of drug regimens because the remaining drugs have more severe side effects resulting in high mortality rates, especially among HIV/AIDS patients. R and H are the two drugs which have the lowest probability of developing resistances of $10^{-8}$ and $10^{-6}$, respectively, and for which the measurement of resistance is the most reliable. Resistance to S, E, Z, and Tb1 is more difficult to measure reliably.

Drug susceptibility tests for mycobacteria are by far the most difficult of laboratory tests to standardize. The universal adoption of standardized methodology would offer guidance to the clinician in the treatment of patients, would enable accurate assessment of the problem of TB drug resistance throughout the world, and would greatly facilitate the interpretation of published data which are, more often than not, difficult to compare.

## DRUG RESISTANCE DETERMINATION

From the bacteriological point of view, drug resistance can be defined as a "decrease in susceptibility of sufficient degree to be reasonably certain that the strain concerned is different from a sample of wild strains of human type that have never come into contact with the drug" (13,14). When such a resistance is demonstrated in the laboratory, clinical response to treatment will usually diminish.

The following three conventional drug-susceptibility testing methods are used throughout the world to measure drug resistance in *M. tuberculosis* (14):

*The Absolute Concentration Method.* Originally described by Meissner (15), this test determines the minimal inhibitory concentrations of H and S by the inoculation of control and drug-containing media with a carefully controlled inoculum of *M. tuberculosis.* Media containing several sequential twofold dilutions of each drug are used and resistance is indicated by the lowest concentration of the drug which will inhibit growth, which is defined as 20 colonies or more at the end of 4 weeks.

A reliable test can only be obtained if the inoculum size and the critical drug concentrations are standardized in each laboratory by reference to "wild"-type strains. Because of the strict inoculum standardization required, this technique cannot be used as a direct test.

*The Resistance Ratio Method.* This method, a variant of the absolute concentration method, was first introduced by Mitchison in 1954 (16) to prevent variation of minimal inhibitory concentrations (MICs) when a strain of *M. tuberculosis* was tested on different batches of medium because of fluctuation in the degree of inactivation of S by the egg components of the medium during inspissation. Intralaboratory and interlaboratory variation between media batches with regard to the drug-susceptibility testing of streptomycin and other drugs was corrected after comparing the resistance of the wild strains with reference strain H37Rv tested in parallel on the same batch of medium. This method was extensively used in cooperative studies carried out by the British Medical Research Council in many countries of Africa and Asia.

Growth is defined as the presence of 20 or more colonies at the end of 4 weeks. The resistance ratio is defined as the minimal concentration inhibiting growth of the test strain divided by the minimal concentration inhibiting growth of the standard susceptible strain in the same set of tests. A culture showing a resistance ratio of 2 or less is defined as susceptible, whereas a ratio of 8 or more denotes resistance.

Acceptable performance of this test depends on the standardization of the inoculum size and the use of a standard strain, but the critical concentration need not be determined because the susceptible control is given by the standard strain. Because the recommended inoculum size ($10^5$ bacilli) can, in fact, contain $10^3$–$10^7$ bacilli, it is easy to see that depending on the relative richness of test versus control inocula, the same strain could be classified as resistant or susceptible. This technique, as the previous one, does not lend itself to direct testing because of the need for inoculum size standardization. It uses slightly more medium than the absolute concentration method and requires retesting of doubtful results more often.

*The Proportion Method.* Originally described by Middlebrook and Cohn in 1958 (17), it was later refined by Canetti et al. (14,18) at the Pasteur Institute in Paris. This technique has gained acceptance in all the countries of the American Region as well as in many other countries throughout the world. In this method, the ratio of the number of colonies growing on drug-containing medium to the number of colonies growing on drug-free medium indicates the proportion of drug-resistant bacilli present in the bacterial population. A high and a low dilution of the inoculum are planted on the media so that isolated (i.e., countable colonies) can be obtained with at least one of the dilutions. From these bacterial colony counts, the proportion of mutants resistant to the drug concentration tested can be determined and expressed as a percentage of the total number of viable colony-

Table 12.1 Drug-resistance criteria for *M. tuberculosis*

| Drug | Concentration[a] (mg/l) | Critical Proportion (%) |
|---|---|---|
| Isoniazid | 0.20 | 1 |
| PAS sodium salt | 0.50 | 1 |
| Dihydrostreptomycin sulfate[b] | 4.00 | 1 |
| Rifampicin | 40.00 | 1 |
| Ethambutol dihydrochloride | 2.00 | 1 |
| Ethionamide | 20.00 | 10 |
| Kanamycin sulfate | 20.00 | 10 |
| Thioacetozone | 2.00 | 10 |

[a]Concentration of drugs in L-J medium prior to inspissation.
[b]Dihydrostreptomycin sulfate must always be used.

Table 12.2 Potency of pure drugs

| Drugs | Microgram of active substance per milligram |
|---|---|
| Isoniazid | 1000 |
| PAS sodium salt | 877 |
| Dihydrostreptomycin sulfate | 800 |
| Rifampicin | 1000 |
| Ethambutol dihydrochloride | 740 |
| Ethionamide | 1000 |
| Kanamycin sulfate | [a] |
| Thioacetzone | 1000 |

[a]Information supplied by manufacturer.

forming units in the population. Below a certain proportion called the critical proportion, a strain is classified as susceptible, above as resistant. The significant resistance proportion levels for the different antituberculous drugs are the levels above which the drugs are no longer clinically useful. These critical proportions are 1% for H, PAS, and R and 10% for the other drugs.

Another key criterion is the critical concentration of the drug, which is the level inhibiting growth of wild-type organisms. The critical concentration criterion is not directly based on achievable blood serum levels nor on the minimal inhibitory concentrations. It is a compromise between these variables and the ability to discriminate between wild-type strains and strains isolated from treatment failure cases.

There are two variants of this method. The simplified variant requires the testing of only one drug concentration and is widely used in most public health settings.

The standard variant uses several drug concentrations and is mainly used in research laboratories.

Standardization of the inoculum size in the proportion method is not crucial as long as isolated colonies appear at one of the two dilutions plated. This circumstance permits the performance of direct drug-susceptibility testing of sputum, which can yield results many weeks earlier than indirect testing.

On the negative side, the proportion of resistant colonies can increase with the degree of clumping of the inoculum. Results can also be biased by calculating the number of bacilli in a suspension of higher concentration from colony counts obtained at a lower concentration. This problem, due to the disintegration of bacterial clumps when the initial suspension is diluted, occurs only when using the proportion method.

Detailed protocols for the performance of these drug susceptibility tests can be found in several laboratory manuals (9,10,14,17). Of the three methods described above, the most widespread and most accessible to the resources of developing countries is the proportion method performed on Loewenstein–Jensen medium.

Cultures no less than 4 weeks old should be used for the test because resistant mutants to H and other drugs can be more dysgonic than susceptible bacilli. Therefore, if younger (not fully developed) cultures are used, it is theoretically possible to classify a culture as susceptible when, in fact, it is resistant. However, it is of doubtful value to run drug-susceptibility tests on very old cultures. When the number of colonies developed on primary isolation is small (for instance, less than 10 colonies), note should be made to the physician that the drug-susceptibility test results might not be representative of the true bacillary population in the lesions.

## PREPARATION OF DRUG-SUSCEPTIBILITY TESTING MEDIA (PROPORTION METHOD)

Culture media with and without drugs are prepared simultaneously and can be used for up to 1 month following the date of preparation, provided they are kept at temperatures ranging from 4°C to 8°C.

The Loewenstein–Jensen (L-J) medium without drugs is prepared as described in Appendix I. Drug-containing medium is prepared by adding the drugs in L-J medium prior to inspissation.

Table 12.1 shows the critical concentrations and critical proportions for the principal antituberculous drugs.

## POTENCY OF THE DRUGS

The true potency of a drug is the number of micrograms ($\mu$g) of active drug per milligram (mg) total weight of the product. Table 12.2 shows potency values for some antituberculous drugs.

Stock solutions should be prepared with pure drugs (see Table 12.3), weighed using an analytical balance, and dissolved in the appropriate solvent on the same day in which the culture medium is prepared. Drug solutions do not require sterilization. If necessary, they can be stored at −20°C for a maximum of 2 months. In that case, it is recommended that the stock solutions be aliquoted in volumes required for single use.

Sterile glassware should always be used and all manipulations should be carried out under standard microbiological aseptic conditions.

## PREPARATION OF WORKING DRUG SOLUTIONS

Working drug dilutions should be prepared on the day of their use. Unused amounts should be discarded. Each dilution should be well mixed by shaking before the preparation of the next one. Table 12.4 shows the stock solution dilutions required to obtain working solutions. One milliliter of working solution added to 500 ml of L-J medium will yield final drug concentrations equivalent to the different critical concentrations.

One milliliter of each working dilution is added to 500 ml of L-J medium and shaken gently by rotating the flask for 5 min. The medium is then dispensed into tubes, inspissated for 45 min, the temperature not to exceed 85°C, and then allowed to cool at room temperature, following which the screw caps are tightened. If the medium has been dispensed in tubes with cotton stoppers, they should be covered with plastic and sealed. All media should be stored in a refrigerator at 4°C and maximum storage time is 1 month. Each new batch of medium should be quality controlled. Table 12.5 shows storage conditions and the shelf life of some antituberculous drugs.

## NEPHELOMETRIC STANDARDS

The following suspension can be used as nephelometric standards for adjusting standard bacillary suspensions:

(a) A suspension of 1 mg/ml of BCG vaccine prepared by reconstituting one ampoule of the freeze-dried vaccine to that concentration. This standard solution can stored at 4°C for up to 1 month. Shake vigorously before use.
(b) A MacFarland barium sulphate ($BaSO_4$) No. 1 standard can be prepared by combining the following solutions in the proportions indicated:

0.1 ml of 1% aqueous barium chloride solution
9.9 ml of 1% aqueous sulphuric acid solution

Table 12.3. Solvents for the preparation of drug stock solutions; solvent volumes to be added to 100 mg of drug

| Drug | Solvent | Volume of solvent for 100 mg of drug | Final concentration (μg/ml) |
|---|---|---|---|
| Ioniazid | Dist. water | 10.0 ml | 10,000 |
| PAS | Dist. water | 8.8 ml | 10,000 |
| Dihydrostreptomycin sulfate | Dist. water | 8.0 ml | 10,000 |
| Rifampicin | Dimethylformamide | 5.0 | 20,000 |
| Ethambutol dihydrochloride | Dist. water | 10.0 ml | 10,000 |
| Ethionamide | Propylene glycol[a] | 10.0 ml | 10,000 |
| Kanamycin sulfate | Dist. water | [b] | 10,000 |
| Thioacetazone | Propylene glycol[c] | 10.0 ml | 10,000 |

[a]Ethylene glycol can be used.
[b]95° Ethyl alcohol can be used.
[c]Ethylene glycol or 2-methoxy ethanol can be used. Heat gently to dissolve.

Table 12.4. Working drug dilutions to be added to L-J medium to obtain critical concentrations

| Drug | Dilution from Stock solution[a] | | Volume of dilution per 500 ml L-J medium |
|---|---|---|---|
| Isoniazid | 1:100 | 1000 $\mu$/ml | 1 ml |
| PAS sodium salt | 1:40 | 250 $\mu$/ml | 1 ml |
| Dihydrostreptomycin sulfate | 1:5 | 2000 $\mu$/ml | 1 ml |
| Rifampicin | Undiluted | | 1 ml |
| Ethambutol dihydrochloride | 1:10 | 1000 $\mu$/ml | 1 ml |
| Ethionamide | Undiluted | | 1 ml |
| Kanamycin sulfate | Undiluted | | 1 ml |
| Thioacetazone | 1:10 | 1000 $\mu$/ml | 1 ml |

[a]Solvent used is distilled water except for thioacetazone, which requires propylene glycol.

The resulting suspension stored in screw-capped tubes which should be shaken before use has an indefinite shelf life if the caps are airtight.

## PREPARATION OF MYCOBACTERIAL SUSPENSIONS

To obtain a suspension representative of the test isolate, a sample is taken with a spatula or a metal loop from as many colonies as possible, or in the case of confluent growth, from as wide an area as possible. Place bacterial growth in a small flat-bottom round flask containing glass beads and five drops of sterile distilled water and shake manually for 1 or 2 min before diluting with approximately 4 ml of sterile distilled water. The suspension is allowed to stand for 30–60 s; the homogeneous portion is removed with a Pasteur pipette and transferred to a sterile tube and visually compared with the standard. Sterile distilled water is added, if needed, until its opacity is adjusted to that of the standard. This is the so-called "standard suspension."

Occasionally, when the growth on primary culture is scarce, the opacity will be adjusted to the $10^{-1}$ standard, which is equivalent to 0.1 mg of bacterial mass per milliliter.

### INOCULATION OF THE BACTERIAL SUSPENSION

Six tenfold dilutions are prepared from the stock suspension using one sterile pipette per tenfold dilution. Of the six dilutions prepared, only the $10^{-3}$, $10^{-5}$,

Table 12.5. Storage and shelf life of pure drug powders

| Drug | Maximum shelf life in dry atmosphere[a] | Storage conditions in airtight bottles |
|---|---|---|
| Isoniazid | 7 years | protect from light |
| PAS | 4 years | protect from light |
| Dihydrostreptomycin sulfate | 2 years | 4°C; protect from moisture |
| Rifampicin | 5 years | <25°C; protect from light, moisture |
| Ethambutol dihydrochloride | 5 years | Protect from moisture |
| Ethionamide | 5 years | 4°C; protect from moisture |
| Kanamycin sulfate | 2 years | <20°C; protect from light |
| Thioacetazone | 4 years | Protect from light |

[a]Dessicator.

Table 12.6. Resistant mutants in wild-type *M. tuberculosis*

| Drug | Concentration (mg/l) | Minimum[a] | Median[a] | Maximum[a] |
|---|---|---|---|---|
| Isoniazid | 0.2 | 0 | 4 | 32 |
| PAS sodium salt | 0.5 | 17 | 900 | 50,000 |
| Dihydrostreptomycin sulfate | 4.0 | 0 | 7 | 300 |
| Ethionamide | 20.0 | 120 | 3500 | 70,000 |
| Kanamycin sulfate | 20.0 | 1 | 80 | 4,000 |
| Rifampicin | 40.0 | 0 | 0.02 | — |
| Ethambutol | 2.0 | 100 | — | 1,000 |

[a]Number of drug resistant bacilli per $10^{-6}$ in *M. tuberculosis* H37Rv strain; six-week readings.

and $10^{-6}$ will be inoculated. Sterile, 1-ml, graduated pipettes are used to inoculate 0.2-ml volumes of each dilution onto the appropriate set of control and drug-containing tubes. The inoculated tubes are gently rotated so that the inoculated suspension will be homogeneously distributed over the whole surface of the medium. The tubes, placed on sloping trays, are incubated at 37°C. If the inoculated tubes have screw caps, they should be left loosely capped until the inoculation liquid evaporates (24–48 h) after which they should be capped tightly. Tubes with cotton stoppers should be rubber-sealed 24–48 h after inoculation.

## READING AND RECORDING OF RESULTS

This standard suspension contains 1 mg milligram of bacterial growth per milliliter. However, the quantity of bacilli contained in this standard suspension varies considerably from test to test; this variation ranges from $10^6$ to $10^8$ germs per milliliter. Therefore, it is necessary to inoculate both dilutions, $10^{-3}$ and $10^{-5}$, on both control and drug-containing L-J medium tubes.

Sometimes it is not possible to count the colonies in the control tube at either of the two dilutions ($10^{-3}$ and $10^{-5}$). In such cases, the colonies inoculated in the $10^{-6}$ dilution series tubes, which are countable, will make it possible to estimate the actual number of colonies in the tubes corresponding to the $10^{-5}$ dilution series and thus establish the results of the test.

In cases in which the number of colonies growing on the $10^{-3}$ control tubes is less than 200, the test should be repeated, but only if it appears that the strain is showing drug susceptibility.

The averaged sum of the colonies counted in the two control tubes indicates the number of bacilli inoculated per tube. The number of colonies growing on the drug-containing tubes of the corresponding dilution series over the number of bacilli inoculated (control tubes) multiplied by 100 gives the percentage of resistance to that drug. This percentage, when compared to the critical proportion established for that drug, will determine whether the strain is susceptible or resistant.

Results are read at 28 days (4 weeks) and 42 days (6 weeks). The reading at 28 days is only definitive for determining resistance on the basis of growth observed on drug-containing media. The 42-day reading yields definitive results for drug susceptibility.

Resistance to para-aminosalicylic acid (PAS), or resistance to several drugs, could indicate the presence of mycobacteria other than tubercle bacilli.

## QUALITY CONTROLS

With each new batch of drug containing L-J medium, a complete DST should be done with the standard strain of *M. tuberculosis* H37Rv. In addition to the usual inoculation of the three dilutions, the undiluted standard suspension and its $10^{-1}$ dilution should also be inoculated on another series of control and drug-containing tubes. Readings should be done at 28 and 42 days. These readings will estimate the number of resistant mutants per $10^6$ bacilli of the standard strain to each drug, which will be compared to the allowable limit values shown on Table 12.6. Should the estimated resistance values exceed normal variation, the following conditions should be investigated:

Quality, purity, and dryness of the drug
Storage conditions of the drug and expiry date
Storage of stock solutions in freezer
Preparation of working drug dilutions
Time and temperature of inspissation of the medium

Each new batch of drug containing L-J medium should be numbered and dated for future reference.

#### SUSCEPTIBILITY OF *M. TUBERCULOSIS* TO PYRAZINAMIDE

Pyrazinamide (PZA) seems to be active only at an acid pH. Hence, to perform the DST by the proportion method, the L-J medium must be acidified to pH 5 before inspissation. At this pH, the growth of *M. tuberculosis* is slow and difficult, and some strains do not grow at all. Moreover, if the pH is above 5.5, pyrazinamide is not active and the drug-susceptibility test would give a false resistance result.

*Mycobacterium tuberculosis* strains susceptible to pyrazinamide possess the pyrazidamidase enzyme (PZase) which metabolizes pyrazinamide into pyrazinoic acid (POA). Determination of pyrazinamidase activity, as described by Wayne (20) (see Appendix I), indicates by a rose-colored band of diffused ferrous salt in the agar medium whether pyrazinoic acid has been formed from PZA. This estimation of pyrazinoic acid can replace the PZA susceptibility test of *M. tuberculosis.*

This test is not useful in determining susceptibility to PZA in nontuberculous mycobacteria, as many of these species show positive results to the test in spite of their PZA resistance.

## 4. Identification of the Mycobacteria

The diagnosis of mycobacterial infection is dominated by the importance of the disease caused by the *M. tuberculosis* complex. However, mycobacteria other than tuberculosis are frequently isolated from all types of clinical and environmental specimens.

In many developing countries, the majority of the strains isolated from human clinical specimens are *M. tuberculosis.* The *M. tuberculosis* complex comprises strains of *M. tuberculosis* (21), *M. bovis, M. africanum,* and *M. microti. M. bovis* (22) is responsible for tuberculosis in cattle, which can cause pulmonary and nonpulmonary disease in humans. *M. africanum* (23), the African variant of *M. tuberculosis,* shares characteristics of *M. tuberculosis* and *M. bovis* and is mainly found in west and east Africa. *M. microti,* the "vole or murine tubercle bacillus" (24), is the etiologic agent of tuberculosis in vole mice, *Microtus agrestis.* All the members of this complex with the exception of *M. microti* can cause human disease. Therefore, it is of practical interest to screen the members of this complex from the other mycobacterial species.

### *4.1 Screening of the* **Mycobacterium tuberculosis** *Complex*

The first step in the screening procedure for the *M. tuberculosis* complex is the microscopic examination of the isolates. The Ziehl–Neelsen staining ensures that the isolate is acid-alcohol-fast and is free of nonmycobacterial contaminants. Although the microscopic morphology of the bacilli is only an indication and not an identification, the presence of serpentine cording should be noted because it is characteristic—albeit not unique—to *M. tuberculosis. M. bovis* and some strains of *M. kansasii* can also show cording.

The next step is the visual morphological examination of the primary isolate. The colonies of *M. tuberculosis* on egg media are characteristically eugonic, rough, buff to yellowish and resemble bread crumbs or cauliflowers. The colonies of *M. bovis* growing on egg media are smaller, flatter, and smoother than those of *M. tuberculosis. M. bovis* isolates grow best in glycerol-free medium and have a preference for pyruvate. *M. africanum,* although similar in appearance to *M. bovis,* are often dysgonic and do not always show pyruvate preference. *M. microti* shows cultural characteristics similar to those of tubercle bacilli.

The above-mentioned characteristics together with a few growth inhibition tests such as the *p*-nitro-$\alpha$-acetylamino-$\beta$-hydroxypropiophenone (NAP) (25,26), the *p*-nitrobenzoic acid (25,27), the hydroxylamine tests (28), and the susceptibility to isonicotinic acid hydrazide (INH), streptomycin, rifampicin, ethambutol, and *p*-aminosalicylate, as well as the catalase production test (29), will provide a good indication of the presence of members of the *M. tuberculosis* complex. Some of these procedures are described in detail in Annex I.

### *4.2. The Differentiation of Tubercle Bacilli*

The aforementioned screening tests separate tubercle bacilli from the other mycobacterial species, but there is still need to differentiate between the species of the *M. tuberculosis* complex. The isolation of *M. africanum* and *M. microti* requires very long incubation periods; the presence of the former should only be suspected in cases related with Africa. *M. africanum* is probably a subspecies or biovar of *M. bovis,* whereas the *M. microti* is extremely rare and could be considered a biovar or pathovar of *M. tuberculosis.* Strains of *M. bovis,* as opposed to those of *M. tuberculosis,* do not secrete niacin in the growth medium (30), are inhibited by thiophene-2-carboxylic acid hydrazide (TCH) (31), do not reduce nitrates (32), and are resistant to PZA (i.e., pyrazinamidase negative).

The stains of Bacille–Calmette–Guérin (BCG) are pathovars of *M. bovis* and are difficult to differentiate bacteriologically from *M. bovis.* However, they are usually more eugonic than strains of *M. bovis;* some show resistance to cycloserine, borderline resistance to TCH, and some are niacin positive. BCG strains are easily differentiated by animal inoculation and should always be suspected if isolated from abscesses following BCG vaccination or from lymph nodes draining the vaccination site. Table 12.8 shows key characteristics of the *M. tuberculosis* complex.

### *4.3. The Differentiation of Mycobacteria Other Than Tubercle Bacilli*

From the clinical and the public health point of view, it is sufficient to determine whether a mycobacterium is *M. tuberculosis* or not. So for most developing countries and within the framework of a NTP, it is enough to apply the screening techniques to separate the *M. tuberculosis* complex from the rest of the mycobacteria and to use some of the key tests described above to identify *M. tuberculosis* and *M. bovis.* Although mycobacteria other than tubercle bacilli can cause mycobacteriosis in man, these species are frequently isolated from the environment. For this reason, special care should be exercised in determining their role as etiological agents. Their clinical significance increases under the following circumstances:

Table 12.7. Classification of "atypical" mycobacteria, according to Runyon (31)

| Group | Description | |
|---|---|---|
| I. | Photochromogens | Slow growth. Colonies not pigmented in the dark. Young colonies acquire yellow color when exposed to light. |
| II. | Scotochromogens | Slow growth. Colonies pigmented yellowish to orange, in both the light and the dark. |
| III. | Not chromogenous | Slow growth. Colonies usually not chromogenous or weakly pigmented. |
| IV. | Fast growers | Form grossly visible mature colonies in less than 7 days at 24°C and 37°C from diluted bacterial suspensions (1 mg/ml). |

Several cultures of the same strain and from the same subject are obtained, especially in the absence of *M. tuberculosis.*
The cultures obtained present abundant growth.
Disease has been confirmed.
A poor response to antituberculosis treatment has been observed.
A pure culture was obtained from a sample taken from a small lesion by sterile manipulation.

Some central laboratories might want to proceed a step further in the identification of mycobacteria, separating the four main groups of mycobacteria other than tubercle bacilli.

In 1954, E. Runyon proposed a classification of mycobacteria more frequently isolated in the laboratory (31), based on easily observed characteristics such as rate of growth and pigmentation of the colonies (see Table 12.7). This first classification continues to be a very useful guide, both for the clinical bacteriologist and the medical doctor. Further identification down to the species level is warranted in very few cases, and unless there is a substantial number of documented cases of mycobacterioses, these isolates should be submitted to specialized centers

Table 12.8. Key characteristics of *M. tuberculosis* complex[a]

| | *M. tuberculosis* | *M. bovis* | *M. bovis* BCG |
|---|---|---|---|
| Growth in 3 days | − | − | − |
| Growth at 30°C and 42°C | − | − | − |
| Growth in presence of NAP, PNB, Hydroxylamine | − | − | − |
| Growth in presence of TCH | + | − | +[b] |
| Growth in presence of: | | | |
| Cycloserine | − | − | ± |
| Pyrazinamidase | + | − | − |
| Nitrate reduction | + | − | − |

[a]Includes *M. africanum* which causes tuberculosis in humans in west and central Africa. Strains of *M. africanum* are unlikely to be found in North American patients; they grow very slowly and possess characteristics common to both *M. tuberculosis* and *M. bovis. M. microti* causes generalized tuberculosis vole mice and represents an intermediate form between *M. tuberculosis* and *M. bovis* and is unlikely to be found in human clinical specimens.
[b]*M. bovis* BCG is partially resistant to TCH.
\+ = more than 84% of strains positive
± = between 50% and 84% of strains positive
∓ = between 16% and 49% of strains positive
− = less than 16% of strains positive.

such as university research labs or to supranational regional reference centers where this kind of identification procedures is done on a routine basis.

## 5. Safety Measures

### *5.1. Biosafety in the Tuberculosis Bacteriology Laboratory*

Laboratory work with infectious specimens always represents a certain risk of infection. Most cases of laboratory-acquired infection result from breathing aerosol particles called droplet nuclei containing viable bacilli, particularly those droplets measuring 3–5 $\mu$m in diameter, which easily reach the lung alveoli when inhaled. **Biosafety measures in tuberculosis bacteriology are a set of commonsense practices strictly observed by a conscientious and well-trained staff.**

The head of the laboratory is responsible for drawing up regulations and standards to ensure the protection of the staff, but each technician is responsible not only for his own safety but also for the safety of other co-workers. The often inferior conditions of premises and equipment which generally prevail in the laboratories of developing countries make it necessary to observe the biosafety measures even more strictly than in those countries with well-equipped laboratories and excellent facilities.

Of all the biosafety measures, the most important one is **to perform each technique with the greatest attention to detail; even the best equipment cannot replace order and care in the performance of the work.**

Biosafety measures concern not only the staff but are also designed to protect the working environment and the safety equipment from possible contamination. They are also designed to regulate the response in case of accident and the "housekeeping" procedures to be carried out upon finishing the work.

#### STAFF

Staff working in the tuberculosis bacteriology laboratory must receive prior technical training, which should include careful teaching of measures designed to protect personal safety as well as the safety of co-workers.

All persons working in a laboratory handling specimens for tuberculosis bacteriology, whether they process these samples or only share the work area, either regularly or occasionally (clerks, cleaning staff, and lab attendants), should be tuberculin positive. Before joining the laboratory, new staff members should have a medical examination, a chest x-ray and an intradermal tuberculin test with PPD 2 TU (Mantoux test). Should the Mantoux test be negative, the individual should be BCG vaccinated. Three months after vaccination, a second tuberculin test is done; if negative, it should be followed with revaccination because it is not advisable for persons with a postvaccinal negative tuberculin test to work in the TB laboratory. Staff members should be submitted to a chest x-ray once a year; the films should be stored in a personal file for future reference.

### *5.2. Biosafety Measures for Laboratories Performing Smear Microscopy*

#### HIGH-RISK OPERATIONS

Laboratory acquired infection is often related to the production of aerosols by the following procedures:

- The opening of specimen containers
- Smear preparation
- Flaming the wire loop (use of wooden applicators recommended whenever possible)

#### GENERAL CONDITIONS

The working environment should be spacious and well ventilated. The place where infected material is handled is considered to be a **CONTAMINATED AREA.** The following precautions should be taken:

It should be located at a point far removed from the laboratory entrance door and well protected from drafts.

Persons not directly connected with the laboratory should not be allowed entry.

While smears are being prepared, the movement of persons working in that area should be limited to a strict minimum.

The work area should only contain the equipment and materials needed for the procedure. Superfluous objects must be avoided. Laboratory notebooks and log books should not be removed from the work area and telephones should not be installed in work areas.

The walls and floors in the **CONTAMINATED** area should be smooth and washable (tiles, mosaics, oil base paint, etc.). The walls and floors of the laboratory should be washed daily with water and detergent at the end of each day's work. Floors should never be swept with a dry broom nor should they be waxed.

The technologist is responsible for the disinfection of the contaminated area before and after each work session. A solution of 5% phenol or 3% cresol is recommended for this purpose; the time of contact of the solution with contaminated areas should be of at least 30 min.

#### PRECAUTIONS WHILE WORKING

Protective clothing with long sleeves made of washable material should be worn. Worn protective clothing should be sterilized in an autoclave before removal from the laboratory for laundering.

Hands should be frequently disinfected, washed with abundant water, soap, and brush, and dried with paper towels.

It is advisable, although not essential, to use disposable masks, made of a material of suitable thickness to serve as a protective barrier.

**Smoking, eating, drinking, and grooming is not authorized in the laboratory.**

#### ACCIDENTS

In case of an accident caused by spillage of infectious materials on the floor or on the bench, pour 5% phenol solution on the specimen, cover with a newspaper, soak thoroughly with the same disinfectant solution, and leave it for at least 30 min before cleaning the area.

### *5.3. Biosafety Measures for Regional and Central Laboratories*

These are laboratories that perform culture, for drug-susceptibility tests and other sophisticated diagnostic procedures.

#### HIGH-RISK OPERATIONS

In addition to the high-risk operations already mentioned, special attention should be devoted to the following:

- Pipetting bacterial suspensions
- Centrifuging infectious liquids
- Decanting of supernatants following centrifugation
- Opening of tubes after centrifuging or shaking
- Manual or mechanical shaking of tubes containing infectious material
- Grinding of tissue samples

#### GENERAL CONDITIONS

The general conditions that apply to laboratories where only smear microscopy is performed also apply to more sophisticated labs. However, the following additional requirements should also be met:

Access will be restricted and warning signs posted on the outside of the doors.

The centrifuge and the mechanical shaker should be located in a well-ventilated area separate from the main body of the laboratory.

The above-mentioned high-risk procedures should be performed in a biosafety cabinet such as the one illustrated in Fig. 12.1.

If no biosafety cabinet is available, it is advisable to install ultraviolet lamps on

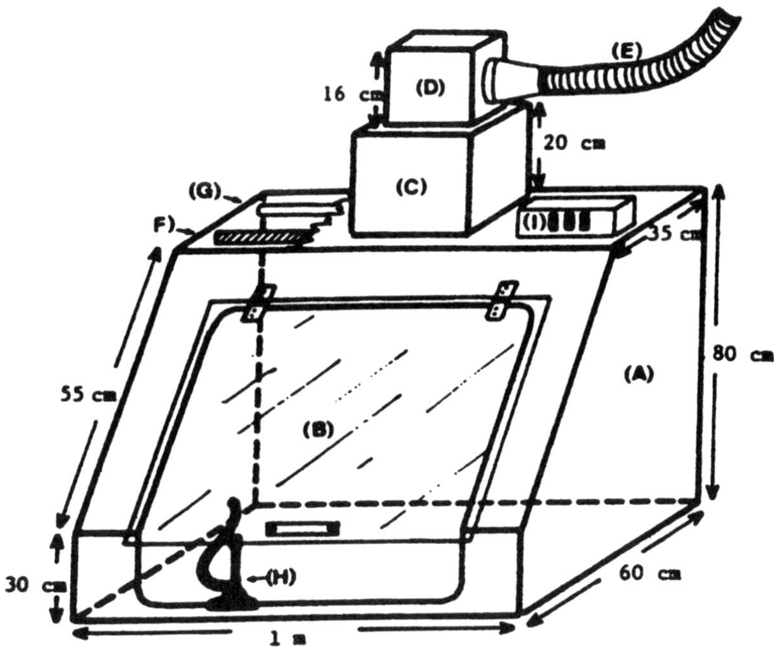


Figure 12.1. Diagram of a simple biosafety cabinet (type 1) which can be made by a craftsman and which has the following characteristics:
**(i)** The walls and roof are made of a sheet of bent aluminum. **(ii)** The front has a sheet of thick glass which opens upward. **(iii)** Between the bench and the lower edge of the glass sheet there should be an opening measuring 1 m × 0.30 m for the technician to put his arms through. **(iv)** An air extractor with metal blades is placed on top of the cabinet, connected with the outside by a vent tube. **(v)** Between the extractor and the roof of the cabinet there should be a biological filter made of glass wool with a minimum thickness of 5 cm. These filters retain dust particles so that the efficiency is gradually reduced. One way of measuring that efficiency is to place the flame of the burner under the extractor. The flame should grow brighter. This simple test must be performed once a month. If this test fails, the filter should be replaced. Before doing so, all biological material should be removed from the area close to the cabinet and gaseous formaldehyde should be passed through the filter to be replaced. Gaseous formaldehyde is produced by pouring formalin (a 40% solution of formaldehyde) over potassium permanganate crystals in the following proportion: 25 ml of formalin and 15 g potassium permanganate for each cubic meter to be disinfected. The potassium permanganate crystals should be placed in a deep metal container. The reaction rapidly produces the release of gas and heat. The formalin should be allowed to act overnight, the extractor being left on all that time. **(vi)** An ultraviolet germicidal lamp emitting 95% radiation of approximately 254 nm in wavelength should be placed inside on the front wall of the cabinet. **(vii)** A fluorescent light source should be placed on the back wall of the cabinet. **(viii)** An inlet in the back wall for a Bunsen burner. **(ix)** Switches for the ultraviolet lamp, the fluorescent lamp, and the air extractor should be placed outside of the hood.

the walls at a distance of not more than 40 cm from the bench surface. It should be noted, however, that the use of such lamps is only a safety measure additional to the chemical disinfection, which is meant to reduce the number of microorganisms in the air and on the lab surfaces. The effectiveness of UV light is reduced by dust accumulation and lamp age; therefore, they should be cleaned weekly with alcohol-soaked cotton. The minimum time of UV light exposure should be 2 h. Moreover, it should be remembered that ultraviolet rays will damage latex, rubber, and plastics.

Biosafety equipment reduces the frequency of accidents due to contamination; however, human error is often the decisive factor.

Each worker should be responsible for maintaining a clean and neat work area.

## PRECAUTIONS WHILE WORKING

*Pipetting of infectious liquids:* Infectious liquids should never be mouth-pipetted. Pipetting aids should replace mouth-pipetting.

*Sterilization of the wire loop or metal spatula:* Wire inoculating loops covered with clumps of bacilli and organic debris should be cleaned by stabbing them repeatedly into flasks containing 95% alcohol with washed sand before flaming. The abrasive action of the sand removes most of the material and the alcohol causes rapid incineration when the loops are flamed.

*Centrifugation of infectious liquids:* The use of sealed centrifuge safety cups is strongly recommended to eliminate aerosol production in case of tube breakage. If such a centrifuge is unavailable, the lid should never be opened while the centrifuge is in operation.

In case of accident, the laboratory manager should be notified immediately and after proper handling of the spill, a record should be made and kept, stating the date of the accident, the names of staff members involved, and the equipment affected.

(a) Breakage or spilling of tubes or flasks containing specimens, or other bacillary suspensions or inoculated cultures: Proceed as indicated above. Wearing the appropriate protective clothing, pick up the debris with forceps, place them in a suitable receptacle, and sterilize them in the autoclave or burn them.

(b) Breakage of tubes or accidental opening of lids inside the centrifuge: If, during centrifugation, a tube is broken, disconnect the centrifuge, leave lid unopened for at least 10 min after complete stoppage, sprinkle the inside of the centrifuge with a 5% phenol solution, pour a larger quantity on the broken tubes, and leave it in contact for 30 min. Pick up the tube and jacket with forceps and place them in a receptacle, clean the inside of the centrifuge with cotton impregnated with 5% phenol solution, discard it in the same receptacle, and sterilize in the autoclave. Proceed in the same manner if the accident occurs during mechanical shaking (i.e., Vortex or Kahn shakers).

(c) Autoinoculation: Wash the inoculated area with soap and water and refer the accident victim to a TB specialist. Coordinate with the physician the follow-up or the specific treatment.

## Appendix

### *Ziehl-Neelsen Method (ZN)*

#### PREPARATION OF REAGENTS

(a) Fuchsin solution
    Basic fuchsin 3 g
    95% Alcohol 100 ml

Add alcohol to the fuchsin in a flask and shake to dissolve. Mix with 55 ml aqueous phenol.* With continuous shaking add enough distilled water sufficient quantity (q.s.) to 1 L. Allow to settle for 24 h and filter.

(b) Methylene Blue counterstain solution
    Methylene Blue 1 g
    95% Alcohol 100 ml

Dissolve by shaking and add distilled water q.s. to 1 L. Allow to settle for 24 h and filter through paper.

(c) Decolorizing solution
    Hydrochloric acid (HCl) 3 ml
    95% Alcohol 970 ml

With a pipette, add HCl slowly into the alcohol and shake gently. A 25% sulphuric acid ($H_2SO_4$) solution in distilled water can be used in place of 3% acid alcohol

*Aqueous phenol is prepared by adding 10 ml distilled water to 100 g of crystallized phenol. It is boiled in a water bath until complete dissolution, and then allowed to cool. Aqueous phenol remains liquid at room temperature.
*See Appendix

solution. To prepare this solution, $H_2SO_4$ analytical reagent is added with a pipette to distilled water with gentle shaking.

Reagents (a) and (b) should be kept in amber-colored flasks and transferred into small glass or plastic flasks prior to staining. Each time the fuchsin solution is poured into these small flasks, it should be filtered. A small funnel with a filter paper can be used for this purpose. This filtration can also be made directly on the smear by placing a strip of filter paper or even plain newspaper over the smear. Fuchsin can precipitate and the crystals formed can be a source of false-positive error in reading slides if solutions are not freshly filtered.

These stains are usually prepared in volumes that will cover the needs for 2 weeks. The date of preparation must be indicated on the bottle. As time elapses, fuchsin or Methylene Blue crystals form spontaneously and the resulting solutions become less concentrated. This could cause false-negative readings or weaken counterstaining.

## LOEWENSTEIN–JENSEN MEDIUM

| | |
|---|---|
| Monopotassium phosphate, anhydrous ($KH_2PO_4$) | 2.4 g |
| Magnesium sulphate ($7H_20$) | 0.24 g |
| Magnesium citrate | 0.60 g |
| Glycerol | 12.0 ml |
| L-Asparagine | 3.6 g |
| Distilled water q.s. to | 600.0 ml |
| Whole eggs | 1000.0 ml |
| 2% Aqueous fresh Malachite Green solution | 20.0 ml |

The mineral salts solution and the freshly prepared Malachite Green solution are autoclaved at 121°C for 15 min. Fresh eggs, no more than a week old, are selected, cleaned with soap and water, rinsed with water, and wiped with a gauze soaked in 70% alcohol. These eggs are broken into a big flask containing a few glass beads or small pieces of glass. The salt base and the Malachite Green solutions are then added. This mixture is homogenized by shaking or blending, filtered through a sterile gauze, and distributed into tubes. The medium is inspissated in a slanted position in the tubes, at 80°C for 45 min. An inspissator or a water bath with well-regulated temperature should be employed. A dry oven can also be used if the temperature is well controlled in the 80–85°C range (8).

## STONEBRINK MEDIUM (12)

| | |
|---|---|
| Monopotassium phosphate, anhydrous ($KH_2PO_4$) | 3.5 g |
| Disodium phosphate ($Na_2HPO_4 \cdot 2H_2O$) | 2.0 g |

| | |
|---|---|
| Sodium pyruvate | 6.3 g |
| Distilled water to | 500.0 ml |
| Whole eggs | 1000.0 ml |
| 2% aqueous Malachite Green solution | 20.0 ml |

The preparation of this medium is similar to the one described above for Loewenstein–Jensen medium.

**FLOCCULATION METHOD**

In methods without centrifugation, two solutions are used:

1. The decontaminant solution—4% NaOH
2. The flocculation solution, composed as follows:

| | |
|---|---|
| Calcium chloride ($CaCl_2 \cdot 2H_2O$) | 3.0 g |
| Distilled water | 50.0 ml |
| Barium chloride ($BaCl_2 \cdot 2H_2O$) | 2.5 g |
| Distilled water | 50.0 ml |

The two solutions are mixed and two drops of the resulting solution are added to each 10 ml of the mixture specimen–decontaminant solution and allowed to stand overnight. The mixture is shaken repeatedly during the first 2 h of contact. The flocculated sediment can be aspirated with a pipette and inoculated onto egg medium (8).

***Niacin Test (Konno, K., 1956 modified)***

(a) PRINCIPLE: Detection of the presence of nicotinic acid in the culture medium. Nicotinic acid is an intermediate in the biosynthesis of NAD (nicotine adenine dinucleotide). *M. tuberculosis* is blocked in this pathway and nicotinic acid is excreted in the culture medium. This test differentiates *M. tuberculosis* from most other mycobacteria.

(b) CULTURES: Test all mycobacterial cultures after 3 weeks of growth on L-J medium.

(c) REAGENTS:

1. 10% (w/v) saturated cyanogen bromide aqueous solution
   5.0 g cyanogen bromide in 50.0 ml distilled water
   **Caution: Cyanogen bromide is tear gas.**
2. 3% (w/v) benzidine solution

3.0 g benzidine in 100.0 ml of 95% ethyl alcohol

**Caution: Benzidine is carcinogenic.** Store solutions in brown bottles at 4°C for a maximum of 4 weeks.

(d) PROCEDURE:

1. Add 8 drops of sterile distilled water to 3-week-old culture on L-J. slant, 16 drops if culture is dry.
2. Autoclave for 30 min at 15.0 lbs pressure, in a slanted position, so that the water covers the bacterial growth.
3. Upon cooling, transfer 8 drops of the water extract into Wassermann test tubes (12 × 75 mm), add 8 drops of cyanogen bromide solution followed by 16 drops of benzidine solution, shake the tubes.

(e) CONTROLS:

1. Reagent control on uninoculated medium (negative)
2. Reagent control of *M. tuberculosis* culture (positive)

(f) RESULTS:

White precipitate = negative reaction
Pink to red precipitate = positive reaction

**CAUTION: This test should always be performed under a safety hood.**

***Nitrate-Reduction (Virtanen, S., 1960)***

(a) PRINCIPLE: This test detects the ability of mycobacteria to reduce nitrates to nitrites. *M. tuberculosis, M. kansasii,* and *M. szulgai* are nitrate reduction positive.

(b) CULTURES: Test all mycobacterial cultures at 3 weeks of growth on L-J medium.

(c) SUBSTRATE AND REAGENTS:

1. Substrate M/100 sodium nitrate in M/45 phosphate buffer.

M/15 pH 7 phosphate buffer solution* 100.0 ml
Distilled water 200.0 ml
$NaNO_3$ 255.0 mg

Dispense 2.0 ml of the $NaNO_3$ substrate solution to each screw-cap tube (13 × 100 mm). Autoclave for 15 min at 15.0 lbs pressure. Stable for 3 weeks at 4°C.

2. Reagents

1. A 1:2 dilution of concentrated HCl (add 50.0 ml HCl to 50.0 ml $H_2O$).
2. Dissolve 0.2 g sulfanilamide in 100.0 ml distilled water.

3. Dissolve 0.1 g *n*-naphthylethylenediamine dihydrochloride in 100.0 ml distilled water.

The reagents are stable for 3 weeks when stored in the dark at 4°C.

(d) PROCEDURE

1. Emulsify one loopful of bacterial growth in the substrate medium.
2. Incubate at 37°C or at optimal temperature of growth for 2 h.
3. Add to each tube in sequence one drop of reagent 1, two drops of reagent 2, and two drops of reagent 3.

(e) CONTROLS:

One uninoculated tube (negative)
One tube inoculated with *M. tuberculosis* (positive)

(f) RESULTS:

1. Development of a red coloration indicates a positive reaction. Results are compared with color standard† and recorded negative (−) to 5 + according to the intensity of the color change.
2. To confirm a negative result, add a small amount of powdered zinc to each tube. If no color change occurs, the reaction was positive (nitrate is reduced beyond nitrite).

### *Thiophen-2-carboxylic Acid Hydrazide (TCH) Susceptibility (Boenicke, R., 1958)*

(a) PRINCIPLE: Selective growth inhibition of *M. bovis*. Differentiates *M. bovis* from other mycobacteria.

(b) CULTURES: Test all mycobacterial cultures after visible growth appears on L-J medium.

(c) MEDIUM: L-J. slopes containing 10.0 $\mu$g/ml thiophen-2-carboxylic acid hydrazide. Medium can be kept at 4°C for 3 weeks.

(d) PROCEDURE:

1. Inoculate slopes with one 3.0-mm loopful of mycobacterial suspension (1.0 mg/ml).
2. Incubate at 37°C.
3. Read after 3 weeks of incubation.

(e) CONTROL: One inoculated drug-free L-J slope.

(f) RESULTS:

† See Appendix

Confluent growth similar to control = resistant
No growth or few discrete colonies = susceptible generalized tuberculosis in vole mice and represents an intermediate form between *M. tuberculosis* and *M. bovis* and is unlikely to be found in human clinical specimens.

2. *M. bovis* BCG is partially resistant to TCH.

\+ = more than 84% of strains positive
± = between 50% and 84% of strains positive
∓ = between 16% and 49% of strains positive
− = less than 16% of strains positive

## Appendix

NITRATE REDUCTION STANDARDS

(Vestal, A.L. 1976)

(a) BUFFER SOLUTION
Mix;

| | |
|---|---|
| M/15 sodium phosphate (dibasic, anhydrous) | 35.0 ml |
| M/15 potassium phosphate (monobasic) | 5.0 ml |
| M/15 trisodium phosphate | 100.0 ml |

(b) INDICATOR SOLUTION

1. 1% phenolphthalein (1.0 g in 100.0 ml 95% ethanol).
2. 1% bromthymol blue (1.0 g in 100.0 ml 95% ethanol).
3. 0.01% bromthymol blue (1.0 ml of solution 2 in 100.0 ml distilled water).

(c) PREPARATION OF STANDARDS

1. Prepare eight 13 × 100 mm screw-cap tubes.
2. Add 2 ml of buffer solution into seven tubes.
3. Add 0.1 ml of indicator solution 1 and 0.2 ml of indicator solution 3 to 10.0 ml buffer solution.
4. Add 2.0 ml of this mixture (step 3) to the 8th tube. This is the 5+ colour standard.
5. To the first tube in the series of seven, add 2.0 ml of mixture (step 3). Make serial dilution by carrying over 2.0 ml, discard 2.0 ml from the last tube.
6. Tube 8 − 5+
   1 = 4+
   2 = 3+
   4 = 2+

5 = 1+
7 = ±

7. Autoclave tubes, store at 4°C. Colours range from pink ± to purple 5+.

**References**


1. Enarson DA (1993) The IUATLD "Model" National Tuberculosis Programmes. In: Acchinelli R, ed. XXV Pan American Congress ULASTER, pp. 190–198. Lima: Libro de Resúmenes.
2. Pan American Health Organization (PAHO/WHO) (1993) Manual of Technical Standards and Procedures for Tuberculosis Bacteriology, Spanish electronic revised edition, Martinez, Argentina: INPPAZ.
3. Pan American Health Organization (1986) Tuberculosis Control: A Manual on Methods and Procedures for Integrated Programs. Washington DC: PAHO/WHO.
4. World Health Organization (1993) Laboratory Biosafety Manual, 2nd ed., pp. 48–54. Geneva: WHO.
5. Canetti G, Grosset J (1969) Techniques et Indications des Examens Bacteriologiques en Tuberculose. St. Mandé: Ed.de la Tourelle.
6. International Union Against Tuberculosis and Lung Diseases (IUATLD) (1993) Tuberculosis Guide for High Prevalence Countries. 2nd ed. Paris: Ed.de l'Aulne.
7. Collins CH, Grange JM, Yates MD (1985) Organization and Practice in Tuberculosis Bacteriology. London: Butterworths.
8. Kleeberg HH, Koornhof HJ, Palmhert H (1980) Laboratory Manual of Tuberculosis Methods, 2nd ed., revised. Pretoria: Tuberculosis Research Institute of the South Africa Medical Research Council.
9. David H, Levy-Frebault V, Thorel MF (1989) Methodes de Laboratoire pour Mycobacteriologie Clinique. Paris: Institut Pasteur.
10. Groothuis DG, Yates MD, eds. (1991) Diagnostic and Public Health Mycobacteriology, revised 2nd ed. Reading: The Eastern Press Ltd.
11. Kantor IN (1979) Bacteriología de la Tuberculosis Humana y Animal, Buenos Aires: OPS/OMS.
12. Stonebrink B, Duoma J, Manten A, Mulder RJ (1969) A comparative investigation on the quality of various culture media as used in the Netherlands for the isolation of Mycobacteria. Selected Papers (The Hague) 12:5–47.
13. Steward SM, Crofton JW (1964) The clinical significance of low degree of drug resistance in pulmonary tuberculosis. Am Rev Tuberc Pulmon Dis 89:811–829.
14. Canetti G, Fox W, Khomenko A et al. 1969. Advances in techniques of testing mycobacterial drug sensitivity and the use of sensitivity tests in tuberculosis control programmes. Bull WHO 41:21–43.
15. Meissner G (1964) The bacteriology of the tubercle bacillus. In Barry VC, ed. Chemotherapy of Tuberculosis, pp. 65–110. London: Butterworths.

16. Mitchison DA (1954) Problems of drug resistance. Br Med Bull 10:115.
17. Middlebrook G, Cohn ML (1958) Bacteriology of tuberculosis laboratory methods. Am J Publ Hlth 48:844–853.
18. Canetti G, Rist N, Grosset J (1963) Mésure de la sensibilité du bacille tuberculeux aux drogues antibacillaires pour la méthode des proportions. Méthodologie, critères de résistance, résultats, interprétation. Rev Tuberc Pheumal 27:217–272.
19. Casal Román M (1989) Microbiologia clinica de las enfermedades por micobacterias. Córdoba: Servicio de Publicaciones Universidad de Córdoba.
20. Wayne LG (1974) Simple pyraminamidase and urease tests for routine identification of mycobacteria. Am Rev Respir Dis 109:147–151.
21. Lehmann KB, Neumann R (1896) Atlas und Grundiss der Bakteriologie und Lehrbuch der speciellen bakteriologschen Diagnostik. 2nd ed., pp. 408–413. Munich: JF Lehmann.
22. Karlson AG, Lessel EF (1970) *Mycobacterium bovis* nom. nov. Int J System Bacteriol 20:273–283.
23. Castets M, Rist N, Boisvert H (1969) La variété africaine du bacille tuberculeux humain. Med Afrique Noive 16:321–322.
24. Redd GB (1957) Family 1 Mycobacteriaceae Chester 1897. In: Breed RS, Murray EGD, Smith NR, eds. Bergey's Manual of Determinative Bacteriology, 7th ed. Baltimore, MD: Williams and Wilkins.
25. Laszlo A, Eidus L (1978) Test for differentiation of *M. tuberculosis* and *M. bovis* for other mycobacteria. Can J Microbiol 24:754–756.
26. Rastogi N, Goh KS, David HL (1989) Selective inhibition of the *Mycobacterium tuberculosis* complex by *p*-nitro-alpha-acetylamino-beta-hydroxypropiophenone (NAP) and *p*-nitrobenzoic acid (PNB) used in 7H11 agar medium. Res Microbial 140(6)419–423.
27. Tsukamura M, Tsukamura S (1964) Differentiation of *Mycobacterium tuberculosis* and *M. bovis* by *p*-nitrobenzoic acid susceptibility. Tubercle 45:64–65.
28. Tsukamura M (1965) Differentiation of mycobacteria by susceptibility to hydroxylamine and 8-azaguanine J Bacteriol 90:556–557.
29. Konuo K (1956) New chemical method to differentiate human-type tubercle bacilli from other mycobacteria. Science 124:985.
30. Bönicke R (1958) Die differenzierung humaner un boviner Tuberkel Bakterien mit Hilfe von Thiophen-2-Carbon Soüre hydazid. Naturwisseuschaften 45:392–393.
31. Timpe A, Runyon EH (1954) The relationship of "atypical" acid-fast bacteria. Am Rev Respir Dis 85:753–754.
32. Virtanen S (1960) A study of nitrate reduction by mycobacteria. Acta Tuberc. Scand 48(Suppl):1–119.

# Index